WJEC GCSE
MATHEMATICS
Higher Student's Book

Wyn Brice, Linda Mason and Tony Timbrell

Second Edition

HODDER
EDUCATION
AN HACHETTE UK COMPANY

This material has been endorsed by WJEC and offers high quality support for the delivery of WJEC qualifications. While this material has been through a WJEC quality assurance process, all responsibility for the content remains with the publisher.

Acknowledgements
The Publishers would like to thank the following for permission to reproduce copyright material:
p.599 Kidshealth.org/The Nemours Foundation

Photo credits: p.50 © Daniel Bosworth/Britain On View

The Publishers would also like to thank WJEC for permission to reproduce past examination questions throughout this book.

Every effort has been made to trace copyright holders. However, if any have been inadvertently overlooked the Publishers would be happy to make the necessary arrangements at the first opportunity.

Hachette UK's policy is to use papers that are natural, renewable and recyclable products and made from wood grown in sustainable forests. The logging and manufacturing processes are expected to conform to the environmental regulations of the country of origin.

Orders: please contact Bookpoint Ltd, 130 Milton Park, Abingdon, Oxon OX14 4SB. Telephone: (44) 01235 827720. Fax: (44) 01235 400454. Lines are open 9.00–5.00, Monday to Saturday, with a 24-hour message answering service.
Visit our website at www.hoddereducation.co.uk

Original edition: Howard Baxter, Michael Handbury, John Jeskins, Jean Matthews, Mark Patmore, Brian Seager 2006
© This edition Wyn Brice, Linda Mason, Tony Timbrell 2006, 2010
First published in 2006 by
Hodder Education, an Hachette UK Company,
338 Euston Road
London NW1 3BH
This second edition published in 2010.

Impression number   5 4 3 2 1
Year   2014 2013 2012 2011 2010

Cover photo © Imagestate Media
Illustrations by Barking Dog Art
Typeset in 10/12pt Times Ten by Tech-Set Ltd, Gateshead, Tyne and Wear
Printed in Italy

A catalogue record for this title is available from the British Library
ISBN 978 1444 114 836

→ CONTENTS

Algebra   131

→ INTRODUCTION

About this book

This book covers the complete specification for the Higher Tier of GCSE Mathematics. It has been written especially for students following WJEC's 2010 Unitised and Linear specifications.

The chapters have been grouped to cover the four main areas of the specifications.

- Chapters 1–9   Number
- Chapters 10–25   Algebra
- Chapters 26–38   Geometry and measure
- Chapters 39–44   Statistics

Two new chapters have been introduced. Chapter 45 consists of questions involving the functional elements of mathematics. Chapter 46 consists of questions involving the solution of mathematical problems.

The work is arranged in the above way to allow centres to plan a scheme of work and introduce topics in a manner which best suits the needs of their students and the requirements of the specification being used. It should provide the flexibility necessary to meet the wide range of needs of students both within a teaching group and across different teaching groups within the same centre.

Each chapter is presented in a way which will help students to understand the mathematics involved, with straightforward explanations and worked examples covering every type of problem.

- 'Check ups' are provided to check understanding of work already covered.
- 'Discoveries' encourage students to find out something for themselves, either from an external source such as the internet, or through a guided activity.
- 'Challenges' are more searching and are designed to make students think mathematically.
- There are a large number of exercises to work through to practise and develop skills.
- Some questions are designed to be done without a calculator in order to prepare students for a non-calculator paper.
- 'Tips' are included which give advice on improving examination performance, direct from experienced examiners who have prepared this book.
- At the end of each chapter there is a 'Mixed exercise', which helps students to revise all the topics covered in that chapter.

Other components in the series

- A Homework Book
  This contains parallel exercises to those in this book and gives more practice.
- A Teacher's Resource Website
  This online teaching resource contains helpful notes on teaching the course and the answers to all the exercises in the Student's and Homework Books.

Top ten tips

Here are some general tips from the examiners who wrote this book to help students to do well in examinations.

Practise

1   **taking time** to read and work through each question carefully.
2   answering questions **without** using a calculator.
3   answering questions which require **explanations**.
4   answering **unstructured** problems.
5   **accurate** drawing and constructions.
6   answering questions which **need a calculator**, trying to use the calculator efficiently.
7   **checking answers**, especially for reasonable size and degree of accuracy.
8   taking care over the **presentation of work**, making sure that all steps are clearly presented and that **the mathematical notation used is correct**.
9   checking that you have **answered the given question**.
10 **rounding numbers**, but only at the appropriate stage.

1 → INTEGERS, POWERS AND ROOTS

Prime numbers and factors

You already know that a **factor** of a number is any number that divides exactly into that number. This includes 1 and the number itself.

Check up 1.1

The factors of 2 are 1 and 2.
The factors of 22 are 1, 2, 11 and 22.

Write down all the factors of these numbers.

a) 14   b) 16   c) 40

Discovery 1.1

a) Write down all the factors of the other numbers from 1 to 20.

b) Write down all the numbers under 20 that have two, and only two, different factors.

The numbers you found in part **b)** of Discovery 1.1 are called **prime numbers**. Notice that 1 is not a prime number as it only has one factor.

> **TIP**
> It is useful to learn the prime numbers up to 50.

Writing a number as a product of its prime factors

When you multiply two or more numbers together, the result is a **product**.

When you write a number as a product of its prime factors, you work out which prime numbers are multiplied together to give the number.

The number 6 written as a product of its prime factors is 2×3.

It is easy to write down the prime factors of 6 because it is a small number. To write a larger number as a product of its prime factors, use this method.

- Try dividing the number by 2.
- If it divides by 2 exactly, try dividing by 2 again.
- Continue dividing by 2 until your answer will not divide by 2.
- Next try dividing by 3.
- Continue dividing by 3 until your answer will not divide by 3.
- Then try dividing by 5.
- Continue dividing by 5 until your answer will not divide by 5.
- Continue to work systematically through the prime numbers.
- Stop when your answer is 1.

EXAMPLE 1.1

a) Write 12 as a product of its prime factors.

b) Write 126 as a product of its prime factors.

Solution

a) $2\overline{)12}$
$2\overline{)6}$
$3\overline{)3}$
1

$12 = 2 \times 2 \times 3$

You already know that you can write 2×2 as 2^2.

So you can write $2 \times 2 \times 3$ in a shorter way as $2^2 \times 3$.

b) 2)126
   3)63
   3)21
   7)7
    1

$126 = 2 \times 3 \times 3 \times 7 = 2 \times 3^2 \times 7$

Remember 3^2 means 3 squared and this is the special name for 3 to the power 2.
The power, 2 in this case, is called the **index**.

> **TIP**
> Check your answer by multiplying the prime factors together. Your answer should be the original number.

EXERCISE 1.1

Write each of these numbers as a product of its prime factors.

1 6	**2** 10	**3** 15	**4** 21	**5** 32
6 36	**7** 140	**8** 250	**9** 315	**10** 420

Challenge 1.1

The factors of 24 are 1, 2, 3, 4, 6, 8, 12, 24.

This is eight different factors. You can write this as F(24) = 8.

24 written as a product of its prime factors is $2 \times 2 \times 2 \times 3 = 2^3 \times 3^1$.

(You do not usually include the index if it is 1 but you need it for this activity.)

Now add 1 to each of the indices: (3 + 1) = 4 and (1 + 1) = 2.
Then multiply these numbers: $4 \times 2 = 8$.

Your answer is the same as F(24), the number of factors of 24.

Here is another example.

The factors of 8 are 1, 2, 4, 8.

This is four different factors so F(8) = 4.

8 written as a product of its prime factors is 2^3.

There is just one power this time.

Add 1 to the index: (3 + 1) = 4.

This is the same as F(8), the number of factors of 8.

a) Try this for 40.

b) Investigate if there is a similar connection between the number of factors and the powers of the prime factors for some other numbers.

Highest common factors and lowest common multiples

The **highest common factor (HCF)** of a set of numbers is the largest number that will divide exactly into each of the numbers.

The largest number that will divide into both 8 and 12 is 4.
So 4 is the highest common factor of 8 and 12.

You can find the highest common factor of 8 and 12 without using any special methods. You list, perhaps mentally, the factors of 8 and 12 and compare the lists to find the largest number that appears in both lists.

When you can give the answer without using any special methods it is called finding **by inspection**.

Check up 1.3

Find, by inspection, the highest common factor (HCF) of each of these pairs of numbers.

a) 12 and 18

b) 27 and 36

c) 48 and 80

> The HCF is never bigger than the smaller of the numbers.

You probably found parts **a)** and **b)** of Check up 1.3 fairly easy but part **c)** more difficult.

This is the method to use when it is not easy to find the highest common factor by inspection.

- Write each number as a product of its prime factors.

- Find the common factors.

- Multiply them together.

This method is shown in the next example.

EXAMPLE 1.2

Find the highest common factor of each of these pairs of numbers.

a) 28 and 72   **b)** 96 and 180

Solution

a) Write each number as the product of its prime factors.

$28 = ② \times ② \times 7 \qquad\qquad = 2^2 \times 7$

$72 = ② \times ② \times 2 \times 3 \times 3 = 2^3 \times 3^2$

The common factors are 2 and 2.

The highest common factor is $2 \times 2 = 2^2 = 4$.

b) Write each number as the product of its prime factors.

$96 = ② \times ② \times 2 \times 2 \times 2 \times ③ = 2^5 \times 3$

$180 = ② \times ② \times ③ \times 3 \times 5 \qquad = 2^2 \times 3^2 \times 5$

The common factors are 2, 2 and 3.

The highest common factor is $2 \times 2 \times 3 = 2^2 \times 3 = 12$.

The **lowest common multiple (LCM)** of a set of numbers is the smallest number into which all the members of the set will divide.

The smallest number into which both 8 and 12 will divide is 24.
So 24 is the lowest common multiple of 8 and 12.

As for the highest common factor, you can find the lowest common multiple of small numbers by inspection. One way is to list the multiples of each of the numbers and compare the lists to find the smallest number that appears in both lists.

Check up 1.4

Find, by inspection, the lowest common multiple (LCM) of each of these pairs of numbers.

a) 3 and 5   **b)** 12 and 16   **c)** 48 and 80

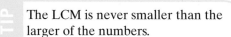

TIP   The LCM is never smaller than the larger of the numbers.

You probably found parts **a)** and **b)** of Check up 1.4 fairly easy but part **c)** more difficult.

This is the method to use when it is not easy to find the lowest common multiple by inspection.

- Write each number as a product of its prime factors.
- Find the highest power of each of the factors that occur in either of the lists.
- Multiply these numbers together.

This method is shown in the next example.

EXAMPLE 1.3

Find the lowest common multiple of each of these pairs of numbers.

a) 28 and 42   **b)** 96 and 180

Solution

a) Write each number as the product of its prime factors.

$28 = 2 \times 2 \times 7 = (2^2) \times 7$

$42 = 2 \times (3) \times (7)$

The highest power of 2 is 2^2.
The highest power of 3 is $3^1 = 3$.
The highest power of 7 is $7^1 = 7$.

The lowest common multiple is $2^2 \times 3 \times 7 = 84$.

Notice that a number can be written as that number to the power 1, for example 3 was written as 3^1. A number to the power 1 is equal to the number itself. $3^1 = 3$.

b) Write each number as the product of its prime factors.

$96 = 2 \times 2 \times 2 \times 2 \times 2 \times 3 = (2^5) \times 3$

$180 = 2 \times 2 \times 3 \times 3 \times 5 = 2^2 \times (3^2) \times (5)$

The highest power of 2 is 2^5.
The highest power of 3 is 3^2.
The highest power of 5 is $5^1 = 5$.

The lowest common multiple is $2^5 \times 3^2 \times 5 = 1440$.

Summary

- To find the highest common factor (HCF), use the prime numbers that appear in *both* lists and use the *lower* power for each prime.
- To find the lowest common multiple (LCM), use all the prime numbers that appear in the lists and use the *higher* power of each prime.

> **TIP**
>
> Check your answers.
>
> Does the HCF divide into both numbers?
>
> Do both numbers divide into the LCM?

For each of these pairs of numbers
- write the numbers as products of their prime factors.
- state the highest common factor.
- state the lowest common multiple.

1 4 and 6   **2** 12 and 16   **3** 10 and 15   **4** 32 and 40   **5** 35 and 45

6 27 and 63   **7** 20 and 50   **8** 48 and 84   **9** 50 and 64   **10** 42 and 49

Challenge 1.2

The students in Year 11 at a school are to be split into groups of equal size.
Two possible sizes for the groups are 16 and 22.

What is the smallest number of students that there can be in Year 11?

Multiplying and dividing by negative numbers

Discovery 1.2

a) Work out this sequence of calculations.

$$5 \times 5 = 25$$
$$5 \times 4 = 20$$
$$5 \times 3 =$$
$$5 \times 2 =$$
$$5 \times 1 =$$
$$5 \times 0 =$$

What is the pattern in the answers?
Use the pattern to continue the sequence.

$$5 \times -1 =$$
$$5 \times -2 =$$
$$5 \times -3 =$$
$$5 \times -4 =$$

b) Work out this sequence of calculations.

$$5 \times 4 =$$
$$4 \times 4 =$$
$$3 \times 4 =$$
$$2 \times 4 =$$
$$1 \times 4 =$$
$$0 \times 4 =$$

Spot the pattern and continue the sequence.

You should have found in Discovery 1.2 that a positive number multiplied by a negative number gives a negative answer.

Discovery 1.3

 Work out this sequence of calculations.

$-3 \times 5 =$
$-3 \times 4 =$
$-3 \times 3 =$
$-3 \times 2 =$
$-3 \times 1 =$
$-3 \times 0 =$

What is the pattern in the answers?
Use the pattern to continue the sequence.

$-3 \times -1 =$
$-3 \times -2 =$
$-3 \times -3 =$
$-3 \times -4 =$
$-3 \times -5 =$

Your answers to Discoveries 1.2 and 1.3 suggest these rules.

$$+ \times - = -$$
and
$$- \times + = -$$

$$+ \times + = +$$
and
$$- \times - = +$$

EXAMPLE 1.4

Work out these.

a) 6×-4 b) -7×-3 c) -5×8

Solution

a) $+ \times - = -$
$6 \times 4 = 24$
So $6 \times -4 = -24$

b) $- \times - = +$
$7 \times 3 = 21$
So $-7 \times -3 = +21 = 21$

c) $- \times + = -$
$5 \times 8 = 40$
So $-5 \times 8 = -40$

Discovery 1.4

$4 \times 3 = 12$ From this calculation you can say that $12 \div 4 = 3$ and $12 \div 3 = 4$.

$10 \times 6 = 60$ From this calculation you can say that $60 \div 6 = 10$ and $60 \div 10 = 6$.

In Example 1.4 you saw that $6 \times -4 = -24$.

In the same way as for the calculations above you can write down these two divisions.

$$-24 \div 6 = -4 \quad \text{and} \quad -24 \div -4 = 6$$

a) Work out 2×-9.
   Then write two divisions in the same way as above.

b) Work out -7×-4.
   Then write two divisions in the same way as above.

Your answers to Discovery 1.4 suggest these rules.

$+ \div - = -$	$+ \div + = +$
and	and
$- \div + = -$	$- \div - = +$

You now have a complete set of rules for multiplying and dividing positive and negative numbers.

$+ \times - = -$	$+ \times + = +$
$+ \div - = -$	$+ \div + = +$
$- \times + = -$	$- \times - = +$
$- \div + = -$	$- \div - = +$

Here is another way of thinking of these rules.

Signs different: answer negative	Signs the same: answer positive

EXAMPLE 1.5

Work out these.

a) 5×-3 **b)** -2×-3 **c)** $-10 \div 2$ **d)** $-15 \div -3$

Solution

First work out the signs. Then work out the numbers.

a) -15 $(+ \times - = -)$ **b)** $+6 = 6$ $(- \times - = +)$

c) -5 $(- \div + = -)$ **d)** $+5 = 5$ $(- \div - = +)$

You can extend the rules to calculations with more than two numbers.

If there is an even number of negative signs the answer is positive.
If there is an odd number of negative signs the answer is negative.

EXAMPLE 1.6

Work out $-2 \times 6 \div -4$.

Solution

You can work this out by taking each part of the calculation in turn.

$$-2 \times 6 = -12 \qquad (- \times + = -)$$
$$-12 \div -4 = 3 \qquad (- \div - = +)$$

Or you can count the number of negative signs and then work out the numbers.

There are two negative signs so the answer is positive.

$$-2 \times 6 \div -4 = 3$$

EXERCISE 1.3

Work out these.

1 4×3	**2** -5×4	**3** -6×-5
4 -9×6	**5** 4×-7	**6** -2×8
7 -3×-6	**8** $24 \div -6$	**9** $-25 \div -5$
10 $-32 \div 4$	**11** $18 \div 6$	**12** $-14 \div -7$
13 $-45 \div 5$	**14** $49 \div -7$	**15** $36 \div -9$
16 $6 \times 10 \div -5$	**17** $-84 \div -12 \times -3$	**18** $4 \times 9 \div -6$
19 $-3 \times -6 \div -2$	**20** $-6 \times 2 \times -5 \div -3$	

Challenge 1.3

Find the value of each of these expressions when $x = -3, y = 4$ and $z = -1$.

a) $5xy$ **b)** $x^2 + 2x$ **c)** $2y^2 - 2yz$

d) $3xz - 2xy + 3yz$ **e)** $4xyz$

Powers and roots

The **square** of a number is the number multiplied by itself.

For example, 2 squared is written 2^2 and equals $2 \times 2 = 4$.

The **cube** of a number is the number $\times$ number $\times$ number.

For example, 2 cubed is written 2^3 and equals $2 \times 2 \times 2 = 8$.

It is useful to know the squares of the numbers 1 to 15 and the cubes of the numbers 1 to 5 and 10.

Check up 1.5

a) What are the squares of the numbers 1 to 15?

b) What are the cubes of the numbers 1 to 5 and 10?

The squares of integers are called **square numbers**.

The cubes of integers are called **cube numbers**.

Because $4^2 = 4 \times 4 = 16$, the **square root** of 16 is 4.
This is written as $\sqrt{16} = 4$.

But $(-4)^2 = -4 \times -4 = 16$.
So the square root of 16 is also -4.

This is often written as $\sqrt{16} = \pm 4$.
Similarly $\sqrt{81} = \pm 9$ and so on.

In many practical problems where the answer is a square root, the negative answer has no meaning. You should state that the negative value has no meaning in the context and then only consider the positive answers.

Because $5^3 = 5 \times 5 \times 5 = 125$, the **cube root** of 125 is 5.
This is written as $\sqrt[3]{125} = 5$.
It can only be positive.

$(-5)^3 = -5 \times -5 \times -5 = -125$.
So $\sqrt[3]{-125} = -5$.

Finding the square root is the reverse operation to squaring. Finding the cube root is the reverse operation to cubing. So you can find the square roots and cube roots of the squares and cubes you know.

Make sure you also know how to use your calculator to work out square roots and cube roots.

TIP

A common error is to think that $1^2 = 2$ rather than 1.

EXAMPLE 1.7

a) Find the square root of 57. Give your answer to 2 decimal places.

b) Find the cube root of 86. Give your answer to 2 decimal places.

Solution

a) $\sqrt{57} = \pm 7.55$ You press $\boxed{\sqrt{}}\,\boxed{5}\,\boxed{7}$ on your calculator.

b) $\sqrt[3]{86} = 4.41$ You press $\boxed{\sqrt[3]{}}\,\boxed{8}\,\boxed{6}$ on your calculator.

◎ EXERCISE 1.4

Do not use your calculator for questions **1** and **2**.

1 Write down the value of each of these.

   a) 7^2 b) 11^2 c) $\sqrt{36}$ d) $\sqrt{144}$

   e) 2^3 f) 10^3 g) $\sqrt[3]{64}$ h) $\sqrt[3]{1}$

2 A square has an area of 36 cm². What is the length of one side?

You may use your calculator for questions **3** to **7**.

3 Find the square of each of these numbers.

   a) 25   b) 40   c) 35   d) 32   e) 1.2

4 Find the cube of each of these numbers.

   a) 12   b) 2.5   c) 6.1   d) 30   e) 5.4

5 Find the square root of each of these numbers.
   Where necessary, give your answer correct to 2 decimal places.

   a) 400   b) 575   c) 1284   d) 3684   e) 15 376

6 Find the cube root of each of these numbers.
   Where necessary, give your answer correct to 2 decimal places.

   a) 512   b) 676   c) 8000   d) 9463   e) 10 000

7 Find two numbers less than 200 which are both a square number and a cube number.

Challenge 1.4

a) The side of a square is 2.2 m long.
What is the area of the square?

b) A cube has edges of length 14 cm.
What is its volume?

c) A square has an area of 29 cm².
What is the length of each side? Give your answer correct to 2 decimal places.

d) The volume of a cube is 96 cm³.
What is the area of one face of the cube? Give your answer correct to 2 decimal places.

Discovery 1.5

$2^2 \times 2^5 = (2 \times 2) \times (2 \times 2 \times 2 \times 2 \times 2) = (2 \times 2 \times 2 \times 2 \times 2 \times 2 \times 2) = 2^7$

$3^5 \div 3^2 = (3 \times 3 \times 3 \times 3 \times 3) \div (3 \times 3) = (3 \times 3 \times 3) = 3^3$

Copy and complete the following.

a) $5^2 \times 5^3 = (5 \times 5) \times (..................) = (..................) = ...$

b) $2^4 \times 2^2 =$ **c)** $6^5 \times 6^3 =$ **d)** $5^5 \div 5^3 =$ **e)** $3^6 \div 3^3 =$ **f)** $7^5 \div 7^2 =$

What do you notice?

Your answers to Discovery 1.5 were examples of these two rules.

$$n^a \times n^b = n^{a+b} \quad \text{and} \quad n^a \div n^b = n^{a-b}$$

You have already met a number with an index of 1 in Example 1.3.
A number to the power 1 equals the number itself.

$n^1 = n$ For example, $3^1 = 3$.

Any number with an index 0 is 1.

$n^0 = 1$ For example, $3^0 = 1$.

> **TIP**
> To confirm this, put $a = b$ in $n^a \div n^b = n^{a-b}$.
> $n^a \div n^a = 1$ and $n^{a-a} = n^0$.

EXAMPLE 1.8

Write each of these as a single power of 3.

a) $3^4 \times 3^2$ **b)** $3^7 \div 3^2$ **c)** $\dfrac{3^5 \times 3}{3^6}$

Solution

a) When you multiply powers you add the indices.
$$3^4 \times 3^2 = 3^{4+2} = 3^6$$

b) When you divide powers you subtract the indices.
$$3^7 \div 3^2 = 3^{7-2} = 3^5$$

c) You can also combine operations.
$$\frac{3^5 \times 3}{3^6} = 3^{5+1-6} = 3^0$$

> **TIP**
>
> $3^0 = 1$ but you have been asked to write your answer as a power of 3 so you leave your answer as 3^0.
>
> If you are asked to simplify the expression, you write $3^0 = 1$.

EXERCISE 1.5

1 Write these in simpler form using indices.
  a) $3 \times 3 \times 3 \times 3 \times 3$
  b) $7 \times 7 \times 7$
  c) $3 \times 3 \times 3 \times 3 \times 5 \times 5$
    Hint: Write the 3s separately from the 5s.

2 Work out these, giving your answers in index form.
  a) $5^2 \times 5^3$ **b)** $10^5 \times 10^2$ **c)** 8×8^3
  d) $3^6 \times 3^4$ **e)** $2^5 \times 2$

3 Work out these, giving your answers in index form.
  a) $5^4 \div 5^2$ **b)** $10^5 \div 10^2$ **c)** $8^6 \div 8^3$
  d) $3^6 \div 3^4$ **e)** $2^3 \div 2^3$

4 Work out these, giving your answers in index form.
  a) $5^4 \times 5^2 \div 5^2$ **b)** $10^7 \times 10^6 \div 10^2$ **c)** $8^4 \times 8 \div 8^3$ **d)** $3^5 \times 3^3 \div 3^4$

5 Work out these, giving your answers in index form.
  a) $\dfrac{2^6 \times 2^3}{2^4}$ **b)** $\dfrac{3^6}{3^2 \times 3^2}$ **c)** $\dfrac{5^3 \times 5^4}{5 \times 5^2}$ **d)** $\dfrac{7^4 \times 7^4}{7^2 \times 7^3}$

Reciprocals

The **reciprocal** of a number is $\dfrac{1}{\text{the number}}$.

For example, the reciprocal of 2 is $\frac{1}{2}$.

The reciprocal of n is $\dfrac{1}{n}$.

The reciprocal of $\dfrac{1}{n}$ is n.

The reciprocal of $\dfrac{a}{b}$ is $\dfrac{b}{a}$.

0 does not have a reciprocal.

To find the reciprocal of a number without a calculator you divide 1 by the number. To find the reciprocal of a number with a calculator you use the $\boxed{x^{-1}}$ button.

EXAMPLE 1.9

Without using a calculator, find the reciprocal of each of these.

a) 5   **b)** $\frac{5}{8}$ **c)** $1\frac{1}{8}$

Solution

a) To find the reciprocal of a number, divide 1 by the number.

The reciprocal of 5 is $\frac{1}{5}$ or 0.2.

b) The reciprocal of $\frac{5}{8}$ is $\frac{8}{5} = 1\frac{3}{5}$.

Note: You should always convert improper fractions to mixed numbers unless you are told not to.

c) First convert $1\frac{1}{8}$ to an improper fraction: $1\frac{1}{8} = \frac{9}{8}$

The reciprocal of $\frac{9}{8}$ is $\frac{8}{9}$.

EXAMPLE 1.10

Use your calculator to find the reciprocal of 1.25.
Give your answer as a decimal.

Solution

This is the sequence of keys to press.

$\boxed{1}\ \boxed{\cdot}\ \boxed{2}\ \boxed{5}\ \boxed{x^{-1}}\ \boxed{=}$

The display should read 0.8.

Write down the reciprocal of each of these numbers.

a) 2   **b)** 5   **c)** 10   **d)** $\frac{3}{5}$

Discovery 1.6

a) Multiply each of the numbers in Check up 1.6 by its reciprocal.
What do you notice about your answers?

b) Now try these products on your calculator.

    (i) 55×2 (press $=$) $\times \frac{1}{2}$ (press $=$)

    (ii) 15×4 (press $=$) $\times \frac{1}{4}$ (press $=$)

    (iii) 8×10 (press $=$) $\times 0.1$ (press $=$)

    What do you notice about your answers?

c) Try some more calculations and explain what is happening.

Challenge 1.5

An **inverse operation** takes you back to the previous number.

Multiplying by a number and multiplying by its reciprocal are inverse operations.

Write down as many operations and their inverse operations as you can.

EXERCISE 1.6

Do not use your calculator for questions **1** to **3**.

1 Write down the reciprocal of each of these numbers.

a) 3   b) 6   c) 49   d) 100   e) 640

2 Write down the numbers of which these are the reciprocals.

a) $\frac{1}{16}$ b) $\frac{1}{9}$ c) $\frac{1}{52}$ d) $\frac{1}{67}$ e) $\frac{1}{1000}$

3 Find the reciprocal of each of these numbers.
Give your answers as fractions or mixed numbers.

a) $\frac{4}{5}$ b) $\frac{3}{8}$ c) $1\frac{3}{5}$ d) $3\frac{1}{3}$ e) $\frac{2}{25}$

You may use your calculator for question **4**.

4 Find the reciprocal of each of these numbers.
Give your answers as decimals.

a) 2.5   b) 0.5   c) 125   d) 0.16   e) 3.2

WHAT YOU HAVE LEARNED

- **A prime number has two factors only, 1 and itself**
- **How to write a number as the product of its prime factors**
- **The highest common factor (HCF) of a set of numbers is the largest number that will divide exactly into each of the numbers**
- **How to find the highest common factor of a pair of numbers using prime factors**
- **The lowest common multiple (LCM) of a set of numbers is the smallest number into which all the members of the set will divide**
- **How to find the lowest common multiple using prime factors**
- **That, when multiplying or dividing positive and negative numbers,**

$$+ \times + = + \qquad - \times - = + \qquad + \times - = - \qquad - \times + = -$$
$$+ \div + = + \qquad - \div - = + \qquad + \div - = - \qquad - \div + = -$$

- **$5^3 = 5 \times 5 \times 5 = 125$, so the cube root of 125 is 5**
- **That, when multiplying and dividing powers,**
$$n^a \times n^b = n^{a+b} \quad \text{and} \quad n^a \div n^b = n^{a-b}$$
- **The reciprocal of a number is 1 divided by the number: the reciprocal of n is $\dfrac{1}{n}$**
- **The reciprocal of $\dfrac{a}{b}$ is $\dfrac{b}{a}$**
- **0 does not have a reciprocal**

1 Write each of these numbers as a product of its prime factors.

  a) 75   **b)** 140   **c)** 420

2 For each of these pairs of numbers
  • write the numbers as products of their prime factors.
  • state the highest common factor.
  • state the lowest common multiple.

  a) 24 and 60   **b)** 100 and 150   **c)** 81 and 135

 Do not use your calculator for questions **3** to **6**.

3 Work out these.

  a) 4×-3 **b)** -2×8 **c)** $-48 \div -6$ **d)** $2 \times -6 \div -4$

4 Write down the square and the cube of each of these numbers.

  a) 4   **b)** 6   **c)** 10

5 Write down the square root of each of these numbers.

  a) 64   **b)** 196

6 Write down the cube root of each of these numbers.

  a) 125   **b)** 27

 You may use your calculator for questions **7** and **8**.

7 Find the square and the cube of each of these numbers.

  a) 4.6   **b)** 21   **c)** 2.9

8 Find the square root and the cube root of each of these numbers.
  Give your answers correct to 2 decimal places.

  a) 89   **b)** 124   **c)** 986

9 Work out these, giving your answers in index form.

  a) $5^5 \times 5^2$ **b)** $10^5 \div 10^2$ **c)** $8^4 \times 8^3 \div 8^5$

  d) $\dfrac{2^4 \times 2^4}{2^2}$ **e)** $\dfrac{3^9}{3^4 \times 3^2}$

10 Find the reciprocal of each of these numbers.

  a) 5   **b)** 8   **c)** $\frac{1}{8}$

  d) 0.1   **e)** 1.6

2 → FRACTIONS, DECIMALS AND PERCENTAGES

Comparing fractions

Sometimes it is obvious which of two fractions is the bigger. If not, the best way is to use equivalent fractions.

To find suitable equivalent fractions, you first need to look for a number that the two denominators (bottom numbers) will divide into exactly.

For example, if you want to compare $\frac{1}{2}$ and $\frac{1}{3}$ you need to find a number that 2 and 3 divide into exactly. Because 2 and 3 are both factors of 6, you can convert both fractions into sixths.

To convert $\frac{1}{2}$ into sixths you need to multiply both the denominator and the numerator by $6 \div 2 = 3$.

To convert $\frac{1}{3}$ into sixths you need to multiply both the denominator and the numerator by $6 \div 3 = 2$.

$$\frac{1 \times 3}{2 \times 3} = \frac{3}{6} \text{ and } \frac{1 \times 2}{3 \times 2} = \frac{2}{6}$$

Now that both fractions are expressed as sixths, you can easily tell which is the bigger by looking at the numerators.

Fractions with the same denominator are said to have a **common denominator**.

EXAMPLE 2.1

Which is the bigger, $\frac{3}{4}$ or $\frac{5}{6}$?

Solution

First find a common denominator, 24 is an obvious one, as $4 \times 6 = 24$, but a smaller one is 12.

12 is the lowest common multiple (LCM) of 4 and 6. You learned how to find LCMs in Chapter 1.

Then convert both fractions into twelfths.

$$\frac{3 \times 3}{4 \times 3} = \frac{9}{12} \qquad \frac{5 \times 2}{6 \times 2} = \frac{10}{12}$$

$\frac{10}{12}$ is bigger than $\frac{9}{12}$, so $\frac{5}{6}$ is bigger than $\frac{3}{4}$.

> **TIP**
> Multiplying the two denominators together will always work to find a common denominator but the LCM is sometimes smaller.

Check up 2.1

Which of these fractions is the bigger?

a) $\frac{3}{4}$ or $\frac{5}{8}$ **b)** $\frac{7}{9}$ or $\frac{5}{6}$ **c)** $\frac{3}{10}$ or $\frac{4}{15}$

> **TIP**
> In each case the LCM is smaller than the number you get by simply multiplying the two denominators together.

Check up 2.2

Put these fractions in order, smallest first.

$\frac{2}{5}$ $\frac{1}{2}$ $\frac{9}{20}$ $\frac{17}{40}$ $\frac{3}{8}$

Adding and subtracting fractions and mixed numbers

You can only add and subtract fractions if they have a common denominator. Sometimes this means you have to find the common denominator first.

Adding and subtracting fractions with a common denominator

This rectangle is divided into twelfths.

$\frac{4}{12}$ of the rectangle is shaded blue and $\frac{3}{12}$ is shaded red. The total fraction shaded is $\frac{7}{12}$.

This shows that $\frac{4}{12} + \frac{3}{12} = \frac{7}{12}$.

To add fractions with a common denominator, simply add the numerators.

Do *not* add the denominators.

You subtract fractions in a similar way.

You may need to cancel the answer, to give the fraction in its lowest terms.

For example, $\frac{7}{12} - \frac{5}{12} = \frac{2}{12} = \frac{1}{6}$.

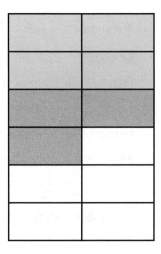

 TIP
Unless you are told not to, always cancel your answer.

Adding and subtracting fractions with different denominators

To add and subtract fractions with different denominators, you use the same method as when comparing fractions to convert to equivalent fractions with a common denominator.

EXAMPLE 2.2

Work out $\frac{3}{8} + \frac{1}{4}$.

Solution

First find the common denominator. The LCM of 4 and 8 is 8.
$\frac{3}{8}$ already has a denominator of 8.

$\frac{1}{4} = \frac{2}{8}$ Multiply the numerator and the denominator by 2.

$\frac{3}{8} + \frac{1}{4} = \frac{3}{8} + \frac{2}{8} = \frac{5}{8}$ Add the numerators only.

TIP
Remember that you add the numerators but not the denominators.

EXAMPLE 2.3

Work out $\frac{2}{3} - \frac{3}{5}$.

Solution

First find the common denominator. The LCM of 3 and 5 is 15.

$\frac{2}{3} = \frac{10}{15}$ Multiply the numerator and the denominator by 5.

$\frac{3}{5} = \frac{9}{15}$ Multiply the numerator and the denominator by 3.

$\frac{10}{15} - \frac{9}{15} = \frac{1}{15}$ Subtract the numerators only.

EXAMPLE 2.4

Work out $\frac{3}{4} + \frac{2}{5}$.

Solution

The lowest common denominator is 20.

$\frac{3}{4} + \frac{2}{5} = \frac{15}{20} + \frac{8}{20}$

$\qquad\quad = \frac{23}{20}$ $\frac{23}{20}$ is an improper ('top-heavy') fraction.

$\qquad\quad = 1\frac{3}{20}$ You need to change it to a mixed number.

Adding and subtracting mixed numbers

To add mixed numbers you add the whole number parts and then add the fraction parts.

EXAMPLE 2.5

Work out $1\frac{1}{4} + 2\frac{1}{2}$.

Solution

$1\frac{1}{4} + 2\frac{1}{2} = 1 + 2 + \frac{1}{4} + \frac{1}{2}$

$\qquad\qquad = 3 + \frac{1}{4} + \frac{1}{2}$ Add the whole numbers first.

$\qquad\qquad = 3 + \frac{1}{4} + \frac{2}{4}$ Change the fractions into equivalent fractions with a common denominator.

$\qquad\qquad = 3\frac{3}{4}$ Add the fractions.

EXAMPLE 2.6

Work out $2\frac{3}{5} + 4\frac{2}{3}$.

Solution

$$2\frac{3}{5} + 4\frac{2}{3} = 6 + \frac{3}{5} + \frac{2}{3} \quad \text{Add the whole numbers first.}$$

$$= 6 + \frac{9}{15} + \frac{10}{15} \quad \text{Change the fractions into equivalent fractions with a common denominator.}$$

$$= 6 + \frac{19}{15} \quad \text{Add the fractions. } \frac{19}{15} \text{ is an improper fraction. You need to change it into a mixed number and add the whole number to the 6 you already have.}$$

$$= 7\frac{4}{15} \quad \frac{19}{15} = 1\frac{4}{15} \text{ and } 6 + 1 = 7.$$

You subtract mixed numbers in a similar way.

EXAMPLE 2.7

Work out $3\frac{3}{4} - 1\frac{1}{3}$.

Solution

$$3\frac{3}{4} - 1\frac{1}{3} = 3 - 1 + \frac{3}{4} - \frac{1}{3} \quad \text{Split the calculation into two parts.}$$

$$= 2 + \frac{3}{4} - \frac{1}{3} \quad \text{Subtract the whole numbers first.}$$

$$= 2 + \frac{9}{12} - \frac{4}{12} \quad \text{Change the fractions into equivalent fractions with a common denominator.}$$

$$= 2\frac{5}{12} \quad \text{Subtract the fractions.}$$

EXAMPLE 2.8

Work out $5\frac{3}{10} - 2\frac{3}{4}$.

Solution

$$5\frac{3}{10} - 2\frac{3}{4} = 5 - 2 + \frac{3}{10} - \frac{3}{4} \quad \text{Split the calculation into two parts.}$$

$$= 3 + \frac{3}{10} - \frac{3}{4} \quad \text{Subtract the whole numbers first.}$$

$$= 3 + \frac{6}{20} - \frac{15}{20} \quad \text{Change the fractions into equivalent fractions with a common denominator.}$$

$$= 2 + \frac{20}{20} + \frac{6}{20} - \frac{15}{20} \quad \frac{6}{20} \text{ is smaller than } \frac{15}{20} \text{ and would give a negative answer. Take 1 of the whole units and change it to } \frac{20}{20}.$$

$$= 2 + \frac{26}{20} - \frac{15}{20} \quad \text{Add it to } \frac{6}{20}.$$

$$= 2\frac{11}{20} \quad \text{Subtract the fractions.}$$

1 For each pair of fractions
- find the common denominator.
- state which is the bigger fraction.

a) $\frac{2}{3}$ or $\frac{7}{9}$ **b)** $\frac{5}{6}$ or $\frac{7}{8}$ **c)** $\frac{3}{8}$ or $\frac{7}{20}$

2 Work out these.

a) $\frac{2}{9} + \frac{5}{9}$ **b)** $\frac{4}{11} + \frac{3}{11}$ **c)** $\frac{5}{12} - \frac{1}{12}$ **d)** $\frac{7}{13} - \frac{2}{13}$

e) $\frac{7}{12} + \frac{3}{12}$ **f)** $\frac{5}{8} + \frac{4}{8}$ **g)** $\frac{8}{9} - \frac{5}{9}$ **h)** $\frac{7}{10} + \frac{9}{10}$

i) $1\frac{5}{12} + 2\frac{1}{12}$ **j)** $3\frac{5}{8} - 1\frac{3}{8}$ **k)** $4\frac{5}{9} - \frac{4}{9}$ **l)** $5\frac{4}{7} - 2\frac{5}{7}$

3 Work out these.

a) $\frac{1}{2} + \frac{3}{8}$ **b)** $\frac{4}{9} + \frac{1}{3}$ **c)** $\frac{5}{6} - \frac{1}{4}$ **d)** $\frac{11}{12} - \frac{2}{3}$

e) $\frac{4}{5} + \frac{1}{2}$ **f)** $\frac{5}{7} + \frac{3}{4}$ **g)** $\frac{8}{9} - \frac{1}{6}$ **h)** $\frac{7}{10} + \frac{4}{5}$

i) $\frac{8}{9} + \frac{5}{6}$ **j)** $\frac{7}{15} + \frac{3}{10}$ **k)** $\frac{4}{9} - \frac{1}{12}$ **l)** $\frac{7}{20} + \frac{5}{8}$

4 Work out these.

a) $3\frac{1}{2} + 2\frac{1}{5}$ **b)** $4\frac{7}{8} - 1\frac{3}{4}$ **c)** $4\frac{2}{7} + \frac{1}{2}$ **d)** $6\frac{5}{12} - 3\frac{1}{3}$

e) $4\frac{3}{4} + 2\frac{5}{8}$ **f)** $5\frac{5}{6} - 1\frac{1}{4}$ **g)** $4\frac{7}{9} + 2\frac{5}{6}$ **h)** $4\frac{7}{13} - 4\frac{1}{2}$

i) $3\frac{5}{7} + 2\frac{1}{3}$ **j)** $7\frac{2}{5} - 1\frac{3}{4}$ **k)** $5\frac{2}{7} - 3\frac{1}{2}$ **l)** $4\frac{1}{12} - 3\frac{1}{4}$

Challenge 2.1

Brittany has some sweets.

$\frac{1}{4}$ of her sweets are red, $\frac{2}{5}$ are yellow and the rest are orange.

What fraction are orange?

Challenge 2.2

Simon says that $\frac{1}{3}$ of his class come to school by car, $\frac{1}{6}$ walk and $\frac{5}{8}$ come on the bus.

Show how you know that he must be wrong.

Challenge 2.3

Find a formula to add these fractions.

$$\frac{a}{b} + \frac{c}{d}$$

Multiplying and dividing fractions and mixed numbers

You already know how to multiply a fraction by a whole number. You can also multiply two fractions together.

Multiplying proper fractions

You know that $\frac{1}{3}$ is the same as $1 \div 3$.

To multiply another fraction, for example $\frac{2}{5}$ by $\frac{1}{3}$, you divide $\frac{2}{5}$ by 3.

The diagram shows $\frac{2}{5}$ divided by 3, which is the same as $\frac{1}{3}$ of $\frac{2}{5}$.

$\frac{1}{3}$ of $\frac{2}{5}$ is $\frac{2}{15}$.

Notice that $1 \times 2 = 2$ (the numerators) and $3 \times 5 = 15$ (the denominators).

So $\frac{1}{3} \times \frac{2}{5} = \frac{1 \times 2}{3 \times 5} = \frac{2}{15}$.

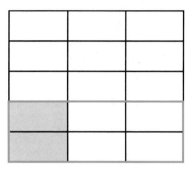

To multiply fractions you

> multiply the numerators and multiply the denominators.

EXAMPLE 2.9

Work out $\frac{2}{3} \times \frac{5}{7}$.

Solution

$$\frac{2}{3} \times \frac{5}{7} = \frac{2 \times 5}{3 \times 7} = \frac{10}{21}$$

EXAMPLE 2.10

Find $\frac{3}{4}$ of $\frac{6}{7}$.

Solution

$\frac{3}{4} \times \frac{6}{7} = \frac{18}{28}$ 'of' means the same as '×'.

$\frac{18}{28} = \frac{9}{14}$ Cancel by dividing the numerator and the denominator by 2.

Look again at Example 2.10. $\frac{3}{4} \times \frac{6}{7}$

The numbers 4 and 6 are both multiples of 2.

This means that you can cancel before you multiply the fractions. This makes the arithmetic easier.

$\frac{3}{\cancel{4}_2} \times \frac{\cancel{6}^3}{7} = \frac{9}{14}$ Divide both 4 and 6 by 2, then multiply the numerators and denominators.

EXAMPLE 2.11

Work out $4 \times \frac{3}{10}$.

Solution

$4 \times \frac{3}{10} = \frac{4}{1} \times \frac{3}{10}$ First write 4 as $\frac{4}{1}$.

$\qquad = \frac{\cancel{4}^2}{1} \times \frac{3}{\cancel{10}_5}$ Cancel by dividing the 4 and the 10 by 2.

$\qquad = \frac{6}{5} = 1\frac{1}{5}$

Dividing proper fractions

When you work out $6 \div 3$, you are finding how many 3s there are in 6.

Finding $6 \div \frac{1}{3}$ is the same as finding how many $\frac{1}{3}$s there are in 6, which is $6 \times 3 = 18$.

So dividing by $\frac{1}{3}$ is the same as multiplying by 3.

Notice that $\frac{1}{3}$ is the reciprocal of 3.

To find $6 \div \frac{2}{3}$, you need to multiply by 3 and also divide by 2, because there will be half as many $\frac{2}{3}$s as there are $\frac{1}{3}$s.
That means you multiply by $\frac{3}{2}$, the reciprocal of $\frac{2}{3}$.

$\qquad 6 \div \frac{2}{3} = \frac{6}{1} \times \frac{3}{2} = \frac{18}{2} = 9$

Dividing by a fraction is the same as multiplying by the reciprocal of the fraction.

TIP

The reciprocal of a fraction is a fraction with the numerator and denominator interchanged. You can think of this as 'turning the fraction upside down'.

EXAMPLE 2.12

Work out $\frac{3}{4} \div \frac{2}{7}$.

Solution

$\frac{3}{4} \div \frac{2}{7} = \frac{3}{4} \times \frac{7}{2}$ The reciprocal of $\frac{2}{7}$ is $\frac{7}{2}$.

$\qquad = \frac{21}{8}$ Multiply the numerators and denominators.

$\qquad = 2\frac{5}{8}$ Change the improper fraction into a mixed number.

EXAMPLE 2.13

Work out $\frac{5}{8} \div \frac{3}{4}$.

Solution

$\frac{5}{8} \div \frac{3}{4} = \frac{5}{8} \times \frac{4}{3}$ The reciprocal of $\frac{3}{4}$ is $\frac{4}{3}$.

$\qquad = \frac{5}{\underset{2}{\cancel{8}}} \times \frac{\cancel{4}^{1}}{3}$ Cancel by dividing the 4 and the 8 by 4.

$\qquad = \frac{5}{6}$

> **TIP**
>
> Never cancel fractions at the divide stage. Wait until it has changed to a multiplication.

Challenge 2.4

a) Calculate the area of this rectangle.

b) Find the perimeter of this rectangle.

Give your answers in their lowest terms.

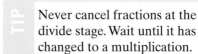

$5\frac{1}{4}$ cm

$3\frac{2}{3}$ cm

Reciprocals

Discovery 2.1

Work out these.

a) $1 \div \frac{3}{4}$ **b)** $1 \div \frac{5}{6}$ **c)** $1 \div \frac{5}{3}$

What do you notice?

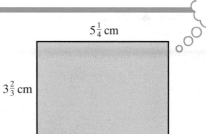

In Chapter 1 you learned that the reciprocal of a number is 1 ÷ the number. You can see now that this definition also applies to fractions.

Your calculator has a reciprocal button. It may be labelled $\boxed{x^{-1}}$.

Use your calculator to try to work out the reciprocal of 0 (zero).

You should get 'error'. This is because you cannot divide by zero. Zero has no reciprocal.

Multiplying and dividing mixed numbers

When multiplying and dividing mixed numbers, you first have to change the mixed numbers into improper fractions.

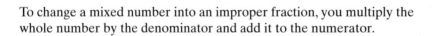

Discovery 2.2

a) (i) How many halves are there in two whole units?
   (ii) How many halves are there in $2\frac{1}{2}$?

b) (i) How many quarters are there in three whole units?
   (ii) How many quarters are there in $3\frac{3}{4}$?

c) (i) How many fifths are there in two whole units?
   (ii) How many fifths are there in $2\frac{4}{5}$?

What do you notice about your answers to part (ii) of these questions?

To change a mixed number into an improper fraction, you multiply the whole number by the denominator and add it to the numerator.

EXAMPLE 2.14

Change $3\frac{2}{3}$ into an improper fraction.

Solution

$3\frac{2}{3} = \dfrac{3 \times 3 + 2}{3}$ Multiply the whole number (3) by the denominator (3) and add it to the numerator (2). This gives you the numerator of the improper fraction. The denominator stays the same.

$= \dfrac{11}{3}$

EXAMPLE 2.15

Change $4\frac{3}{5}$ to an improper fraction.

Solution

$4\frac{3}{5} = \dfrac{4 \times 5 + 3}{5}$

$= \dfrac{23}{5}$

Multiplying and dividing mixed numbers is the same as multiplying and dividing fractions, once you have changed the mixed numbers to improper fractions.

EXAMPLE 2.16

Work out $2\frac{1}{2} \times 4\frac{3}{5}$.

Solution

$2\frac{1}{2} \times 4\frac{3}{5} = \frac{5}{2} \times \frac{23}{5}$ First change the mixed numbers into improper fractions.

$\qquad = \frac{1\cancel{5}}{2} \times \frac{23}{\cancel{5}_1}$ Cancel out the two 5s. This makes the arithmetic much easier.

$\qquad = \frac{23}{2}$ Multiply the numerator and the denominator.

$\qquad = 11\frac{1}{2}$ Give your answer as a mixed number.

EXAMPLE 2.17

Work out $2\frac{3}{4} \div 1\frac{5}{8}$.

Solution

$2\frac{3}{4} \div 1\frac{5}{8} = \frac{11}{4} \div \frac{13}{8}$ Change the mixed numbers into improper fractions. You must do this before turning the calculation into a multiplication.

$\qquad = \frac{11}{1\cancel{4}} \times \frac{\cancel{8}^2}{13}$ The reciprocal of $\frac{13}{8}$ is $\frac{8}{13}$. The numbers 4 and 8 are both multiples of 4.

$\qquad = \frac{22}{13}$

$\qquad = 1\frac{9}{13}$ Give your answer as a mixed number.

> **TIP**
> If you are multiplying or dividing by a whole number, for example 6, you can write it as $\frac{6}{1}$.

 EXERCISE 2.2

1   Change these mixed numbers to improper fractions.

   a) $4\frac{3}{4}$ b) $5\frac{2}{3}$ c) $6\frac{1}{2}$ d) $2\frac{5}{8}$

   e) $3\frac{2}{7}$ f) $1\frac{5}{12}$ g) $2\frac{5}{6}$ h) $5\frac{7}{11}$

2 Work out these.
Write your answers as proper fractions or mixed numbers in their lowest terms.

a) $\frac{3}{5} \times 4$ b) $\frac{3}{4} \times 6$ c) $\frac{2}{3} \div 5$

d) $7 \times \frac{5}{8}$ e) $\frac{5}{7} \div 3$ f) $6 \div \frac{2}{3}$

3 Work out these.
Write your answers as proper fractions or mixed numbers in their lowest terms.

a) $\frac{1}{2} \times \frac{3}{8}$ b) $\frac{4}{9} \times \frac{1}{3}$ c) $\frac{5}{6} \times \frac{1}{4}$ d) $\frac{11}{12} \div \frac{2}{3}$

e) $\frac{4}{5} \div \frac{1}{2}$ f) $\frac{5}{7} \times \frac{3}{4}$ g) $\frac{8}{9} \times \frac{1}{6}$ h) $\frac{7}{10} \div \frac{4}{5}$

i) $\frac{8}{9} \times \frac{5}{6}$ j) $\frac{7}{15} \div \frac{3}{10}$ k) $\frac{4}{9} \div \frac{1}{12}$ l) $\frac{7}{20} \times \frac{5}{8}$

4 Work out these.
Write your answers as proper fractions or mixed numbers in their lowest terms.

a) $3\frac{1}{2} \times 2\frac{1}{5}$ b) $4\frac{2}{7} \times \frac{1}{2}$ c) $2\frac{3}{4} \div 1\frac{3}{4}$ d) $1\frac{5}{12} \div 3\frac{1}{3}$

e) $3\frac{1}{5} \times 2\frac{5}{8}$ f) $2\frac{7}{8} \div 1\frac{3}{4}$ g) $2\frac{7}{9} \times 3\frac{3}{5}$ h) $5\frac{5}{6} \div 1\frac{3}{4}$

i) $3\frac{5}{7} \times 2\frac{1}{13}$ j) $5\frac{2}{5} \div 2\frac{1}{4}$ k) $5\frac{2}{7} \times 3\frac{1}{2}$ l) $4\frac{1}{12} \div 3\frac{1}{4}$

Challenge 2.5

Work out these.

a) $\left(3\frac{1}{2} + 2\frac{4}{5}\right) \times 2\frac{1}{12}$ b) $5\frac{1}{3} \div 3\frac{3}{5} + 2\frac{1}{3}$

c) $4\frac{2}{3} \times 2\frac{2}{7} - 4\frac{7}{8}$ d) $\left(2\frac{4}{5} + 3\frac{1}{4}\right) \div \left(3\frac{1}{3} - 2\frac{3}{4}\right)$

Fractions on your calculator

You need to be able to calculate with fractions without a calculator.
However, when a calculator is allowed, you can use the fraction button.

The fraction button looks like this $\boxed{a^b/c}$.

To enter a fraction such as $\frac{2}{5}$ into your calculator you press $\boxed{2}$ $\boxed{a^b/c}$ $\boxed{5}$ $\boxed{=}$.

Your display will look like this. $\boxed{2 \lrcorner 5}$

This is the calculator's way of showing the fraction $\frac{2}{5}$.

Discovery 2.3

Some calculators may have the ⌐ symbol a different way round.

Check now what you see when you press $\boxed{2}$ $\boxed{a^b/c}$ $\boxed{5}$ $\boxed{=}$.

To do a calculation like $\frac{2}{5} + \frac{1}{2}$, the sequence of buttons is

[2] [a$^{b/c}$] [5] [+] [1] [a$^{b/c}$] [2] [=].

This is what you should see on your display.   [9⌐10]

You must, of course, write this down as $\frac{9}{10}$ for your answer.

EXAMPLE 2.18

Use your calculator to work out $\frac{3}{4} + \frac{5}{6}$.

Solution

This is the sequence of buttons to press.

[3] [a$^{b/c}$] [4] [+] [5] [a$^{b/c}$] [6] [=]

The display on your calculator should look like this.   [1⌐7⌐12]

This is the calculator's way of showing the mixed number $1\frac{7}{12}$.

So the answer is $1\frac{7}{12}$.

To enter a mixed number such as $2\frac{3}{5}$ into your calculator you press

[2] [a$^{b/c}$] [3] [a$^{b/c}$] [5] [=].

Your display will look like this.   [2⌐3⌐5]

EXAMPLE 2.19

Use your calculator to work out these.

a) $2\frac{3}{5} - 1\frac{1}{4}$ **b)** $2\frac{2}{3} \times 3\frac{3}{4}$

Solution

a) This is the sequence of buttons to press.

[2] [a$^{b/c}$] [3] [a$^{b/c}$] [5] [−] [1] [a$^{b/c}$] [1] [a$^{b/c}$] [4] [=]

The display on your calculator should look like this.   [1⌐7⌐20]

So the answer is $1\frac{7}{20}$.

b) This is the sequence of buttons to press.

[2] [a$^{b/c}$] [2] [a$^{b/c}$] [3] [×] [3] [a$^{b/c}$] [3] [a$^{b/c}$] [4] [=]

The answer is 10.

Cancelling fractions

To **cancel** fractions to their **lowest terms** you divide the numerator and the denominator by the same number.

For example $\frac{8}{12} = \frac{2}{3}$ (by dividing both the numerator and the denominator by 4).

You can also do this on a calculator.

When you press 8 a^b/c 1 2, you should see $8 \lrcorner 12$.

When you press =, the display changes to $2 \lrcorner 3$, meaning $\frac{2}{3}$.

When you do calculations with fractions on your calculator, it will automatically give the answer as a fraction in its lowest terms.

If you do a calculation which is a mixture of fractions and decimals, your calculator will give the answer as a decimal.

EXAMPLE 2.20

Use your calculator to work out $2\frac{3}{4} \times 1.5$.

Solution

This is the sequence of buttons to press.

2 a^b/c 3 a^b/c 4 $\times$ 1 $\cdot$ 5 =

The answer is 4.125.

Improper fractions

If you enter an improper fraction into your calculator and press the = button, the calculator will automatically change it to a mixed number.

EXAMPLE 2.21

Use your calculator to change $\frac{187}{25}$ to a mixed number.

Solution

This is the sequence of buttons to press.

1 8 7 a^b/c 2 5 =

The display on your calculator should look like this. $7 \lrcorner 12 \lrcorner 25$

So the answer is $7\frac{12}{25}$.

1 Work out these.

a) $\frac{2}{7} + \frac{1}{3}$ **b)** $\frac{3}{4} - \frac{2}{5}$ **c)** $\frac{5}{8} \times \frac{4}{11}$ **d)** $\frac{11}{12} \div \frac{5}{8}$

e) $2\frac{3}{7} + 3\frac{1}{2}$ **f)** $5\frac{2}{3} - 3\frac{3}{4}$ **g)** $4\frac{2}{7} \times 3$ **h)** $5\frac{7}{8} \div 1\frac{5}{6}$

2 Write these fractions in their lowest terms.

a) $\frac{24}{60}$ **b)** $\frac{35}{56}$ **c)** $\frac{84}{180}$ **d)** $\frac{175}{400}$ **e)** $\frac{18}{162}$

3 Write these improper fractions as mixed numbers.

a) $\frac{124}{60}$ **b)** $\frac{130}{17}$ **c)** $\frac{73}{15}$ **d)** $\frac{168}{35}$ **e)** $\frac{107}{13}$

4 Calculate

a) the perimeter of this rectangle.

b) the area of this rectangle.

$6\frac{3}{4}$ cm

$3\frac{2}{3}$ cm

Changing fractions to decimals

Since a fraction like $\frac{5}{8}$ means the same as $5 \div 8$, you can use division to change a fraction into a decimal.

EXAMPLE 2.22

Convert $\frac{5}{8}$ to a decimal.

Solution

First write 5 as 5.000.   You may need more or fewer zeros depending on the fraction.

Now work out $5.000 \div 8$.

$$\begin{array}{r} 0.6\ 2\ 5 \\ 8\overline{)5.0^20^40} \end{array}$$

If the division is not exact, you may need to round your answer to a given number of decimal places.

Challenge 2.6

Use the method of Example 2.22 to convert $\frac{1}{3}$ into a decimal.

How is this different from the example?

Some fractions, such as $\frac{5}{8}$, convert to decimals which stop. These are **terminating decimals**. Others, such as $\frac{1}{3}$, just keep going. These are called **recurring decimals**.

Discovery 2.4

Convert the following fractions into decimals.

a) $\frac{1}{2}$ **b)** $\frac{1}{3}$ **c)** $\frac{3}{4}$ **d)** $\frac{2}{5}$

e) $\frac{5}{6}$ **f)** $\frac{2}{7}$ **g)** $\frac{7}{8}$ **h)** $\frac{8}{9}$

Try some more conversions of your own.

What can you say about the numbers in the denominators of the fractions that give terminating decimals?

EXAMPLE 2.23

State whether each of these fractions gives a terminating or a recurring decimal.

a) $\frac{1}{6}$ **b)** $\frac{1}{5}$ **c)** $\frac{1}{7}$ **d)** $\frac{1}{11}$

Solution

a) $\frac{1}{6}$ is a recurring decimal $1 \div 6 = 0.166\,666\,\ldots$

b) $\frac{1}{5}$ is a terminating decimal $1 \div 5 = 0.2$

c) $\frac{1}{7}$ is a recurring decimal $1 \div 7 = 0.142\,857\,142\,\ldots$

d) $\frac{1}{11}$ is a recurring decimal $1 \div 11 = 0.090\,909\,\ldots$

EXERCISE 2.4

1 Change each of these fractions to a decimal.
If necessary, give your answer to 3 decimal places.

  a) $\frac{4}{5}$ **b)** $\frac{3}{8}$ **c)** $\frac{2}{11}$ **d)** $\frac{1}{9}$ **e)** $\frac{9}{20}$

2 State whether each of these fractions gives a recurring or a terminating decimal.
Give a reason for each answer.

  a) $\frac{3}{5}$ **b)** $\frac{2}{3}$ **c)** $\frac{4}{9}$ **d)** $\frac{1}{16}$ **e)** $\frac{3}{7}$

3 **a)** Find the recurring decimal equivalent to $\frac{5}{7}$.

  b) How many digits are there in the repeating pattern?

Challenge 2.7

a) In Exercise 2.4 questions **1** and **2**, you found these.

$\frac{1}{9} = 0.111\ 111\ 111...$ $\qquad\qquad$ $\frac{4}{9} = 0.444\ 444\ 444...$

Write down the decimal equivalent of these without using your calculator.

$\frac{2}{9}$ $\qquad$ $\frac{3}{9}$ $\qquad$ $\frac{5}{9}$ $\qquad$ $\frac{6}{9}$ $\qquad$ $\frac{7}{9}$ $\qquad$ $\frac{8}{9}$

b) In Example 2.23 you found that $\frac{1}{11} = 0.090\ 909\ 090...$.

In addition, $\frac{2}{11} = 0.181\ 818\ 181...$ and $\frac{5}{11} = 0.454\ 545\ 454...$.

Write down the decimal equivalent of these without using your calculator.

$\frac{3}{11}$ $\qquad$ $\frac{4}{11}$ $\qquad$ $\frac{6}{11}$ $\qquad$ $\frac{7}{11}$ $\qquad$ $\frac{8}{11}$ $\qquad$ $\frac{9}{11}$ $\qquad$ $\frac{10}{11}$

Mental arithmetic with decimals

You should be able to add and subtract simple decimals in your head. It is similar to adding and subtracting whole numbers.

For example, you can do $63 + 24$ by adding 20 to get 83 and then adding 4 to get 87.

In the same way, you can do $6.3 + 2.4$ by adding 2 to get 8.3 and then adding 0.4 to get 8.7.

Subtraction can also be done in stages.

EXAMPLE 2.24

Work out these.

a) $5.8 + 7.3$ $\qquad\qquad$ **b)** $8.5 - 3.7$

Solution

a) $5.8 + 7 = 12.8$ $\qquad$ Add the units first.

$12.8 + 0.3 = 13.1$ $\qquad$ Then add the tenths.

b) $8.5 - 3 = 5.5$ $\qquad$ Subtract the units first.

$5.5 - 0.5 = 5$ $\qquad$ You need to subtract 7 tenths. Subtract 5 tenths first.

$5 - 0.2 = 4.8$ $\qquad$ Then subtract the remaining 2 tenths.

EXERCISE 2.5

Work out these. As far as possible, write down only your final answer.

1	4.2 + 3.5	**2**	5.1 + 2.8	**3**	7.8 − 4.2	**4**	5.6 − 3.4
5	5.8 + 1.3	**6**	4.6 + 3.5	**7**	6.5 − 0.8	**8**	6.4 − 2.6
9	7.9 + 4.3	**10**	7.8 + 8.7	**11**	7.8 − 6.9	**12**	7.6 − 1.8

Multiplying and dividing decimals

This section shows you how to extend the techniques you used for
multiplying simple decimals to multiplying any decimals.

Look again at your answers to Discovery 2.5. These are the steps you take to multiply decimals.

1   Carry out the multiplication ignoring the decimal points. The digits in the answer will be the same as the digits in the final answer.
2   Count the total number of decimal places in the two numbers to be multiplied.
3   Put the decimal point in the answer you got in step 1 so that the final answer has the same number of decimal places as you found in step 2.

EXAMPLE 2.25

Work out 8×0.7.

Solution

1   First do $8 \times 7 = 56$.
2   The total number of decimal places in 8 and 0.7 is $0 + 1 = 1$.
3   The answer is 5.6.

> **TIP** Notice that when you multiply by a number between 0 and 1, such as 0.7, you decrease the original number (8 to 5.6).

EXAMPLE 2.26

Work out 8.3×3.4.

Solution

1   First do 83×34.

$$
\begin{array}{r}
8\ 3 \\
\times\ \ \ 3\ 4 \\
\hline
2\ 4\ 9\ 0 \\
3\ 3\ 2 \\
\hline
2\ 8\ 2\ 2 \\
\end{array}
$$

The method used here is the traditional long multiplication. You may have learnt another method.

2   The total number of decimal places in 8.3 and 3.4 is $1 + 1 = 2$.
3   The answer is 28.22.

EXAMPLE 2.27

Work out 8.32×2.6.

Solution

1   First do 832×26.

$$
\begin{array}{r}
8\ 3\ 2 \\
\times\ \ \ \ \ 2\ 6 \\
\hline
1\ 6\ 6\ 4\ 0 \\
4\ 9\ 9\ 2 \\
\hline
2\ 1\ 6\ 3\ 2 \\
\end{array}
$$

2   The total number of decimal places in 8.32 and 2.6 is $2 + 1 = 3$.
3   The answer is 21.632.

Discovery 2.6

a) Do these calculations on your calculator.
   (i) $26 \div 1.3$ **(ii)** $260 \div 13$
b) What do you notice?
c) Now do these calculations on your calculator.
   (i) $5.92 \div 3.7$ **(ii)** $59.2 \div 37$ **(iii)** $3.995 \div 2.35$ **(iv)** $399.5 \div 235$
d) Can you explain your results?

The result of a division is unchanged when you multiply both numbers by 10 (i.e. move the decimal point one place in both numbers).

The result is also unchanged when you multiply both numbers by 100 (i.e. move the decimal point two places in both numbers).

This rule is exactly the same as when you are writing equivalent fractions.

For example, $\frac{3}{5} = \frac{30}{50} = \frac{300}{500}$.

You use this rule when you are dividing decimals.

EXAMPLE 2.28

Work out $6 \div 0.3$.

Solution

First multiply both numbers by 10, so that the number you are dividing by is a whole number.

The calculation becomes $60 \div 3$.

$60 \div 3 = 20$

so $6 \div 0.3$ is also 20.

> **TIP**
> Notice that when you divide by a number between 0 and 1, such as 0.3, you increase the original number (6 to 20).

EXAMPLE 2.29

Work out $4.68 \div 0.4$.

Solution

First multiply both numbers by 10 (move the decimal point one place).

The calculation becomes $46.8 \div 4$.

$$\begin{array}{r} 11.7 \\ 4\overline{)46.^28} \end{array}$$ The decimal point in the answer goes above the decimal point in 46.8.

$4.68 \div 0.4$ is also 11.7.

EXAMPLE 2.30

Work out 3.64 ÷ 1.3.

Solution

First multiply both numbers by 10 (move the decimal point one place).

The calculation becomes 36.4 ÷ 13.

$$\begin{array}{r} 2.8 \\ 13\overline{)36.^{10}4} \end{array}$$ You may have been taught to do this by long division rather than by short division.

3.64 ÷ 1.3 is also 2.8.

EXERCISE 2.6

1 Work out these.

a) 4×0.3 b) 0.5×7 c) 3×0.6

d) 0.8×9 e) 0.6×0.4 f) 0.8×0.6

g) 40×0.3 h) 0.5×70 i) 0.3×0.2

j) 0.8×0.1 k) $(0.7)^2$ l) $(0.3)^2$

2 Work out these.

a) $8 \div 0.2$ b) $1.2 \div 0.3$ c) $2.8 \div 0.7$

d) $3.6 \div 0.4$ e) $24 \div 1.2$ f) $50 \div 2.5$

g) $9 \div 0.3$ h) $15 \div 0.3$ i) $16 \div 0.2$

j) $24 \div 0.8$ k) $1.55 \div 0.5$ l) $48.8 \div 0.4$

3 Work out these.

a) 4.2×1.5 b) 6.2×2.3 c) 5.9×6.1

d) 7.2×2.7 e) 63×1.8 f) 72×5.4

g) 5.6×8.9 h) 10.9×2.4 i) 12.7×0.4

j) 2.34×0.8 k) 5.46×0.7 l) 6.23×1.6

4 Work out these.

a) $14.7 \div 0.3$ b) $13.6 \div 0.8$ c) $14.4 \div 0.6$

d) $22.4 \div 0.7$ e) $47.7 \div 0.9$ f) $85.8 \div 1.1$

g) $3.42 \div 0.6$ h) $1.96 \div 0.4$ i) $1.45 \div 0.5$

j) $3.51 \div 1.3$ k) $5.55 \div 1.5$ l) $6.3 \div 1.4$

Challenge 2.8

In a 4 by 400-metres relay race, the four members of a team run the following times.

44.5 seconds	45.6 seconds	45.8 seconds	43.9 seconds

What was their average time?

Challenge 2.9

a) Calculate the area of this rectangle.

6.3 cm

2.6 cm

b) This rectangle has the same area as the one in part **a)**.
Calculate the length of this rectangle.

3.9 cm

Percentage increase and decrease

You already know how to find a percentage of a quantity.

Percentage increase

To increase £240 by 23% you first work out 23% of £240. $240 \times 0.23 = £55.20$
Then you add £55.20 to £240. $240 + 55.20 = £295.20$

There is a quicker way to do the same calculation.

To increase a quantity by 23% you need to find the original quantity plus 23%.

This means that to increase £240 by 23% you need to find 100% of £240 + 23% of £240 = 123% of £240.

The decimal equivalent of 123% is 1.23.

The calculation can therefore be done in one stage: $240 \times 1.23 = £295.20$

The number that you multiply the original quantity by (here 1.23) is called the **multiplier**.

EXAMPLE 2.31

Amir's salary is £17 000 per year. He receives a 3% increase.
Find his new salary.

Solution

Amir's new salary is 103% of his original salary. So the multiplier is 1.03.

£17 000 × 1.03 = £17 510

This method is much quicker when repeated calculations are needed.

EXAMPLE 2.32

Invest now and receive a guaranteed 6% compound interest over 5 years

Compound interest means that interest is paid on the total amount in the account. It is different from simple interest, when interest is paid only on the original amount invested.

Jane invests £1500 for the full 5 years.
What will her investment be worth at the end of the 5 years?

Solution

At the end of year 1 the investment will be worth £1500 × 1.06 = £1590.00

At the end of year 2 the investment will be worth £1590 × 1.06 = £1685.40
This is the same as £1500 × 1.06 × 1.06 = £1685.40
or £1500 × 1.06^2 = £1685.40

At the end of year 3 the investment will be worth £1685.40 × 1.06 = £1786.524
This is the same as £1500 × 1.06 × 1.06 × 1.06 = £1786.524
or £1500 × 1.06^3 = £1786.524

At the end of year 4 the investment will be worth £1786.524 × 1.06 = £1893.7154
This is the same as £1500 × 1.06 × 1.06 × 1.06 × 1.06 = £1893.7154
or £1500 × 1.06^4 = £1893.7154

At the end of year 5 the investment will be worth £1893.7154 × 1.06 = £2007.34
This is the same as £1500 × 1.06 × 1.06 × 1.06 × 1.06 × 1.06 = £2007.34
or £1500 × 1.06^5 = £2007.34
(to the nearest penny)

Notice that at the end of year n you multiply £1500 by 1.06^n.

TIP Use the power button (1 or x^y or y^x) on your calculator.

Percentage decrease

Percentage decrease can be done in a similar way.

EXAMPLE 2.33

In a sale there is 15% off everything.

Kieran buys a DVD recorder in the sale. The original price was £225.
Calculate the sale price.

Solution

£225 × 0.85 = £191.25   A percentage decrease of 15% is the same as
100% − 15% = 85%. So the multiplier is 0.85.

Again, this method is very useful for repeated calculations.

EXAMPLE 2.34

The value of a car decreases by 12% every year.
Zara's car cost £9000 when new.
Calculate its value 4 years later. Give your answer to the nearest pound.

Solution

100% − 12% = 88%

Value after 4 years = £9000 × 0.88^4 At the end of year 4 you multiply £9000 by 0.88^4.
= £5397.26
= £5397 to the nearest pound

EXERCISE 2.7

1 Write down the multiplier that will increase an amount by
   a) 13%.   b) 20%.   c) 68%.   d) 8%.
   e) 2%.   f) 17.5%.   g) 100%.   h) 150%.

2 Write down the multiplier that will decrease an amount by
   a) 14%.   b) 20%.   c) 45%.   d) 7%.
   e) 3%.   f) 23%.   g) 86%.   h) 16.5%.

3 Sanjay earns £4.60 per hour from his Saturday job.
   If he receives a 4% increase, how much will he earn?
   Give your answer to the nearest penny.

4 In a sale, all items were reduced by 30%. Abi bought a pair of shoes.
   The original price was £42. What was the sale price?

5 Mark invested £2400 at 5% compound interest.
What was the investment worth at the end of 4 years?
Give your answer to the nearest pound.

6 This painting was worth £15 000 in 2004.
The painting increased in value by 15% every year for 6 years.
How much was it worth at the end of the 6 years?
Give your answer to the nearest pound.

7 The value of a car decreased by 9% per year.
When it was new it was worth £14 000.
What was its value after 5 years?
Give your answer to the nearest pound.

8 House prices rose by 12% in 2006, 11% in 2007 and 7% in 2008.
At the start of 2006 the price of a house was £120 000.
What was the price at the end of 2008?
Give your answer to the nearest pound.

9 The value of an investment rose by 8% in 2008 and fell by 8% in 2009. If the value of the
investment was £3000 at the start of 2008, what was the value at the end of 2009?

> ## WHAT YOU HAVE LEARNED
>
> - **To compare fractions you convert them to equivalent fractions with a common denominator**
> - **To add and subtract fractions you use a common denominator**
> - **To add or subtract mixed numbers you deal with the whole numbers first and then the fraction parts**
> - **To multiply fractions you multiply the numerators and multiply the denominators**
> - **Sometimes you can cancel before doing the multiplication**
> - **To divide fractions you find the reciprocal of the second fraction (turn it upside down) and then multiply**
> - **To multiply and divide mixed numbers, you must change the mixed numbers to improper fractions first**
> - **How to work with fractions and mixed numbers on your calculator using the [aᵇ/c] button**
> - **To change a fraction to a decimal you divide the numerator by the denominator**
> - **To multiply decimals, you multiply the numbers without the decimal point and then count the total number of decimal places in the two numbers**
> - **To divide by a decimal with one decimal place, you multiply both numbers by 10 (move the decimal point to the right one place) and then do the division**
> - **A quick way of increasing by e.g. 12% or 7% is to multiply by 1.12 or 1.07**
> - **A quick way of decreasing by e.g. 15% or 8% is to multiply by 0.85 or 0.92**

MIXED EXERCISE 2

Do not use your calculator for questions **1** to **6**.

1 For each pair of fractions
- find the common denominator.
- state which is the bigger fraction.

a) $\frac{4}{5}$ or $\frac{5}{6}$ b) $\frac{1}{3}$ or $\frac{2}{7}$ c) $\frac{13}{20}$ or $\frac{5}{8}$

2 Work out these.

a) $\frac{3}{5} + \frac{4}{5}$ b) $\frac{3}{7} + \frac{2}{3}$ c) $\frac{5}{8} - \frac{1}{6}$ d) $\frac{7}{10} + \frac{2}{15}$ e) $\frac{11}{12} - \frac{3}{8}$

f) $3\frac{1}{4} + 2\frac{1}{6}$ g) $4\frac{3}{4} - 1\frac{2}{5}$ h) $5\frac{1}{2} + 2\frac{7}{8}$ i) $3\frac{5}{6} + 2\frac{2}{9}$ j) $4\frac{1}{4} - 2\frac{3}{5}$

k) $\frac{3}{5} \times \frac{2}{3}$ l) $\frac{4}{7} \times \frac{5}{6}$ m) $\frac{5}{8} \div \frac{2}{3}$ n) $\frac{9}{10} \div \frac{3}{7}$ o) $\frac{15}{16} \times \frac{12}{25}$

p) $1\frac{2}{3} \times 2\frac{1}{5}$ q) $2\frac{5}{6} \div 1\frac{3}{4}$ r) $2\frac{5}{8} \times 1\frac{3}{7}$ s) $1\frac{7}{10} \div 4\frac{2}{5}$ t) $2\frac{3}{4} \times 3\frac{3}{7}$

3 Change each of these fractions to a decimal.
Where necessary, give your answer correct to 3 decimal places.

a) $\frac{1}{8}$ b) $\frac{2}{9}$ c) $\frac{5}{7}$ d) $\frac{3}{11}$

4 Work out these.

a) $4.3 + 5.4$ b) $9.6 - 4.3$ c) $5.8 + 2.9$ d) $6.4 - 1.8$

5 Work out these.

a) 5×0.4 b) 0.7×0.1 c) 0.9×0.8

d) 1.8×6 e) 2.7×3.4 f) 5.2×3.6

6 Work out these.

a) $9 \div 0.3$ b) $3.2 \div 0.4$ c) $6.9 \div 2.3$

d) $56 \div 0.7$ e) $86.9 \div 1.1$ f) $5.22 \div 0.6$

You may use your calculator for questions **7** to **9**.

7 Use your calculator to work out these.

a) $\frac{2}{11} + \frac{5}{6}$ b) $\frac{7}{8} - \frac{3}{5}$ c) $2\frac{2}{7} \times 1\frac{3}{8}$ d) $8\frac{2}{5} \div 2\frac{7}{10}$

8 Sam invested £3500 at 6% compound interest.
What was the investment worth at the end of 7 years?
Give your answer to the nearest pound.

9 In a sale, prices were reduced by 10% every day.
A pair of jeans originally cost £45.
Nicola bought a pair of jeans on the fourth day of the sale.
How much did she pay for them? Give your answer to the nearest penny.

3 → RATIO AND PROPORTION

THIS CHAPTER IS ABOUT

- **Understanding ratio and its notation**
- **Writing a ratio in its lowest terms**
- **Writing a ratio in the form 1 : *n***
- **Using ratios in proportion calculations**
- **Dividing a quantity in a given ratio**
- **Comparing proportions**

YOU SHOULD ALREADY KNOW

- **How to multiply and divide without a calculator**
- **How to find common factors**
- **How to simplify fractions**
- **What is meant by an enlargement**
- **How to change between metric units**

What is a ratio?

A ratio is used to compare two or more quantities.

If you have three sweets and decide to keep one and give two to your best friend, you and your friend have sweets in the ratio 1:2. You say this as '1 to 2'.

Larger numbers can also be compared in a ratio.

If you have six sweets and decide to keep two and give four to your best friend, you and your friend have sweets in the ratio 2:4.

You already know how to give a fraction in its **lowest terms**, by **cancelling**. You can do the same with ratios.

2:4 = 1:2   2 and 4 are both multiples of 2, so you can divide each part of the ratio by 2.

EXAMPLE 3.1

The salaries of three people are £16 000, £20 000 and £32 000. Write this as a ratio in its lowest terms.

Solution

$$
\begin{array}{llll}
& 16\,000 : 20\,000 : 32\,000 & \text{First write the salaries as a ratio.} \\
= & 16 \ : \ 20 \ : \ 32 & \text{Divide each part of the ratio by 1000.} \\
= & 8 \ : \ 10 \ : \ 16 & \text{Divide each part by 2.} \\
= & 4 \ : \ 5 \ : \ 8 & \text{Divide each part by 2.}
\end{array}
$$

Notice that your answer should not include units. £4 : £5 : £8 would be wrong. See Example 3.3.

To write a ratio in its lowest terms in one step, find the highest common factor (HCF) of the numbers in the ratio and then divide each part of the ratio by the HCF.

EXAMPLE 3.2

Write these ratios in their lowest terms.

a) $20:50$ **b)** $16:24$ **c)** $9:27:54$

Solution

a) $20:50 = 2:5$ Divide each part by 10.

b) $16:24 = 2:3$ Divide each part by 8.

c) $9:27:54 = 1:3:6$ Divide each part by 9.

Check up 3.1

a) Jane is 4 years old and Petra is 8 years old.
Write the ratio of their ages in its lowest terms.

b) A recipe uses 500 g of flour, 300 g of sugar and 400 g of raisins.
Write the ratio of these amounts in its lowest terms.

Sometimes you have to change the units of one part of the ratio first.

EXAMPLE 3.3

Write each of these ratios in its lowest terms.

a) 1 millilitre : 1 litre   **b)** 1 kilogram : 200 grams

Solution

a) 1 millilitre : 1 litre = 1 millilitre : 1000 millilitres   Write each part in the same units.

$= 1:1000$ When the units are the same, you do not include them in the ratio.

b) 1 kilogram : 200 grams = 1000 grams : 200 grams   Write each part in the same units.

$= 5:1$ Divide each part by 200.

EXAMPLE 3.4

 Write each of these ratios in its lowest terms.

a) 50p : £2   **b)** 2 cm : 6 mm   **c)** 600 g : 2 kg : 750 g

Solution

a) 50p : £2 = 50p : 200p   Write each part in the same units.
           = 1 : 4   Divide each part by 50.

b) 2 cm : 6 mm = 20 mm : 6 mm   Write each part in the same units.
              = 10 : 3   Divide each part by 2.

c) 600 g : 2 kg : 750 g = 600 g : 2000 g : 750 g   Write each part in the same units.
                 = 12 : 40 : 15   Divide each part by 50.

EXERCISE 3.1

1 Write each of these ratios in its lowest terms.

    a) 6 : 3   **b)** 25 : 75   **c)** 30 : 6

    d) 5 : 15 : 25   **e)** 6 : 12 : 8

2 Write each of these ratios in its lowest terms.

    a) 50 g : 1000 g   **b)** 30p : £2   **c)** 2 minutes : 30 seconds

    d) 4 m : 75 cm   **e)** 300 ml : 2 litres

3 At a concert there are 350 men and 420 women.
    Write the ratio of men to women in its lowest terms.

4 Al, Peta and Dave invest £500, £800 and £1000 respectively in a business.
    Write the ratio of their investments in its lowest terms.

5 A recipe for vegetable soup uses 1 kg of potatoes, 500 g of leeks and 750 g of celery.
    Write the ratio of the ingredients in its lowest terms.

Challenge 3.1

    a) Explain why the ratio 20 minutes : 1 hour is not 20 : 1.

    b) What should it be?

Writing a ratio in the form 1 : n

It is sometimes useful to have a ratio with 1 on the left.
A common scale for a scale model is 1 : 24.
The scale of a map or enlargement is often given as 1 : n.

To change a ratio to this form, divide both numbers by the one on the left.
This can be written in a general form as 1 : n.

EXAMPLE 3.5

Write these ratios in the form 1 : n.

a) 2 : 5   **b)** 8 mm : 3 cm   **c)** 25 mm : 1.25 km

Solution

a) 2 : 5 = 1 : 2.5   Divide each part by 2.

b) 8 mm : 3 cm = 8 mm : 30 mm   Write each part in the same units.
                   = 1 : 3.75   Divide each part by 8.

c) 25 mm : 1.25 km = 25 : 1 250 000   Write each part in the same units.
                     = 1 : 50 000   Divide each part by 25.

1 : 50 000 is a common map scale. It means that 1 cm on the map represents 50 000 cm, or 500 m, on the ground.

> **TIP**
> Use a calculator if necessary to convert the ratio to the form 1 : n.

 ## EXERCISE 3.2

1 Write each of these ratios in the form 1 : n.

    a) 2 : 6   **b)** 3 : 15   **c)** 6 : 15   **d)** 4 : 7

    e) 20p : £1.50   **f)** 4 cm : 5 m   **g)** 10 : 2   **h)** 2 mm : 1 km

2 On a map a distance of 8 mm represents a distance of 2 km.
    What is the scale of the map in the form 1 : n?

3 A negative for a photograph is 35 mm long. An enlargement is 21 cm long.
    What is the ratio of the negative to the enlargement in the form 1 : n?

Using ratios

Sometimes you know one of the quantities in the ratio, but not the other.

If the ratio is in the form $1:n$, you can work out the second quantity by multiplying the first by n.

You can work out the first quantity by dividing the second quantity by n.

EXAMPLE 3.6

a) A negative is enlarged in the ratio $1:20$ to make a picture.
The negative measures 36 mm by 24 mm.
What size is the enlargement?

b) Another $1:20$ enlargement measures 1000 mm $\times$ 1000 mm.
What size is the negative?

Solution

a) $36 \times 20 = 720$ The enlargement will be 20 times bigger than
$24 \times 20 = 480$ the negative, so multiply both dimensions by 20.

The enlargement measures 720 mm by 480 mm.

b) $1000 \div 20 = 50$ The negative will be 20 times smaller than the negative,
so divide the dimensions by 20.

The negative measures 50 mm $\times$ 50 mm.

EXAMPLE 3.7

A map is drawn to a scale of 1 cm : 2 km.

a) On the map, the distance between Amhope and Didburn is 5.4 cm.
What is the actual distance in kilometres?

b) The length of a straight railway track between two stations is 7.8 km.
How long is this track on the map in centimetres?

Solution

a) $2 \times 5.4 = 10.8$ The actual distance, in kilometres, is
Real distance = 10.8 km   twice as large as the map distance, in
centimetres. So multiply by 2.

b) $7.8 \div 2 = 3.9$ The map distance, in centimetres, is
Map distance = 3.9 cm   half as large as the actual distance, in
kilometres. So divide by 2.

What would the answer to part **a)** of Example 3.7 be in centimetres?

What ratio could you use to work this out?

Sometimes you have to work out quantities using a ratio that is not in the form $1:n$.

To work out an unknown quantity, you multiply each part of the ratio by the same number to get an equivalent ratio which contains the quantity you know. This number is called the **multiplier**.

EXAMPLE 3.8

To make jam, fruit and sugar are mixed in the ratio $2:3$.
This means that if you have 2 kg of fruit, you need 3 kg of sugar; if you have 4 kg of fruit, you need 6 kg of sugar.

How much sugar do you need if your fruit weighs

a) 6 kg? **b)** 10 kg? **c)** 500 g?

Solution

a) $6 \div 2 = 3$ Divide the quantity of fruit by the fruit part of the ratio to find the multiplier.

$2:3 = 6:9$ Multiply each part of the ratio by the multiplier, 3.
9 kg of sugar

b) $10 \div 2 = 5$ Divide the quantity of fruit by the fruit part of the ratio to find the multiplier.

$2:3 = 10:15$ Multiply each part of the ratio by the multiplier, 5.
15 kg of sugar

c) $500 \div 2 = 250$ Divide the quantity of fruit by the fruit part of the ratio to find the multiplier.

$2:3 = 500:750$ Multiply each part of the ratio by the multiplier, 250.
750 g of sugar

EXAMPLE 3.9

5 cm

Two photos are in the ratio $2:5$.

a) What is the height of the larger photo?

b) What is the width of the smaller photo?

9 cm

Solution

a) $5 \div 2 = 2.5$ Divide the height of the smaller photo by the smaller part of the ratio to find the multiplier.

$2:5 = 5:12.5$ Multiply each part of the ratio by the multiplier, 2.5.

Height of the larger photo $= 12.5$ cm

b) $9 \div 5 = 1.8$ Divide the width of the larger photo by the larger part of the ratio to find the multiplier.

$2:5 = 3.6:9$ Multiply each part of the ratio by the multiplier, 1.8.

Width of the smaller photo $= 3.6$ cm

EXAMPLE 3.10

To make grey paint, white paint and black paint are mixed in the ratio $5:2$.

a) How much black paint is mixed with 800 ml of white paint?

b) How much white paint is mixed with 300 ml of black paint?

Solution

A table is often useful for this sort of question.

Paint	White	Black
Ratio	5	2
a) Amount	800 ml	$2 \times 160 = 320$ ml
Multiplier	$800 \div 5 = 160$	
b) Amount	$5 \times 150 = 750$ ml	300 ml
Multiplier		$300 \div 2 = 150$

> **TIP** Make sure you haven't made a silly mistake by checking that the bigger side of the ratio has the bigger quantity.

a) Black paint $= 320$ ml   **b)** White paint $= 750$ ml

EXAMPLE 3.11

To make stew for four people, a recipe uses 1.6 kg of beef.
How much beef is needed using the recipe for six people?

Solution

The ratio of people is $4:6$.

$4:6 = 2:3$ Write the ratio in its lowest terms.

$1.6 \div 2 = 0.8$ Divide the quantity of beef needed for four people by the first part of the ratio to find the multiplier.

$0.8 \times 3 = 2.4$ Multiply the second part of the ratio by the multiplier, 0.8.

Beef needed for six people $= 2.4$ kg

1 The ratio of the lengths of two squares is 1 : 6.
   a) The length of the side of the small square is 2 cm.
   What is the length of the side of the large square?
   b) The length of the side of the large square is 21 cm.
   What is the length of the side of the small square?

2 The ratio of helpers to babies in a crèche must be 1 : 4.
   a) There are six helpers on a Tuesday.
   How many babies can there be?
   b) There are 36 babies on a Thursday.
   How many helpers must there be?

3 Sanjay is mixing pink paint.
   To get the shade he wants, he mixes red and white paint in the ratio 1 : 3.
   a) How much white paint should he mix with 2 litres of red paint?
   b) How much red paint should he mix with 12 litres of white paint?

4 The negative of a photo is 35 mm long. An enlargement of 1 : 4 is made.
   What is the length of the enlargement?

5 A road atlas of Great Britain is to a scale of 1 inch to 4 miles.
   a) On the map, the distance between Forfar and Montrose is 7 inches.
   What is the actual distance between the two towns in miles?
   b) It is 40 miles from Newcastle to Middlesbrough. How far is this on the map?

6 For a recipe, Chelsy mixes water and lemon curd in the ratio 2 : 3.
   a) How much lemon curd should she mix with 20 ml of water?
   b) How much water should she mix with 15 teaspoons of lemon curd?

7 To make a solution of a chemical, a scientist mixes 3 parts chemical with 20 parts water.
   a) How much water should he mix with 15 ml of chemical?
   b) How much chemical should he mix with 240 ml of water?

8 An alloy is made by mixing 2 parts silver with 5 parts nickel.
   a) How much nickel must be mixed with 60 g of silver?
   b) How much silver must be mixed with 120 g of nickel?

9 Sachin and Rehan share a flat. They agree to share the rent in the same ratio as their wages.
   Sachin earns £600 a month and Rehan earns £800 a month.
   If Sachin pays £90, how much does Rehan pay?

10 A recipe for hotpot uses onions, carrots and stewing steak in the ratio, by mass, of 1 : 2 : 5.
   a) What quantity of steak is needed if 100 g of onion is used?
   b) What quantity of carrots is needed if 450 g of steak is used?

Dividing a quantity in a given ratio

Discovery 3.1

Maya has an evening job making up party bags for a children's party organiser.
She shares out lemon sweets and raspberry sweets in the ratio 2 : 3.
Each bag contains 5 sweets.

a) On Monday Maya makes up 10 party bags.
    (i)   How many sweets does she use in total?
    (ii) How many lemon sweets does she use?
    (iii) How many raspberry sweets does she use?

b) On Tuesday Maya makes up 15 party bags.
    (i)   How many sweets does she use in total?
    (ii) How many lemon sweets does she use?
    (iii) How many raspberry sweets does she use?

What do you notice?

A ratio represents the number of shares in which a quantity is divided.
The total quantity divided in a ratio is found by adding the parts of the
ratio together.

To find the quantities shared in a ratio:
- Find the total number of shares.
- Divide the total quantity by the total number of shares to find the
  multiplier.
- Multiply each part of the ratio by the multiplier.

TIP
The multiplier may not be a whole number. Work with the
decimal or fraction and round the final answer if necessary.

EXAMPLE 3.12

To make fruit punch, orange juice and grapefruit juice are mixed in the ratio 5 : 3.
Jo wants to make 1 litre of punch.

  a) How much orange juice does she need in millilitres?

  b) How much grapefruit juice does she need in millilitres?

Solution

$5 + 3 = 8$ First work out the total number of shares.

$1000 \div 8 = 125$ Convert 1 litre to millilitres and divide by 8 to find the multiplier.

A table is often helpful for this sort of question.

Punch	Orange	Grapefruit
Ratio	5	3
Amount	$5 \times 125 = 625$ ml	$3 \times 125 = 375$ ml

a) Orange juice = 625 ml   **b)** Grapefruit juice = 375 ml

EXERCISE 3.4

Do not use your calculator for questions **1** to **5**.

1   Share £20 between Dave and Sam in the ratio $2:3$.

2   Paint is mixed in the ratio 3 parts red to 5 parts white to make 40 litres of pink paint.
     a) How much red paint is used?   **b)** How much white paint is used?

3   Asif is making mortar by mixing sand and cement in the ratio $5:1$.
     How much sand is needed to make 36 kg of mortar?

4   To make a solution of a chemical, a scientist mixes 1 part chemical with 5 parts water.
     She makes 300 ml of the solution.
     a) How much chemical does she use?   **b)** How much water does she use?

5   Amit, Bree and Chris share £1600 between them in the ratio $2:5:3$.
     How much does each receive?

You may use your calculator for questions **6** to **8**.

6   In a local election, 5720 people vote.
     They vote for Conservative, Labour and other parties in the ratio $6:3:2$.
     How many people vote Labour?

7   St Anthony's College Summer Fayre raised £1750. The governors decided to share
    the money between the college and a local charity in the ratio 5 to 1.
    How much did the local charity receive? Give your answer correct to the nearest pound.

8   Sally makes breakfast cereal by mixing bran, currants and wheatgerm in the ratio 8 : 3 : 1
    by mass.
    a) How much bran does she use to make 600 g of the cereal?
    b) One day, she has only 20 g of currants.
        How much cereal can she make? She has plenty of bran and wheatgerm.

Challenge 3.3

Okera has a photo measuring 13 cm by 17 cm. He wants to have it enlarged.
Supa Print offer two sizes: 24 inches by 32 inches and 20 inches by 26.5 inches.
He wants to keep the same proportions, or as near as possible.

a) Which of the two enlargements should he choose? Show how you make your decision.

b) Why might he choose the other one?

Best value

Discovery 3.2

Two packets of cornflakes are
available at a supermarket.

Which is the better value for money?

To compare value, you need to compare either

● how much you get for a certain amount of money or
● how much a certain quantity (for example, volume or mass) costs.

In each case you are comparing **proportions**, either of size or of cost.

The better value item is the one with the **lower unit cost** or the **greater
number of units per penny** (or pound).

EXAMPLE 3.13

Sunflower oil is sold in 700 ml bottles for 95p and in 2 litre bottles for £2.45.
Show which bottle is the better value.

Solution

Method 1

Work out the price per millilitre for each bottle.

Size	Small	Large
Capacity	700 ml	2 litre = 2000 ml
Price	95p	£2.45 = 245p
Price per ml	95p ÷ 700 = 0.14p	245p ÷ 2000 = 0.12p

Use the same units for each bottle.

Round your answers to 2 decimal places if necessary.

The price per ml of the 2 litre bottle is lower. It has the lower unit cost.
In this case the unit is a millilitre.

The 2 litre bottle is the better value.

Method 2

Work out the amount per penny for each bottle.

Size	Small	Large
Capacity	700 ml	2 litre = 2000 ml
Price	95p	£2.45 = 245p
Amount per penny	700 ml ÷ 95 = 7.37 ml	2000 ml ÷ 245 = 8.16 ml

Again, use the same units for each bottle.

Round your answers to 2 decimal places if necessary.

The amount per penny is greater for the 2 litre bottle. It has the greater number of units per penny.

The 2 litre bottle is the better value.

> **TIP**
> Make it clear whether you are working out the cost per unit or the amount per penny, and include the units in your answers. Always show your working.

1 A 420 g bag of Choco bars costs £1.59 and a 325 g bag of Choco bars costs £1.09.
   Which is the better value for money?

2 Spa water is sold in 2 litre bottles for 85p and in 5 litre bottles for £1.79.
   Show which is the better value.

3 Wallace bought two packs of cheese: a 680 g pack for £3.20 and a 1.4 kg pack for £5.40.
   Which was the better value?

4 One-inch nails are sold in packets of 50 for £1.25 and in packets of 144 for £3.80.
   Which packet is the better value?

5 Toilet rolls are sold in packs of 12 for £1.79 and in packs of 50 for £7.20.
   Show which is the better value.

6 Brillo white toothpaste is sold in 80 ml tubes for £2.79 and in 150 ml tubes for £5.00.
   Which tube is the better value?

7 A supermarket sells cola in three different sized bottles: a 3 litre bottle costs £1.99,
   a 2 litre bottle costs £1.35 and a 1 litre bottle costs 57p.
   Which bottle gives the best value?

8 Crispy cornflakes are sold in three sizes: 750 g for £1.79, 1.4 kg for £3.20 and 2 kg for £4.89.
   Which packet gives the best value?

WHAT YOU HAVE LEARNED

- **To write a ratio in its lowest terms, divide all parts of the ratio by their highest common factor (HCF)**
- **To write a ratio in the form 1 : *n*, divide both numbers by the one on the left**
- **If the ratio is in the form 1 : *n*, you can work out the second quantity by multiplying the first by *n*, and you can work out the first quantity by dividing the second quantity by *n***
- **To find an unknown quantity, each part of the ratio must be multiplied by the same number, called the multiplier**
- **To find the quantities shared in a given ratio, first find the total number of shares, then divide the total quantity by the total number of shares to find the multiplier, then multiply each part of the ratio by the multiplier**
- **To compare value, work out the cost per unit or the number of units per penny (or pound). The better value item is the one with the lower cost per unit or the greater number of units per penny (or pound)**

1 Write each ratio in its simplest form.
   a) 50 : 35   b) 30 : 72   c) 1 minute : 20 seconds
   d) 45 cm : 1 m   e) 600 ml : 1 litre

2 Write these ratios in the form 1 : n.
   a) 2 : 8   b) 5 : 12   c) 2 mm : 10 cm
   d) 2 cm : 5 km   e) 100 : 40

3 A notice is enlarged in the ratio 1 : 20.
   a) The original is 3 cm wide.
       How wide is the enlargement?
   b) The enlargement is 100 cm long.
       How long is the original?

4 To make 12 scones Maureen uses 150 g of flour.
   How much flour does she use to make 20 scones?

5 To make a fruit and nut mixture, raisins and nuts are mixed in the ratio 5 : 3, by mass.
   a) What mass of nuts is mixed with 100 g of raisins?
   b) What mass of raisins is mixed with 150 g of nuts?

6 Panache made a fruit punch by mixing orange, lemon and grapefruit juice in the ratio
   5 : 1 : 2.
   a) He made a 2 litre bowl of fruit punch.
       How many millilitres of grapefruit juice did he use?
   b) How much fruit punch could he make with 150 ml of orange juice?

7 Show which is the better buy: 5 litres of oil for £18.50 or 2 litres of oil for £7.00.

8 Supershop sells milk in pints at 43p and in litres at 75p.
   A pint is equal to 568 ml.
   Which is the better buy?

4 → MENTAL METHODS

Mental strategies

You can develop your mental skills by practising them and by being open to new and better ideas.

Discovery 4.1

How many ways can you find to work out mentally the answer to each of the following calculations?

Make a note of the methods you use.

Which methods were most efficient?

a) $39 + 47$

b) $126 \div 3$

c) $290 \div 5$

d) $164 - 37$

e) 23×16

f) 21×19

g) 13×13

h) $10 - 1.7$

i) $14.6 + 2.9$

j) 3.6×30

k) $-6 + (-4)$

l) $-10 - (-7)$

m) $0.7 + 9.3$

n) 4×3.7

o) $12 \div 0.4$

p) 15% of £176

Compare your results and methods with the rest of the class.

Did anyone have ideas which you hadn't thought of and which you think work well?

Which calculations could you easily do completely in your head?

In which calculations did you want to make a note on paper of intermediate answers?

For adding and subtracting, there are a number of strategies you can use.

- Use number bonds that you know.
- Count forwards or backwards from one number.
- Use compensation: add or subtract too much, then compensate. For example, to add 9, first add 10 and then subtract 1.
- Use your knowledge of place value to help with adding or subtracting decimals.
- Use partitioning. For example, to subtract 63, first subtract 60 and then subtract 3.
- Jot a number line down on paper.

For multiplying and dividing, there are other strategies.

- Use factors. For example, to multiply by 20, first multiply by 2 and then multiply the result by 10.
- Use partitioning. For example, to multiply by 13, first multiply the number by 10, then multiply the number by 3 and finally add the results.
- Use your knowledge of place value to help with multiplying and dividing decimals.
- Recognise special cases where doubling and halving can be used.
- Use the relationship between multiplication and division.
- Recall percentage and fraction relationships. For example, $25\% = \frac{1}{4}$.

 TIP Check your answers by working them out again, using a different strategy.

Square and cube numbers

You learned about square and cube numbers in Chapter 1.

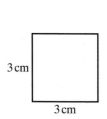

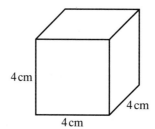

Area of square $= 3 \times 3$
$\qquad = 3^2 = 9$ cm^2

Volume of cube $= 4 \times 4 \times 4$
$\qquad = 4^3 = 64$ cm^3

Numbers such as $9(= 3^2)$ are called **square numbers**.
Numbers such as $64(= 4^3)$ are called **cube numbers**.

You learned in Chapter 1 that because $4^2 = 4 \times 4 = 16$, the **square root** of 16 is 4. This is written as $\sqrt{16} = 4$.

You should learn the squares of the numbers 1 to 15 and the cubes of the numbers 1 to 5 and 10 by heart.

Knowing the square numbers up to 15^2 will help you when you need to work out a square root.
For example, if you know that $7^2 = 49$, you also know that $\sqrt{49} = 7$.

These facts can also be used in other calculations.

EXAMPLE 4.1

Work out 50^2 mentally.

Solution

$50^2 = (5 \times 10)^2$ $50 = 5 \times 10$.
$ = 5^2 \times 10^2$ Square each term inside the brackets.
$ = 25 \times 100 = 2500$ You know that $5^2 = 25$ and $10^2 = 100$.

This is one possible strategy. You may think of another.

EXERCISE 4.1

Work these out mentally. As far as possible, write down only the final answer.

1 **a)** $9 + 17$ **b)** $0.6 + 0.9$ **c)** $13 + 45$ **d)** $143 + 57$ **e)** $72 + 8.4$
   f) $13.6 + 6.5$ **g)** $614 + 47$ **h)** $6.2 + 3.9$ **i)** $246 + 37$ **j)** $92 + 183$

2 **a)** $24 - 8$ **b)** $1.5 - 0.6$ **c)** $132 - 45$ **d)** $76 - 18$ **e)** $78 - 8.4$
   f) $102 - 37$ **g)** $165 - 96$ **h)** $403 - 126$ **i)** $98 - 12.3$ **j)** $1200 - 204$

3 **a)** 9×8 **b)** 13×4 **c)** 0.6×4 **d)** 32×5 **e)** 0.8×1000
   f) 21×16 **g)** 37×5 **h)** 130×4 **i)** 125×8 **j)** 31×25

4 **a)** $28 \div 7$ **b)** $160 \div 2$ **c)** $65 \div 5$ **d)** $128 \div 8$ **e)** $156 \div 12$
   f) $96 \div 24$ **g)** $8 \div 100$ **h)** $3 \div 0.5$ **i)** $4 \div 0.2$ **j)** $1.8 \div 0.6$

5 **a)** $5 + (-1)$ **b)** $-6 + 2$ **c)** $-2 + (-5)$ **d)** $-10 + 16$ **e)** $12 + (-14)$
   f) $6 - (-4)$ **g)** $-7 - (-1)$ **h)** $-10 - (-6)$ **i)** $12 - (-12)$ **j)** $-8 - (-8)$

6 **a)** 4×-2 **b)** -6×2 **c)** -3×-4 **d)** 7×-5 **e)** -4×-10
   f) $6 \div -2$ **g)** $-20 \div 5$ **h)** $-12 \div -4$ **i)** $18 \div -9$ **j)** $-32 \div -2$

7 Write down the square of each of these numbers.
   a) 6 **b)** 5 **c)** 11 **d)** 10 **e)** 13
   f) 20 **g)** 300 **h)** 0.4 **i)** 0.7 **j)** 0.3

8 Write down the square root of each of these numbers.

    a) 16   **b)** 9   **c)** 49   **d)** 169   **e)** 225

9 Write down the cube of each of these numbers.

    a) 1   **b)** 5   **c)** 2   **d)** 40   **e)** 0.3

10 Find 2% of £460.

11 A square has area 64 cm². How long is its side?

12 Jim spends £34.72. How much change does he get from £50?

13 A bottle contains 750 ml of water. Jo pours 330 ml into a glass. How much water is left in the bottle?

14 A rectangle has sides 4.5 cm and 4.0 cm. Work out

    a) the perimeter of the rectangle.   **b)** the area of the rectangle.

15 Find two numbers the sum of which is 13 and the product 40.

Rounding to 1 significant figure

We often use rounded numbers instead of exact ones.

Check up 4.1

In the following statements, which of these numbers are likely to be exact and which have been rounded?

a) Yesterday, I spent £14.62.

b) My height is 180 cm.

c) Her new dress cost £40.

d) The attendance at the Arsenal match was 32 000.

e) The cost of building the new school is £27 million.

f) The value of π is 3.142.

g) The Olympic games were in Athens in 2004.

h) There were 87 people at the meeting.

Discovery 4.2

Look at a newspaper.

Find five articles or advertisements where exact numbers have been used.

Find five articles or advertisements where rounded numbers have been used.

When estimating the answers to calculations, rounding to 1 significant figure is usually sufficient.

This means giving just one non-zero figure, with zeros as placeholders to make the number the correct size.

For example, 87 is 90 to 1 significant figure. It is between 80 and 90 but is nearer 90.

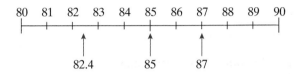

82.4 is 80 to 1 significant figure. It is between 80 and 90 but is nearer 80.
85 is 90 to 1 significant figure. It is halfway between 80 and 90. To avoid confusion, 5 is always rounded up.

The first significant figure is the first non-zero digit.

For example, the first significant figure in 6072 is 6.

So to round to 1 significant figure:

• Find the first non-zero digit. Look at the digit after it.
  If it is less than 5, leave the first non-zero digit as it is.
  If it is 5 or more, add 1 to the first non-zero digit.
• Then look at the place value of the first non-zero digit and add zeros as placeholders, if necessary, to make the number the correct size.

EXAMPLE 4.2

Round each of these numbers to 1 significant figure.

a) £29.95 **b)** 48 235 **c)** 0.072

Solution

a) £29.95 = £30 to 1 s.f. The second non-zero digit is 9 so round 2 up to 3.
Looking at place value, the 2 is 20, so the 3 should be 30.

b) 48 235 = 50 000 to 1 s.f. The second non-zero digit is 8 so round 4 up to 5.
Looking at place value, the 4 is 40 000, so the 5 should be 50 000.

c) 0.072 = 0.07 to 1 s.f. The second non-zero digit is 2 so 7 stays as it is.
Looking at place value, the 7 is 0.07, which stays as it is.

To estimate answers to problems, round each number to 1 significant figure.

Use mental or pencil and paper strategies to help you do the calculation.

EXAMPLE 4.3

Estimate the cost of four CDs at £7.95 each.

Solution

Cost ≈ £8 × 4   £7.95 rounded to 1 significant figure is £8.
      = £32

In practical situations, it is often useful to know whether your estimate is too big or too small. Here, £32 is larger than the exact answer, because £7.95 has been rounded up.

EXAMPLE 4.4

Estimate the answer to this calculation.
$$\frac{4.62 \times 0.61}{52}$$

Solution

$$\frac{4.62 \times 0.61}{52} \approx \frac{5 \times 0.6}{50}$$ Round each number in the calculation to 1 significant figure.

$$= \frac{{}^{1}\!\!\not5 \times 0.6}{\not50_{10}}$$ Cancel by dividing both 5 and 50 by 5.

$$= \frac{0.6}{10}$$

$$= 0.06$$

This is one possible strategy. You may think of another.

Rounding to a given number of significant figures

Rounding to a given number of significant figures involves using a similar method to rounding to 1 significant figure: just look at the size of the first digit which is not required.

For instance, to round to 3 significant figures, start counting from the first non-zero digit and look at the size of the fourth figure.

EXAMPLE 4.5

a) Round 52 617 to 2 significant figures.

b) Round 0.072 618 to 3 significant figures.

c) Round 17 082 to 3 significant figures.

> **TIP** Always state the accuracy of your answers, when you have rounded them.

Solution

a) 52¦617 = 53 000 to 2 s.f.

To round to 2 significant figures, look at the third figure. It is 6, so the second figure changes from 2 to 3. Remember to add zeros for placeholders.

b) 0.072 6¦18 = 0.0726 to 3 s.f.

The first significant figure is 7. To round to 3 significant figures, look at the fourth significant figure. It is 1, so the third figure is unchanged.

c) 170¦82 = 17 100 to 3 s.f.

The 0 in the middle here is a significant figure. To round to 3 significant figures, look at the fourth figure. It is 8, so the third figure changes from 0 to 1. Remember to add zeros for placeholders.

◎ EXERCISE 4.2

1 Round each of these numbers to 1 significant figure.

a) 8.2	b) 6.9	c) 17	d) 25.1
e) 493	f) 7.0	g) 967	h) 0.43
i) 0.68	j) 3812	k) 4199	l) 3.09

2 Round each of these numbers to 1 significant figure.

a) 14.9	b) 167	c) 21.2	d) 794
e) 6027	f) 0.013	g) 0.58	h) 0.037
i) 1.0042	j) 20 053	k) 0.069	l) 1942

3 Round each of these numbers to 2 significant figures.

a) 17.6	b) 184.2	c) 5672	d) 97 520
e) 50.43	f) 0.172	g) 0.0387	h) 0.006 12
i) 0.0307	j) 0.994		

4 Round each of these numbers to 3 significant figures.

a) 8.261	b) 69.77	c) 16 285	d) 207.51
e) 12 524	f) 7.103	g) 50.87	h) 0.4162
i) 0.038 62	j) 3.141 59		

For questions **5** to **12**, round the numbers in your calculations to 1 significant figure. Show your working.

5 At the school fête, Tony sold 245 ice-creams at 85p each. Estimate his takings.

6 Kate had £30. How many CDs, at £7.99 each, could she buy?

7 A rectangle measures 5.8 cm by 9.4 cm. Estimate its area.

8 A circle has diameter 6.7 cm. Estimate its circumference. $\pi = 3.142 \ldots$.

9 A new car is priced at £14 995 excluding VAT. VAT at 17.5% must be paid on it. Estimate the amount of VAT to be paid.

10 A cube has side 3.7 cm. Estimate its volume.

11 Pedro drove 415 miles in 7 hours 51 minutes. Estimate his average speed.

12 Estimate the answers to these calculations.

  a) 46×82 **b)** $\sqrt{84}$ **c)** $\dfrac{1083}{8.2}$ **d)** 7.05^2

  e) $43.7 \times 18.9 \times 29.3$ **f)** $\dfrac{2.46}{18.5}$ **g)** $\dfrac{29}{41.6}$ **h)** 917×38

  i) $\dfrac{283 \times 97}{724}$ **j)** $\dfrac{614 \times 0.83}{3.7 \times 2.18}$ **k)** $\dfrac{6.72}{0.051 \times 39.7}$ **l)** $\sqrt{39 \times 80}$

Challenge 4.1

Write a number which will round to 500 to 1 significant figure.
Write a number which will round to 500 to 2 significant figures.
Write a number which will round to 500 to 3 significant figures.
Compare your results with your classmates. What do you notice?

Using π without a calculator

When finding the area and circumference of a circle, you need to use π. Since $\pi = 3.141\ 592 \ldots$, you often round it to 1 significant figure when working without a calculator.

This is what you did in question **8** of Exercise 4.2.

An alternative is to give an exact answer by leaving π in the answer.

EXAMPLE 4.6

Find the area of a circle of radius 5 cm, leaving π in your answer.

Solution

Area $= \pi r^2$ The formula for the area of a circle is given in Chapter 31.
$\phantom{\text{Area}} = \pi \times 5^2$
$\phantom{\text{Area}} = \pi \times 25 = 25\pi \text{ cm}^2$

EXAMPLE 4.7

A circular pond of radius 3 m is surrounded
by a path 2 m wide.
Find the area of the path.
Give your answer as a multiple of π.

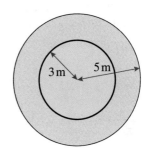

Solution

Area of path $=$ area of large circle $-$ area of small circle
$\phantom{\text{Area of path }} = \pi \times 5^2 - \pi \times 3^2$
$\phantom{\text{Area of path }} = 25\pi - 9\pi$ You already know that you can **collect like terms**.
$\phantom{\text{Area of path }} = 16\pi \text{ m}^2$ You can treat π in the same way.

EXERCISE 4.3

Give your answers to these questions as simply as possible.
Leave π in your answers where appropriate.

1
  a) $2 \times 4 \times \pi$ **b)** $\pi \times 8^2$ **c)** $\pi \times 6^2$
  d) $2 \times 13 \times \pi$ **e)** $\pi \times 9^2$ **f)** $2 \times \pi \times 3.5$

2
  a) $4\pi + 10\pi$ **b)** $\pi \times 8^2 + \pi \times 4^2$ **c)** $\pi \times 6^2 - \pi \times 2^2$
  d) $2 \times 25\pi$ **e)** $\dfrac{24\pi}{6\pi}$ **f)** $2 \times \pi \times 5 + 2 \times \pi \times 3$

3   The circumferences of two circles are in the ratio $10\pi : 4\pi$. Simplify this ratio.

4   Find the area of a circle with radius 15 cm.

5   A circular hole of radius 2 cm is drilled in a square of side 8 cm.
     Find the area that is left.

Deriving unknown facts from those you know

Earlier in this chapter, you saw that knowing a square number means that you also know the corresponding square root. For example, if you know that $11^2 = 121$, you also know that $\sqrt{121} = 11$.

In a similar way, knowing addition facts means that you also know the corresponding subtraction facts. For example, if you know that $91 + 9 = 100$, you also know that $100 - 9 = 91$ and $100 - 91 = 9$.

Knowing multiplication facts means that you also know the corresponding division facts.

For example, if you know that $42 \times 87 = 3654$, you also know that $3654 \div 42 = 87$ and $3654 \div 87 = 42$.

Your knowledge of place value, and of multiplying and dividing by powers of 10, means that you can answer other problems as well.

EXAMPLE 4.8

Given that $73 \times 45 = 3285$, work out these.

a) 730×45

b) $\dfrac{32.85}{450}$

Solution

a) $730 \times 45 = 10 \times 73 \times 45$ You can break 730 down to 10×73.
$= 10 \times 3285 = 32\,850$ You know that $73 \times 45 = 3285$.

b) $\dfrac{32.85}{450} = \dfrac{3285}{45\,000}$ Multiply the numerator and denominator by 100 so that the numerator is 3285.

$= \dfrac{3285}{1000 \times 45}$ You can break $45\,000$ down to 1000×45.

$= \dfrac{3285}{45} \div 1000$ You know that $73 \times 45 = 3285$, so you also know that $3285 \div 45 = 73$.

$= 73 \div 1000 = 0.073$

> **TIP**
> Check your answer by estimating. $\dfrac{32.85}{450}$ is a bit less than $\dfrac{45}{450}$ so the answer will be a bit less than 0.1.

Challenge 4.2

$352 \times 185 = 65\,120$

Write five other multiplication statements, with their answers, using this result.
Write five division statements, with their answers, using this result.

1 Work out these.
a) 0.7×7000 b) 0.06×0.6 c) 0.8×0.05 d) $(0.04)^2$
e) $(0.2)^2$ f) 600×50 g) 70×8000 h) 5.6×200
i) 40.1×3000 j) 4.52×2000 k) 0.15×0.8 l) 0.05×1.2

2 Work out these.
a) $500 \div 20$ b) $10 \div 200$ c) $2.6 \div 20$ d) $35 \div 0.5$
e) $2.4 \div 400$ f) $2.7 \div 0.03$ g) $0.06 \div 0.002$ h) $7 \div 0.2$
i) $600 \div 0.04$ j) $80 \div 0.02$ k) $0.52 \div 40$ l) $70 \div 0.07$

3 Given that $1.6 \times 13.5 = 21.6$, work out these.
a) 16×135 b) $21.6 \div 16$ c) $2160 \div 135$
d) 0.16×0.0135 e) $216 \div 0.135$ f) 160×1.35

4 Given that $988 \div 26 = 38$, work out these.
a) 380×26 b) $98.8 \div 26$ c) $9880 \div 38$
d) $98.8 \div 2.6$ e) $9.88 \div 260$ f) $98.8 \div 3.8$

5 Given that $153 \times 267 = 40\,851$, work out these.
a) 15.3×26.7 b) $15\,300 \times 2.67$ c) $40\,851 \div 26.7$
d) 0.153×26.7 e) $408.51 \div 15.3$ f) $408\,510 \div 26.7$

WHAT YOU HAVE LEARNED

- **How to use different mental strategies for adding and subtracting, and multiplying and dividing**
- **To round to a given number of significant figures, look at the size of the first digit which is not required**
- **When using π without a calculator, round it to 1 significant figure, or give an exact answer by leaving π in the answer, simplifying the other numbers**
- **Knowing a square number means you also know the corresponding square root**
- **Knowing addition facts means you also know the corresponding subtraction facts**
- **Knowing multiplication facts means that you also know the corresponding division facts**

 MIXED EXERCISE 4

1 Work these out mentally. As far as possible, write down only your final answer.
a) $16 + 76$ b) $0.7 + 0.9$ c) $135 + 6.9$ d) $12.3 + 8.8$ e) $196 + 245$
f) $13.6 - 6.4$ g) $205 - 47$ h) $12 - 3.9$ i) $601 - 218$ j) $15.2 - 8.3$

2 Work these out mentally. As far as possible, write down only your final answer.
   a) 17×9 **b)** 0.6×0.9 **c)** 0.71×1000 **d)** 23×15 **e)** 41×25
   f) $85 \div 5$ **g)** $0.7 \div 2$ **h)** $82 \div 100$ **i)** $1.8 \div 0.2$ **j)** $6 \div 0.5$

3 Work these out mentally. As far as possible, write down only your final answer.
   a) $2 + (-6)$ **b)** $-3 + 9$ **c)** $15 - (-2)$ **d)** $-8 + (-1)$ **e)** $-3 - (-4)$
   f) 7×-4 **g)** -6×-5 **h)** $18 \div -2$ **i)** $-50 \div -10$ **j)** $-12 \div 4$

4 Write down the square of each of these numbers.
   a) 7   **b)** 0.9   **c)** 12   **d)** 100   **e)** 14

5 Jackie spends £84.59. How much change does she get from £100?

6 Round each of these numbers to 1 significant figure.
   a) 9.2   **b)** 3.9   **c)** 26   **d)** 34.9   **e)** 582
   f) 6.0   **g)** 985   **h)** 0.32   **i)** 0.57   **j)** 45 218

7 A rectangle measures 3.9 cm by 8.1 cm. Estimate its area.

8 Pam drove for 2 hours 5 minutes and travelled 106 miles. Estimate her average speed.

9 Estimate the answer to each of these calculations.
   a) 46×82 **b)** $\sqrt{107}$ **c)** $\dfrac{983}{5.2}$ **d)** 6.09^2 **e)** $72.7 \times 19.6 \times 3.3$
   f) $\dfrac{2.46}{18.5}$ **g)** $\dfrac{59}{1.96}$ **h)** 307×51 **i)** $\dfrac{586 \times 97}{187}$ **j)** $\dfrac{318 \times 0.72}{5.1 \times 2.09}$

10 Round each of these numbers to 2 significant figures.
   a) 9.16   **b)** 4.72   **c)** 0.0137   **d)** 164 600   **e)** 507

11 Round each of these numbers to 3 significant figures.
   a) 1482   **b)** 10.16   **c)** 0.021 85   **d)** 20.952   **e)** 0.005 619

12 Simplify each of these calculations, leaving π in your answers.
   a) $2 \times 5 \times \pi$ **b)** $\pi \times 7^2$ **c)** $14\pi + 8\pi$
   d) $\pi \times 5^2 - \pi \times 4^2$ **e)** $\pi \times 9^2 + \pi \times 2^2$ **f)** $\pi \times 8^2 - \pi \times 1^2$

13 Work out these.
   a) 500×30 **b)** 0.2×400 **c)** 2.4×20 **d)** 0.3^2 **e)** 5.13×300
   f) $600 \div 30$ **g)** $3.2 \div 20$ **h)** $2.1 \div 0.03$ **i)** $90 \div 0.02$ **j)** $600 \div 0.05$

14 Given that $1.9 \times 23.4 = 44.46$, work out these.
   a) 19×234 **b)** $44.46 \div 19$ **c)** $4446 \div 234$
   d) 0.19×0.0234 **e)** $444.6 \div 0.234$ **f)** 190×0.0234

15 Given that $126 \times 307 = 38\,682$, work out these.
   a) 12.6×3.07 **b)** $12\,600 \times 3.07$ **c)** $38\,682 \div 30.7$
   d) 0.126×30.7 **e)** $386.82 \div 12.6$ **f)** $38.682 \div 30.7$

5 → SOLVING PROBLEMS

THIS CHAPTER IS ABOUT

- **Understanding the order in which your calculator does calculations**
- **Using your calculator efficiently to do more difficult calculations**
- **Estimating and checking your answers**
- **Selecting suitable strategies and techniques in problem-solving**
- **Applying your knowledge to solving problems using ratio, proportion and percentages and to compound measures such as speed and density**
- **Conversion between measures such as time**
- **Interpreting social statistics such as the retail price index**

YOU SHOULD ALREADY KNOW

- **How to work with ratios**
- **How to use proportion**
- **How to find and use percentages**
- **How to use speed and density**
- **How to round numbers, for example to 1 significant figure**

Order of operations

Your calculator always follows the correct order of operations. This means that it does brackets first, then powers (such as squares), then multiplication and division and lastly addition and subtraction.

If you want to change the normal order of doing things you need to give your calculator different instructions.

Sometimes the simplest way of doing this is to press the $=$ button in the middle of a calculation. This is shown in the following example.

EXAMPLE 5.1

Work out $\dfrac{5.9 + 3.4}{3.1}$.

Solution

You need to work out the addition first.

Press $\boxed{5}\boxed{\cdot}\boxed{9}\boxed{+}\boxed{3}\boxed{\cdot}\boxed{4}\boxed{=}$.

You should see 9.3.

Now press $\boxed{\div}\boxed{3}\boxed{\cdot}\boxed{1}\boxed{=}$.

The answer is 3.

Using brackets

Sometimes you need other ways of changing the order of operations.

For example, in the calculation $\dfrac{5.52 + 3.34}{2.3 + 1.6}$, you need to add 5.52 + 3.45, and then add 2.3 + 1.6, before doing the division.

One way to do this is to write down the answers to the two addition sums and then do the division.

$$\frac{5.52 + 3.34}{2.3 + 1.6} = \frac{8.98}{3.9} = 2.3$$

A more efficient way to do it is to use brackets.

You do the calculation as $(5.52 + 3.45) \div (2.3 + 1.6)$.

This is the sequence of keys to press.

$\boxed{(}\boxed{5}\boxed{\cdot}\boxed{5}\boxed{2}\boxed{+}\boxed{3}\boxed{\cdot}\boxed{4}\boxed{5}\boxed{)}\boxed{\div}\boxed{(}\boxed{2}\boxed{\cdot}\boxed{3}\boxed{+}\boxed{1}\boxed{\cdot}\boxed{6}\boxed{)}\boxed{=}$

Check up 5.1

Enter the sequence above in your calculator and check that you get 2.3.

EXAMPLE 5.2

Use your calculator to work out these calculations without writing down the answers to middle stages.

a) $\sqrt{5.2 + 2.7}$

b) $\dfrac{5.2}{3.7 \times 2.8}$

Solution

a) You need to work out 5.2 + 2.7 before finding the square root.
You use brackets so that the addition is done first.
$\sqrt{(5.2 + 2.7)} = 2.811$ correct to 3 decimal places.

b) You need to work out 3.7 × 2.8 before doing the division.
You use brackets so that the multiplication is done first.
$5.2 \div (3.7 \times 2.8) = 0.502$ correct to 3 decimal places.

EXERCISE 5.1

Work these out on your calculator without writing down the answers to middle stages.
If the answers are not exact, give them correct to 2 decimal places.

1 $\dfrac{5.2 + 10.3}{3.1}$

2 $\dfrac{127 - 31}{25}$

3 $\dfrac{9.3 + 12.3}{8.2 - 3.4}$

4 $\sqrt{15.7 - 3.8}$

5 $6.2 + \dfrac{7.2}{2.4}$

6 $(6.2 + 1.7)^2$

7 $\dfrac{5.3}{2.6 \times 1.7}$

8 $\dfrac{2.6^2}{1.7 + 0.82}$

9 $2.8 \times (5.2 - 3.6)$

10 $\dfrac{6.2 \times 3.8}{22.7 - 13.8}$

11 $\dfrac{5.3}{\sqrt{6.2 + 2.7}}$

12 $\dfrac{5 + \sqrt{25 + 12}}{6}$

Estimating and checking

Discovery 5.1

Without working them out, write down whether or not each of these calculations
is correct.

Give your reason in each case.

a) 1975 × 43 = 84 920

b) 697 × 0.72 = 5018.4

c) 3864 ÷ 84 = 4.6

d) 19 × 37 = 705

e) 306 ÷ 0.6 = 51

f) 6127 × 893 = 54 714.11

Compare your reasons with the rest of the class.

Did you all use the same reasons each time?

Did anyone have ideas which you hadn't thought of and which you think work well?

There are a number of facts that you can use to check a calculation.

- Odd $\times$ odd = odd, even $\times$ odd = even, even $\times$ even = even.
- A number multiplied by 5 will end in 0 or 5.
- The last digit in a multiplication comes from multiplying the last digits of the numbers.
- Multiplying by a number between 0 and 1 makes the original number smaller.
- Dividing by a number greater than 1 makes the original number smaller.
- Calculating an estimate by rounding the numbers to 1 significant figure shows whether the answer is the correct size.

When checking a calculation, there are three main strategies that you can use.

- Common sense
- Estimates
- Inverse operations

Using **common sense** is often a good first check: is your answer about the size you expected?

You may already use **estimates** when you are shopping, to make sure you have enough money and to check you are given the correct change.

> Get into the habit of checking your answers to calculations when solving problems, to see if your answer is sensible.

EXAMPLE 5.3

Estimate the cost of five CDs at £5.99 and two DVDs at £14.99.

Solution

CDs: $5 \times 6 = 30$ Split the calculation into two parts and round
DVDs: $2 \times 15 = 30$ the numbers to 1 significant figure.

Total = 30 + 30
     = £60

EXAMPLE 5.4

Estimate the answer to $\dfrac{\sqrt{394} \times 3.7}{49.2}$.

Solution

$$\frac{\sqrt{394} \times 3.7}{49.2} \approx \frac{\sqrt{400} \times 4}{50}$$

$$= \frac{20 \times 4}{50}$$

$$= \frac{80}{50}$$

$$= \frac{8}{5}$$

$$= 1.6$$

TIP $\approx$ means 'is approximately equal to'.

Inverse operations can be particularly useful when you are working with a calculator and want to check that you pressed the correct keys first time.

For example, to check the calculation $920 \div 64 = 14.375$, you can do $14.375 \times 64 = 920$.

Accuracy of answers

Sometimes, you are asked to give your answers to a **given degree of accuracy**. For example, you may be asked to round your answer to 3 decimal places.

Sometimes you are asked to give the answer to a **sensible degree of accuracy**. In Chapter 29 you will learn that your final answer should not be given to a greater degree of accuracy than any of the values used in the calculation.

EXAMPLE 5.5

Calculate the hypotenuse of a right-angled triangle, given that the other two sides are 4.2 cm and 5.8 cm.

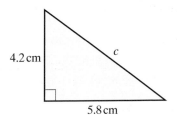

4.2 cm

c

5.8 cm

Solution

Using Pythagoras' theorem,

$c^2 = 4.2^2 + 5.8^2$

$c^2 = 51.28$

$c = \sqrt{51.28}$

$c = 7.161\ 005\ ...$

$c = 7.2$ cm (to 1 decimal place)

Challenge 5.1

In Example 5.5, the answer was given to 1 decimal place, which is to the nearest millimetre.

Think of a situation where it would be most appropriate to give the answer correct to

a) 2 decimal places.   **b)** the nearest 100.

EXERCISE 5.2

Do not use your calculator for questions **1** to **4**.

1 These calculations are all wrong. This can be spotted quickly without working them out. For each one, give a reason why it is wrong.

    a) $6.3 \times -5.1 = 32.13$ **b)** $8.7 \times 0.34 = 29.58$

    c) $3.7 \times 60 = 22.2$ **d)** $\sqrt{62.41} = 8.9$

2 These calculations are all wrong. This can be spotted quickly without working them out. For each one, give a reason why it is wrong.

    a) $5.4 \div 0.9 = 60$ **b)** $-7.2 \div -0.8 = -9$

    c) $5.7^2 = 44.89$ **d)** $13.8 + 9.3 = 22.4$

3 Estimate the answer to each of these calculations. Show your working.

    a) 972×18 **b)** 0.39^2 **c)** $-19.6 \div 5.2$

4 Estimate the answer to each of these calculations. Show your working.

    a) The cost of 7 CDs at £8.99.

    b) The cost of 29 theatre tickets at £14.50.

    c) The cost of 3 meals at £5.99 and 3 drinks at £1.95.

You may use your calculator for questions **5** to **9**.

5 Use inverse operations to check these calculations. Write down the operations you use.

    a) $762.5 \times 81.4 = 62\,067.5$ **b)** $38.3^2 = 1466.89$

    c) $66.88 \div 3.8 = 17.6$ **d)** $69.1 \times 4.3 - 18.2 = 278.93$

6 Work out these. Round your answers to 2 decimal places.

    a) $(48.2 - 19.5) \times 16.32$ **b)** $\dfrac{14.6 + 17.3}{13.8 \times 0.34}$

7 Work out these. Round your answers to 3 decimal places.

    a) $\dfrac{47.3}{6.9 - 3.16}$ **b)** $\dfrac{17.6^3 \times 94.1}{572}$

8 Work out these. Round your answers to 1 decimal place.

    a) 6.3×9.7 **b)** 57×0.085

9 a) Use rounding to 1 significant figure to estimate the answer to each of these calculations. Show your working.

    (i) 39.2^3 **(ii)** 18.4×0.19

    (iii) $\sqrt{7.1^2 - 3.9^2}$ **(iv)** $\dfrac{11.6 + 30.2}{0.081}$

b) Use your calculator to find the correct answer to each of the calculations in part **a)**. Where appropriate, round your answer to a sensible degree of accuracy.

Compound measures

In Chapter 29 you will learn about two **compound measures**: **speed** and **density**. Speed is touched on in this section along with another compound measure: **population density**. This measures the number of people living in a certain area.

EXAMPLE 5.6

a) Figures published in 2007 showed that approximately 118 400 people were living in the county of Gwynedd. The area of Gwynedd is 2 535 km². Calculate the population density of Gwynedd in 2007.

b) The corresponding figures for The Vale of Glamorgan were approximately 124 000 people and an area of 331 km². Calculate the population density of The Vale of Glamorgan in 2007.

c) Comment on your results for parts **a)** and **b)**.

Solution

a) Population density $= \dfrac{\text{population}}{\text{area}}$

    Population density of Gwynedd $= \dfrac{118\,400}{2535}$

                                   $= 46.706\ldots$

                                   $= 47$ people per km²
                                   (to the nearest whole number)

b) Population density of The Vale of Glamorgan $= \dfrac{124\,000}{331}$

                                         $= 374.622\ldots$

                                         $= 375$ people per km²
                                         (to the nearest whole number)

c) The population density for The Vale of Glamorgan is nearly 8 times that of Gwynedd. This reflects the general urban nature of The Vale of Glamorgan compared to the more rural one of Gwynedd.

    Do not forget that statistics like population density are average figures. Both areas have their well populated towns, but overall The Vale of Glamorgan is more densely populated than Gwynedd.

Challenge 5.2

From the internet, or elsewhere, find the population and area of your town or village.

Calculate the population density for your area.

Compare the population densities of two different places locally – perhaps a town and a village, or either of these and a city.

Time

When solving a problem using measures, make sure you check which units you are using. You may need to convert between units.

EXAMPLE 5.7

A train is travelling at 18 metres per second.
Calculate its speed in kilometres per hour.

Solution

18 metres per second	$= 18 \times 60$ metres per minute	1 minute = 60 seconds
	$= 18 \times 60 \times 60$ metres per hour	1 hour = 60 minutes
	$= 64\,800$ metres per hour	
	$= 64.8$ km/h	1 km = 1000 m

EXAMPLE 5.8

Penny's journey to her holiday destination has three stages.
The stages take 3 hours 43 minutes, 1 hour 29 minutes and 4 hours 17 minutes.
How long does her journey take altogether?

Solution

Add the hours and the minutes separately.
$3 + 1 + 4 = 8$ hours
$43 + 29 + 17 = 89$ minutes
89 minutes = 1 hour 29 minutes   There are 60 minutes in 1 hour.
8 hours + 1 hour 29 minutes = 9 hours 29 minutes

TIP

Take care when using your calculator for time problems.
For example, 3 hours 43 minutes is not 3.43 hours, since there are 60 minutes in an hour, not 100.
It is safer to add the minutes separately, as in Example 5.8.

EXAMPLE 5.9

Pali travels 48 miles at an average speed of 30 miles per hour. How long does his journey take? Give your answer in hours and minutes.

Solution

$$\text{Time} = \frac{\text{distance}}{\text{speed}}$$

$$= \frac{48}{30} = 1.6 \text{ hours}$$

$$0.6 \text{ hours} = 0.6 \times 60 \text{ minutes}$$

$$= 36 \text{ minutes}$$

So Pali's journey takes 1 hour 36 minutes.

Discovery 5.2

Some calculators have a $\boxed{\,'\,'\,''\,}$ button. You can use this for working with time on your calculator.

When you do a calculation and the answer is a time in hours but your calculator shows a decimal, you change it to hours and minutes by pressing $\boxed{\,'\,'\,''\,}$ and then the $\boxed{=}$ button.

To change a time in hours and minutes back to a decimal time, use the $\boxed{\text{SHIFT}}$ key.

The $\boxed{\text{SHIFT}}$ key is usually the top left button on your calculator, but it might be called something else.

To enter a time of 8 hours 32 minutes on your calculator, press this sequence of keys.

$\boxed{8}\ \boxed{\,'\,'\,''\,}\ \boxed{3}\ \boxed{2}\ \boxed{\,'\,'\,''\,}\ \boxed{=}$

The display should look like this. $\boxed{8°32°0}$

You may wish to experiment with this button and learn how to use it to enter and convert times.

Solving problems

When solving a problem, break it down into steps.

Read the question carefully and then ask yourself these questions.

- What am I asked to find?
- What information have I been given?
- What methods can I apply?

If you can't see how to find what you need straight away, ask yourself what you can find with the information you are given. Then, knowing that information, ask yourself what you can find next that is relevant.

Many of the complex problems that you meet in everyday life concern money. For example, people have to pay **income tax**. This is calculated as a percentage of what you earn.

Everyone is entitled to a personal allowance (income that is not taxed). For the tax year 2009–2010 this was £6475.

Income in excess of the personal allowance is known as taxable income and is taxed at different rates. For the tax year 2009–2010 the rates were as follows.

Tax bands		Taxable income (£)
Starting rate	10%*	0–2440
Basic rate	22%	0–37 400
Higher rate	40%	over 37 400

*Savings income only.

EXAMPLE 5.10

In the tax year 2009–2010, Stacey earned £48 080.

Calculate how much tax she had to pay.

Solution

Taxable income = £48 080 − £6475
            = £41 605

First subtract the personal allowance from Stacey's total income to find her taxable income.

Tax payable at basic rate = 22% of £37 400
            = 0.22 × £37 400
            = £8228

Stacey's taxable income is more than £37 400. So she pays basic-rate tax on £37 400.

Income to be taxed at higher rate
    = £41 605 − £37 400
    = £4205

To calculate the amount to be taxed at this rate, subtract the £37 400 taxed at the basic rate from Stacey's total taxable income.

Tax payable at higher rate = 0.40 × £4205
            = £1682

Calculate the tax Stacey must pay on the remaining £4205 of her taxable income.

Total tax payable = £8228 + £1682
            = £9910

Finally, add the two lots of tax together to find the total Stacey must pay.

Index numbers

The **Retail Price Index** (RPI) is used by the government to help keep track of how much certain basic items cost. It helps to show how much your money is worth year on year.

The system started in the 1940s and the base price was reset to 100 in January 1987. You can think of this base RPI number as being 100% of the price at the time.

In October 2009 the RPI for all items was 216.0. This showed that there had been an 116% increase in the price of these items since January 1987.

However, the RPI for all items excluding housing costs was 198.8, showing that there had been a smaller increase of 98.8% if housing costs were not included.

The RPI is often referred to in the media, when monthly figures are published: people need price increases to be kept small, otherwise they will, in effect, be poorer unless their income increases.

You can find more information about this and other index numbers on the government's statistics website, www.statistics.gov.uk.

> TIP
>
> This may seem complicated, but index numbers are really just percentages. You learned about percentage increase and decrease in Chapter 2.

EXAMPLE 5.11

In October 2008, the RPI for all items excluding mortgage interest payments was 211.1.

In October 2009, this same RPI was 215.1.

Calculate the percentage increase during that 12 month period.

Solution

Increase in RPI during year = 215.1 − 211.1   First work out the increase in the RPI.
$$= 4$$

Percentage increase $= \dfrac{\text{increase}}{\text{original price}} \times 100$ Then calculate the percentage increase in relation to the 2008 figure.

$$= \dfrac{4}{211.1} \times 100 = 1.89\% \text{ (to 2 decimal places)}$$

1   Write each of these times in hours and minutes.
    a) 2.85 hours   b) 0.15 hours

2   Write each of these times as a decimal.
    a) 1 hour 27 minutes   b) 54 minutes

3   Jason took part in a three-stage race.
    His times for the three stages were 43 minutes, 58 minutes and 1 hour 34 minutes.
    What was his total time for the race? Give your answer in hours and minutes.

4   A courier travelled from Barnsley to Rotherham and then from Rotherham to Sheffield,
    before driving straight back from Sheffield to Barnsley.

    The journey from Barnsley to Rotherham took 37 minutes.
    The journey from Rotherham to Sheffield took 29 minutes.
    The journey from Sheffield to Barnsley took 42 minutes.

    How long did the courier spend travelling in total?
    Give your answer in hours and minutes.

5   Pierre bought 680 g of cheese at £7.25 a kilogram. He also bought some peppers at 69p
    each. The total cost was £8.38. How many peppers did he buy?

6   Two families share the cost of a meal in the ratio 3 : 2. They spend £38.40 on food and
    £13.80 on drinks. How much do the families each pay for the meal?

7   A recipe for four people uses 200 ml of milk.
    Janna makes this recipe for six people. She uses milk from a full 1 litre carton.
    How much milk is left after she has made the recipe?

8   At the beginning of a journey, the mileometer in Steve's car read 18 174.
    At the end of the journey it read 18 309.
    His journey took 2 hours 30 minutes.
    Calculate his average speed.

9   Mr Brown's electricity bill showed that he had used 2316 units of electricity at 7.3p per
    unit. He also pays a standing charge of £12.95. VAT on the total bill was at the rate of 5%.
    Calculate the total bill including VAT.

10 The population density of a Welsh town was 832 people/km^2 in 2009.
    The area of the town is 73 km^2. How many people lived in the town in 2009?

11 In January 2009 the Retail Price Index (RPI) excluding housing was 190.6.
    During the next 12 months it increased by 5.35%.
    What was this RPI in January 2010?

12 A wooden toy brick is a cuboid measuring 2 cm by 3 cm by 5 cm.
Its mass is 66 g. Calculate the density of the wood.

13 A cylindrical water jug has a base radius of 5.6 cm.
Calculate the depth of the water when the jug contains 1.5 litres.

Challenge 5.3

Work in pairs.

Use data from a newspaper or your own experience to write a money problem.

Write the problem on one side of a sheet of paper then, on the other side, solve your problem.

Swap problems with your partner and solve each other's problems.

Check your answers against the solutions provided and discuss instances where you have used different methods.

If you have time, repeat this activity with another problem: perhaps one involving speed, or where you need to change the units.

WHAT YOU HAVE LEARNED

- **How to change the order of operations by using the $=$ button and brackets**
- **Three good strategies for checking answers are common sense, estimates and using inverse operations**
- **Compound measures include speed, density and population density**
- **When solving a problem, break it down into steps. Consider what you have been asked to find, what information you have been given and what methods you can apply. If you can't immediately see how to find what you need, consider what you can find with the given information and then look at the problem again**
- **To check you have used the correct units**

MIXED EXERCISE 5

1 Work out these calculations without writing down the answers to any middle stages.

a) $\dfrac{7.83 - 3.24}{1.53}$

b) $\dfrac{22.61}{1.7 \times 3.8}$

2 Work out $\sqrt{5.6^2 - 4 \times 1.3 \times 5}$.

Give your answer correct to 2 decimal places.

3 These calculations are all wrong.
This can be spotted quickly without working them out.
For each one, give a reason why it is wrong.

a) $7.8^2 = 40.64$ **b)** $2.4 \times 0.65 = 15.6$

c) $58\,800 \div 49 = 120$ **d)** $-6.3 \times 8.7 = 2.4$

4 Estimate the answers to these calculations. Show your working.

a) 894×34 **b)** 0.58^2 **c)** $-48.2 \div 6.1$

5 Work out these using your calculator. Round your answers to 2 decimal places.

a) $(721.5 - 132.6) \times 2.157$ **b)** $\dfrac{19.8 + 31.2}{47.8 \times 0.37}$

6 a) Use rounding to 1 significant figure to estimate the answer to each of these calculations. Show your working.

(i) 21.4^3 **(ii)** 26.7×0.29 **(iii)** $\sqrt{8.1^2 - 4.2^2}$ **(iv)** $\dfrac{31.9 + 48.2}{0.039}$

b) Use your calculator to find the correct answer to each of the calculations in part **a)**. Where appropriate, round your answer to a sensible degree of accuracy.

You may use a calculator for questions **7** to **12**.

7 Ken bought 400 g of meat at £6.95 a kilogram. He also bought some melons at £1.40 each. He paid £6.98. How many melons did he buy?

8 Write each of these times in hours and minutes.

a) 3.7 hours **b)** 2.75 hours

c) 0.8 hours **d)** 0.85 hours

9 Stefan's journey to work took him 42 minutes. He travelled 24 miles. Calculate his average speed in miles per hour.

10 The population density of Anglesey was 94 people/km^2 in 2007. The area of Anglesey is 711 km^2. How many people lived on Anglesey in 2007?

11 Mrs Singh's electricity bill showed that she had used 1054 units of electricity at 7.5p per unit. She also had to pay a standing charge of £13.25. VAT on the total bill was at the rate of 5%. Calculate the total bill including VAT.

12 A glass contains 300 ml of water. It is a cylinder with depth of water 11 cm. Find the radius of the cylinder.

6 → FURTHER PERCENTAGES

THIS CHAPTER IS ABOUT

- **Repeated percentage and proportional change**
- **Reverse percentage and fraction problems**

YOU SHOULD ALREADY KNOW

- **How to use fraction, decimal and percentage notation**
- **How to increase and decrease a quantity by a given percentage**
- **How to find a fraction of a quantity**
- **How to multiply and divide by fractions and mixed numbers**

Repeated percentage change

In Chapter 2 you learned how to increase and decrease a quantity by a given percentage.

Check up 6.1

What must you multiply an amount by to
a) increase it by 5%?
b) increase it by 2.5%?
c) decrease it by 5%?
d) decrease it by 2.5%?
e) increase it by x%?
f) decrease it by x%?

In Chapter 2 you also met repeated percentage change.

A common use of repeated percentage increase is compound interest. This differs from simple interest. When compound interest is paid it is added to the account. The interest is worked out each year on the total in the account, rather than just the original amount.

A common use of repeated percentage decrease is depreciation.

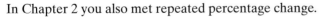

Repeated fractional change

In the same way as you increase a quantity by a given percentage, you increase a quantity by a given fraction by multiplying it by 1 + the fraction. You learned how to multiply by mixed numbers in Chapter 2. To decrease a quantity by a given fraction you divide by 1 − the fraction.

EXAMPLE 6.1

a) Increase £45 by $\frac{1}{5}$.

b) Decrease £63 by $\frac{1}{3}$.

Solution

a) The multiplier is $1 + \frac{1}{5} = \frac{6}{5}$.

Do not convert $\frac{6}{5}$ to a mixed number.

£45 increased by $\frac{1}{5} = 45 \times \frac{6}{5} = £54$.

b) The multiplier is $1 - \frac{1}{3} = \frac{2}{3}$.

£63 decreased by $\frac{1}{3} = 63 \times \frac{2}{3} = £42$.

You use powers to work out repeated fractional change in the same way that you do to work out repeated percentage change.

EXAMPLE 6.2

Andrew said that he would increase his donation to a charity by $\frac{1}{25}$ each year.
His first donation was £120.
How much was his donation after 5 years?

Solution

The multiplier is $1 + \frac{1}{25} = \frac{26}{25}$.

After 5 years his donation is $120 \times \left(\frac{26}{25}\right)^5 = £146.00$ (to the nearest penny).

These are the keys to press on your calculator.

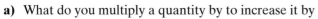

1 2 0 × (2 6 ÷ 2 5) ^ 5 =

EXAMPLE 6.3

The distance that Patricia can walk in a day is reducing by $\frac{1}{15}$ each year.
This year she can walk 12 miles in a day.
How far will she be able to walk in 5 years' time?
Give your answer to 2 decimal places.

Solution

The multiplier is $1 - \frac{1}{15} = \frac{14}{15}$.

In five years the distance is $12 \times \left(\frac{14}{15}\right)^5 = 8.50$ miles to 2 decimal places.

1 Calculate how much these items are worth if they increase by the given fraction each year for the given number of years.

Give your answers to the nearest penny.

	Original value	Fractional increase	Number of years
a)	£280	$\frac{1}{5}$	5
b)	£3500	$\frac{4}{15}$	7
c)	£1400	$\frac{2}{9}$	4

2 Calculate how much these items are worth if they decrease by the given fraction each year for the given number of years.

Give your answers to the nearest penny.

	Original value	Fractional decrease	Number of years
a)	£280	$\frac{1}{5}$	5
b)	£3500	$\frac{4}{15}$	7
c)	£1400	$\frac{2}{9}$	4

3 An investment company claims it will add $\frac{1}{5}$ to your savings each year.

Jasmine invests £3000.

How much should her savings be worth after 10 years?

Give your answer to the nearest pound.

4 In a sale, a clothes shop reduced the price of goods by $\frac{1}{3}$ each day until they were sold.

A coat was originally priced at £60.

What was its price after 3 days, to the nearest penny?

5 It is claimed that the number of rabbits in Freeshire is increasing by $\frac{1}{12}$ each year.

It is estimated there are currently 1700 rabbits.

How many rabbits will there be after 4 years if the statement is true?

Give your answer correct to 3 significant figures.

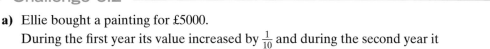

Challenge 6.2

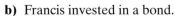

a) Ellie bought a painting for £5000.

During the first year its value increased by $\frac{1}{10}$ and during the second year it increased by $\frac{1}{8}$ of its new value.

What was its value after the two years?

b) Francis invested in a bond.

The value increased by $\frac{1}{20}$ in the first year, but then dropped by $\frac{1}{15}$ of the new value in the second year.

What was the overall fractional increase or decrease of the original value over the two years?

Reverse percentage and fraction problems

If a quantity is increased by a percentage, $x\%$, the new amount is $(100 + x)\%$ of the original amount. The multiplier is $\left(1 + \frac{x}{100}\right)$.

To find the original amount when you are given the new amount, you divide by the multiplier, $\left(1 + \frac{x}{100}\right)$.

If a quantity is decreased by a percentage, $x\%$, the new amount is $(100 - x)\%$ of the original amount. The multiplier is $\left(1 - \frac{x}{100}\right)$.

To find the original amount when you are given the new amount, you divide by the multiplier, $\left(1 - \frac{x}{100}\right)$.

EXAMPLE 6.4

Ben received a 20% increase in his salary.
After the increase his salary was £31 260.
What was his salary before the increase?

Solution

New salary = old salary × 1.2   The multiplier is 1.2.

Old salary = new salary ÷ 1.2   To reverse the process, divide by the multiplier.
$\quad\quad\quad\quad$ = 31 260 ÷ 1.2
$\quad\quad\quad\quad$ = £26 050

EXAMPLE 6.5

In a sale, everything was reduced by 7.5%.

a) David bought a jacket in the sale.
It was originally priced at £85.
What was the sale price?

b) Irene paid £38.70 for a skirt in the sale.
What was the original price of the skirt?

Give your answers to a sensible degree of accuracy.

Solution

Since 1p is the smallest coin, a sensible degree of accuracy is to the nearest penny.

a) Sale price = original price × 0.925   The multiplier is $1 - 0.075 = 0.925$.
Sale price = 85×0.925
$= £78.63$ (to the nearest penny)

b) Original price = sale price ÷ 0.925
$= £38.70 \div 0.925$
$= £41.84$ (to the nearest penny)

> **TIP**
>
> When you are asked to give an answer to a sensible degree of accuracy, look at how accurate the figures are in the question.

Just as with percentages, if a quantity is increased by a fraction, the new amount is (1 + fraction) × the original amount.
To find the original amount when you are given the new amount, you divide by the multiplier, (1 + fraction).

If a quantity is decreased by a fraction, the new amount is (1 − fraction) × the original amount.
To find the original amount when you are given the new amount, you divide by the multiplier, (1 − fraction).

You learned how to divide by a fraction in Chapter 2.

Check up 6.4

Work out these.

a) $45 \div \frac{5}{6}$ **b)** $26 \div \frac{4}{3}$ **c)** $31.5 \div \frac{9}{7}$ **d)** $46.2 \div \frac{7}{4}$ **e)** $297 \div \frac{2}{3}$

EXAMPLE 6.6

On a jar of coffee it says 'Now contains $\frac{1}{5}$ more'.
The new size contains 564 g.
How much did a jar contain before the increase?

Solution

New amount = old amount $\times \frac{6}{5}$

Old amount = new amount $\div \frac{6}{5}$

$$= 564 \div \frac{6}{5}$$

$$= 564 \times \frac{5}{6} \qquad \text{To divide by } \tfrac{6}{5} \text{ you multiply by } \tfrac{5}{6}.$$

$$= 470 \text{ g}$$

EXERCISE 6.2

Give your answers to a sensible degree of accuracy.

1   Copy and complete this table.

	Original value	Percentage increase	New value
a)	£80	15%	
b)		12%	£52.64
c)	£185	4%	
d)		7.5%	£385.50

2   Copy and complete this table.

	Original value	Percentage decrease	New value
a)	£800	13%	
b)		8%	£113.16
c)	£227	2.5%	
d)		4.5%	£374.50

3   Alan invested in a bond that increased by 7.5% each year. After one year his investment was worth £875. How much did he invest?

4   A coat originally cost £79. In a sale, it was reduced by 5%. What was the sale price?

5 Mrs Dale made a profit of £13 250 this year. This was 6% more than her profit last year. What was her profit last year?

6 In a sale, all items were reduced by 2.5%. A computer cost £740 in the sale. What was the pre-sale price of the computer?

7 Henry was given a pay rise of 7%. His salary after the rise was £36 850. What was his salary before the rise?

8 In a local election last year, Labour received 1375 votes. The number of votes they received this year was 12% greater.
How many votes did Labour receive this year?

9 A holiday cost £510 including VAT at 20%.
What was the cost before VAT?

10 At Jane's café all the prices were increased by 5% (to the nearest penny).
   a) A cup of tea cost 75p before the increase. What is the new price?
   b) The new price of a cup of coffee is £1.30. What was the old price?

11 During a special offer, a train company reduces all its fares by a third.
   a) A journey usually costs £84. What is the cost of the journey during the special offer?
   b) During the special offer, a journey costs £36.50. What is the usual cost of the journey?

12 A computer servicing company increased all charges by $\frac{1}{8}$. After the increase, Sam paid £94.50 to have his computer serviced.
How much would it have cost before the increase?

Challenge 6.3

   a) A jumper was reduced by 12.5% in a sale.
   The reduction was £7.50.
   What was the sale price?

   b) Sarah invested in a bond which increased by 4.5% each year for 5 years.
   At the end of that time her bond was worth £10 530.24.
   How much did she invest?

   c) There are 36 more girls than boys in a school.
   54% of the students at the school are girls.
   How many girls are there at the school?

continues ...

d) Joe invests £1000 in a savings account that pays 4.5% compound interest annually.
He invests a further £1000 at the end of each year for 4 years.
He then leaves his money in the account for a further 3 years without adding to it.
How much does he have in the account at the end of this time?

e) Maria has £12 000 in a savings account that pays 4.75% compound interest annually.
Each year she takes £1000 out of the account immediately after the interest has been added.
   (i)   After how many years does she first have less than £5000 in her account?
   (ii) How much is there then in the account?

f) The value of a car decreased by $\frac{1}{5}$ each year for 5 years. At the end of that time the car was worth £6144. How much was it worth when new?

g) A canal boat increased in value by $\frac{1}{15}$ each year for 2 years.
The overall increase was £3100. What was the original value of the boat?

WHAT YOU HAVE LEARNED

- When a quantity is increased by $x\%$ for n years, the multiplier is $\left(1 + \dfrac{x}{100}\right)^n$

- When a quantity is decreased by $x\%$ for n years, the multiplier is $\left(1 - \dfrac{x}{100}\right)^n$

- The multiplier for fractional increase is (1 + fraction)
- The multiplier for fractional decrease is (1 − fraction)
- To find the original amount when given the new amount and the increase, you divide by the multiplier

---

MIXED EXERCISE 6

Give your answers to a sensible degree of accuracy.

1 A population of bacteria is growing at a rate of 5% a day. There are 1450 bacteria on Tuesday. How many are there 3 days later?

2 A newspaper reported that the number of people taking their main holiday in Britain had reduced by 10% a year over the last 5 years. Five years ago, 560 people from a small town took their main holiday in Britain. If the report is true, how many of them would you expect to take their main holiday in Britain now?

3 Copy and complete this table.

	Original value	Increase or decrease	New value
a)	£700	Increase of 3.5%	
b)		Increase of 4.5%	£28.54
c)	£365	Decrease of 5%	
d)		Decrease of 7%	£840
e)		Increase of $\frac{2}{9}$	£92.40

4 Damien sold his bicycle for £286, at a loss of 45% on what he paid for it. How much did he pay?

5 A magazine sold 10 240 copies this month. This was an increase of $\frac{1}{9}$ on the number sold last month. How many were sold last month?

6 At the Star theatre, the seating plan was changed and the number of seats in the stalls was increased by a third. There are now 312 seats in the stalls. How many were there before the increase?

7 Calculate how much £5000 invested with compound interest would be worth in each of the following cases.

    a) 4.5% for 3 years   **b)** 3.1% for 5 years   **c)** 3.8% for 8 years

8 A bond pays 4% compound interest annually for the first 2 years and 6% each year for the next three years. Jenny invested £2000 in the bond. What was the bond worth

    a) after 2 years?   **b)** after 5 years?

9 In a sale, all the prices were reduced by 10%.

    a) A coat cost £125 before the sale. What was its sale price?

    b) The sale price of a suit was £156.42. What the pre-sale price?

10 A car manufacturer claims that for its latest model, the number of miles per gallon has increased by $\frac{1}{5}$. The new model travels 48 miles per gallon.
How many miles per gallon did the old model travel?

7 → INDICES AND STANDARD FORM

Raising a power to another power

In Chapter 1 you learned that $a^m \times a^n = a^{m+n}$. From this rule for multiplication you can work out the following.

$$(a^4)^3 = a^4 \times a^4 \times a^4$$
$$= a^{4+4+4}$$
$$= a^{3 \times 4}$$
$$= a^{12}$$

This result can be written in the general form

$$(a^m)^n = a^{m \times n}$$

Other powers

Discovery 7.1

a) Copy and complete this table for $y = 2^x$.

x	5	4	3	2	1
$y = 2^x$			8		2

b) What is the rule for getting from one term to the next along the bottom row?

c) Use this rule to find the last four entries in this table.

x	5	4	3	2	1	0	-1	-2	-3
$y = 2^x$			8		2				

Two to the power zero is 1.

$$2^0 = 1$$

This fits with the rules for multiplication and division. For example

$2^0 \times 2^3 = 2^{0+3} = 2^3$ which means that $2^0 = 1$.

$2^4 \div 2^4 = 2^{4-4} = 2^0$ which again means $2^0 = 1$ because $2^4 \div 2^4 = 1$.

Discovery 7.2

Use the multiplication and division rules to show that $3^0 = 1$ and $4^0 = 1$.

This is the general rule.

$a^0 = 1$, whatever the value of a.

Here is the completed table from Discovery 7.1.

x	5	4	3	2	1	0	-1	-2	-3
$y = 2^x$	32	16	8	4	2	1	$\frac{1}{2}$	$\frac{1}{4}$	$\frac{1}{8}$

From the table you can see that

$$2^{-1} = \frac{1}{2}, \quad 2^{-2} = \frac{1}{4} = \frac{1}{2^2} \quad \text{and} \quad 2^{-3} = \frac{1}{8} = \frac{1}{2^3}.$$

These fit with the division rule. For example

$$2^2 \div 2^5 = 2^{2-5} = 2^{-3} \quad \text{and} \quad \frac{2^2}{2^5} = \frac{4}{32} = \frac{1}{8} = \frac{1}{2^3}.$$

Discovery 7.3

Use the division rule to show that $3^{-2} = \frac{1}{3^2}$.

This is the general rule for negative powers.

$$a^{-n} = \frac{1}{a^n}$$

Discovery 7.4

Here again is a table for $y = 2^x$.

x	0	1	2	3	4
$y = 2^x$	1	2	4	8	16

a) Plot a graph of $y = 2^x$.

b) Use your graph to suggest values for $2^{\frac{1}{2}}, 2^{\frac{1}{3}}$ and $2^{\frac{3}{2}}$.
Use the rule $(a^m)^n = a^{m \times n}$ to find the value of $(2^{\frac{1}{2}})^2$.
What does this suggest that $2^{\frac{1}{2}}$ means?
Does this agree with the answer you read off the graph?

c) Similarly use the rule $(a^m)^n = a^{m \times n}$ to find the value of $(2^{\frac{1}{3}})^3$.
What does this suggest that $2^{\frac{1}{3}}$ means?
Does this agree with the value you read off the graph?

d) Again use the rule $(a^m)^n = a^{m \times n}$ to find the value of $(2^{\frac{3}{2}})^2$.
What does this suggest that $2^{\frac{3}{2}}$ means?
Does this agree with the value you read off the graph?

Discovery 7.5

Use the rule $(a^m)^n = a^{m \times n}$ to find the meaning of $3^{\frac{1}{2}}$ and $4^{\frac{1}{3}}$.

This is the general rule for fractional powers with 1 as the numerator.

$$a^{\frac{1}{n}} = \sqrt[n]{a}$$

This is the general rule for any fractional power.

$$a^{\frac{m}{n}} = \sqrt[n]{a^m}$$

Now $8^{\frac{2}{3}} = \sqrt[3]{8^2} = \sqrt[3]{64} = 4$.

But also $(\sqrt[3]{8})^2 = 2^2 = 4$.

This suggests that you can do the root and the power in either order: $\sqrt[3]{8^2}$ or $(\sqrt[3]{8})^2$.

So the general rule above becomes as follows.

$$a^{\frac{m}{n}} = \sqrt[n]{a^m} = (\sqrt[n]{a})^m$$

EXAMPLE 7.1

Write the following in index form.

a) The cube root of x

b) The reciprocal of x^2

c) $\sqrt[5]{x^2}$

Solution

a) $x^{\frac{1}{3}}$ b) x^{-2} c) $x^{\frac{2}{5}}$

EXAMPLE 7.2

Work out the following.

a) $49^{\frac{1}{2}}$ b) 5^{-2} c) $125^{\frac{1}{3}}$

d) $4^{\frac{5}{2}}$ e) $\left(\frac{1}{3}\right)^{-2}$

Solution

a) $49^{\frac{1}{2}} = \sqrt{49} = 7$ b) $5^{-2} = \dfrac{1}{5^2} = \dfrac{1}{25}$

c) $125^{\frac{1}{3}} = \sqrt[3]{125} = 5$ d) $4^{\frac{5}{2}} = (\sqrt{4})^5 = 2^5 = 32$

e) $\left(\frac{1}{3}\right)^{-2} = \dfrac{1}{\left(\frac{1}{3}\right)^2} = \dfrac{1}{\left(\frac{1}{9}\right)} = 9$

TIP

If you have to work out the square root of the cube of a number it is usually easier to find the square root first.

If you want a fraction to a negative power, you can use the reciprocal of the fraction to the positive power.

$$\left(\frac{a}{b}\right)^{-n} = \left(\frac{b}{a}\right)^{n}$$

Check up 7.1

Match each number in index form to an ordinary number.

You may not need all of the numbers but you may need some more than once.

2^3 $\left(\frac{1}{2}\right)^{-2}$ $1000^{\frac{2}{3}}$ $64^{\frac{1}{2}}$ 10^2 $64^{\frac{1}{3}}$ $(0.5)^{-1}$ $8^{\frac{2}{3}}$ $4^{\frac{1}{2}}$ $64^{\frac{2}{3}}$ $64^{\frac{1}{6}}$ 7^0 $36^{\frac{1}{2}}$ $64^{\frac{5}{6}}$

0   1   2   4   6   8   9   10   16   20   32   64   100   200

EXERCISE 7.1

1 Write these in index form.

a) The square root of x **b)** $\dfrac{1}{x^4}$ **c)** $\sqrt{x^5}$

2 Work out these. Give your answers as whole numbers or fractions.

a) 4^{-1} **b)** $4^{\frac{1}{2}}$ **c)** 4^0 **d)** 4^{-2} **e)** $4^{\frac{3}{2}}$

f) $100^{\frac{3}{2}}$ **g)** $64^{\frac{2}{3}}$ **h)** $81^{-\frac{1}{2}}$ **i)** 12^0 **j)** $32^{\frac{4}{5}}$

k) $8^{\frac{1}{3}}$ **l)** 8^{-1} **m)** $8^{\frac{4}{3}}$ **n)** $\left(\frac{1}{8}\right)^{-1}$ **o)** $8^{-\frac{1}{3}}$

3 Work out these. Give your answers as whole numbers or fractions.

a) $3^2 \times 4^{\frac{1}{2}}$ **b)** $3^4 \times 9^{\frac{1}{2}}$ **c)** $125^{\frac{1}{3}} \times 5^{-2}$

d) $16^{\frac{1}{2}} \times 6^2 \times 2^{-3}$ **e)** $2^3 + 4^0 + 49^{\frac{1}{2}}$ **f)** $\left(\frac{1}{2}\right)^{-3} \times 27^{\frac{2}{3}}$

g) $6^2 \div 25^{\frac{1}{2}}$ **h)** $5^2 + 8^{\frac{1}{3}} - 7^0$

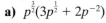 **Challenge 7.1**

Expand and simplify these.

a) $p^{\frac{3}{2}}(3p^{\frac{1}{2}} + 2p^{-2})$ **b)** $(3a^{\frac{1}{2}} - 2)(5a^{\frac{1}{2}} + 1)$

Using a calculator

The power button on a calculator is usually labelled $\boxed{\wedge}$ or $\boxed{x^y}$ or $\boxed{y^x}$.
You can use your calculator to work out any number to any power.

EXAMPLE 7.3

Work out $3.1^{2.4}$.

Solution

This is the sequence of keys to press.

$\boxed{3}\boxed{.}\boxed{1}\boxed{\wedge}\boxed{2}\boxed{.}\boxed{4}\boxed{=}$

You should get 15.110…

Even if the power button on your calculator is labelled $\boxed{x^y}$ or $\boxed{y^x}$ rather
than $\boxed{\wedge}$, you may see the symbol $\wedge$ in your display.

You can combine the use of the power button with the fraction button $\boxed{a^{b}/c}$ for powers such as $\frac{3}{4}$.

Above the power button you may see $\sqrt[x]{}$ or $\sqrt[x]{y}$ or $y^{1/x}$. These are usually in yellow. They allow you to find any root of a number. The yellow functions are operated using the $\boxed{\text{SHIFT}}$ or $\boxed{\text{INV}}$ or $\boxed{\text{2nd F}}$ button (usually yellow) before the main button.

EXAMPLE 7.4

Find the cube root of 27.

Solution

The cube root of 27 in index form is $27^{\frac{1}{3}}$.

This is the sequence of keys to press.

$\boxed{3}\ \boxed{\text{SHIFT}}\ \boxed{\land}\ \boxed{2}\ \boxed{7}\ \boxed{=}$

You should get 3.

EXAMPLE 7.5

Work out these. Give your answers exactly or to 5 significant figures.

a) 4.7^4 b) 2.3^5 c) 2.1^{-2}

d) $17.8^{\frac{1}{4}}$ e) $729^{\frac{5}{6}}$

Solution

a) $4.7^4 = 487.968\,1 = 487.97$ **b)** $2.3^5 = 64.363\,43 = 64.363$

c) $2.1^{-2} = 0.226\,757... = 0.226\,76$ **d)** $17.8^{\frac{1}{4}} = 2.054\,021... = 2.0540$

e) $729^{\frac{5}{6}} = 243$

TIP

When you find a root using a calculator, it is very helpful to check by working backwards.

For example, in part **d)** of Example 7.5, work out $2.054\,021^4 = 17.799\,98...$ to confirm your answer.

EXERCISE 7.2

1 Work out these. Give your answers exactly or to 5 significant figures.

  a) 4.2^3 **b)** 0.52^4 **c)** 2.01^6 **d)** 3.24^{-3}

  e) $16\,807^{\frac{1}{5}}$ **f)** $5.32^{\frac{1}{4}}$ **g)** $\sqrt[3]{23}$ **h)** $243^{\frac{3}{5}}$

2 Work out these. Give your answers exactly or to 5 significant figures.

  a) 200×1.03^4 **b)** $1.3^5 \times 3.2^4$ **c)** $2.5^5 \div 1.3^{-4}$ **d)** $(5.6 \times 2.3^3)^{\frac{1}{4}}$

  e) $2.35 + 1.2^6$ **f)** $2.3^{\frac{1}{3}} - 1.9^{\frac{1}{5}}$ **g)** $5.2^4 + 0.3^{-3}$ **h)** $15^3 - 225^{\frac{3}{2}}$

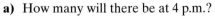

Challenge 7.2

The number of bacteria in a petri dish doubles every 2 hours.
There are 400 bacteria in the dish at midday.

a) How many will there be at 4 p.m.?

b) How many will there be at midnight?

  Write your answer in the form 400×2^t.

  Use the $\boxed{\wedge}$ button on your calculator to work out the answer.

Using the rules of indices with numbers and letters

The rules you have learned can be used with numbers or letters.

Here are the rules again.

- $a^m \times a^n = a^{m+n}$
- $a^m \div a^n = a^{m-n}$
- $(a^m)^n = a^{m \times n}$

> **TIP**
>
> A common mistake is trying to simplify an addition or subtraction.
>
> There is no rule for simplifying $a^x + a^y$ or $a^x - a^y$.
>
> Also, there is no rule for simplifying $a^x \times b^y$ or $a^x \div b^y$ because the **base** is different. This is another way of saying that they are unlike terms.

EXAMPLE 7.6

Where possible, write these as single powers of 2.

a) $2\sqrt{2}$
b) $(\sqrt[3]{2})^2$
c) $2^3 \div 2^{\frac{1}{2}}$
d) $2^3 + 2^4$
e) $8^{\frac{3}{4}}$
f) $2^3 \times 4^{\frac{3}{2}}$
g) $2^n \times 4^3$

Solution

a) $2\sqrt{2} = 2^1 \times 2^{\frac{1}{2}} = 2^{\frac{3}{2}}$

b) $(\sqrt[3]{2})^2 = (2^{\frac{1}{3}})^2 = 2^{\frac{2}{3}}$

c) $2^3 \div 2^{\frac{1}{2}} = 2^{3-\frac{1}{2}} = 2^{2\frac{1}{2}} = 2^{\frac{5}{2}}$

d) $2^3 + 2^4$ (these cannot be added)

e) $8^{\frac{3}{4}} = (2^3)^{\frac{3}{4}} = 2^{3 \times \frac{3}{4}} = 2^{\frac{9}{4}}$

f) $2^3 \times 4^{\frac{3}{2}} = 2^3 \times (2^2)^{\frac{3}{2}} = 2^3 \times 2^3 = 2^6$

g) $2^n \times 4^3 = 2^n \times (2^2)^3 = 2^n \times 2^6 = 2^{n+6}$

 ## EXERCISE 7.3

1 Write these as powers of 3 as simply as possible.

a) 81
b) $\frac{1}{3}$
c) $3 \times \sqrt{3}$
d) $3^4 \times 9^{-1}$
e) 9^n
f) 27^{3n}
g) $9^n \times 27^{3n}$

2 Where possible, write these as powers of 5 as simply as possible.

a) 0.2
b) 125
c) 25^2
d) $125 \times 5^{-2} \times 25^2$
e) $5^4 - 5^3$
f) $25^{3n} \times 125^{\frac{n}{3}}$

3 Write these in the form $2^a \times 3^b$.

a) 18
b) 72
c) $18^{\frac{1}{3}}$
d) $\frac{4}{9}$
e) $13\frac{1}{2}$

4 Write each of these numbers as a product of prime numbers. Use indices where possible.

For example, $\dfrac{8}{\sqrt{3}} = 8 \times \dfrac{1}{\sqrt{3}} = 2^3 \times 3^{-\frac{1}{2}}$

a) 75
b) 288
c) 500
d) 3240

Challenge 7.3

Simplify these.

a) $x^2 \times x^{\frac{1}{2}}$

b) $x^{\frac{1}{2}} \div x^2$

c) $x^3 \times x^{-2}$

d) $x^4 \div x^{-2}$

e) $\sqrt{\dfrac{8a^3}{2a^2}}$

f) $\sqrt[3]{x^6}$

Challenge 7.4

The numbers of a certain species of animal are reducing by 15% every 10 years.

In 1960 there were 30 000.

How many will there be

a) in 2010?

b) in 2050?

c) n years after 1960?

Standard form

Standard form is a very important use of indices. It is a way of making very large numbers and very small numbers easy to deal with.

In standard form, numbers are written as a number between 1 and 10 multiplied by a power of 10.

Large numbers

EXAMPLE 7.7

Write these numbers in standard form.

a) 500 000

b) 6 300 000

c) 45 600

Solution

a) $500\,000 = 5 \times 100\,000 = 5 \times 10^5$

b) $6\,300\,000 = 6.3 \times 1\,000\,000 = 6.3 \times 10^6$

c) $45\,600 = 4.56 \times 10\,000 = 4.56 \times 10^4$

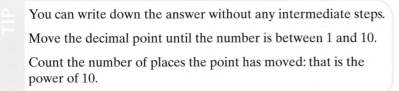

You can write down the answer without any intermediate steps.

Move the decimal point until the number is between 1 and 10.

Count the number of places the point has moved: that is the power of 10.

Small numbers

Write these numbers in standard form.

a) 0.000 003   **b)** 0.000 056   **c)** 0.000 726

Solution

a) $0.000\,003 = \dfrac{3}{1\,000\,000} = 3 \times \dfrac{1}{1\,000\,000} = 3 \times \dfrac{1}{10^6} = 3 \times 10^{-6}$

b) $0.000\,056 = \dfrac{5.6}{100\,000} = 5.6 \times \dfrac{1}{100\,000} = 5.6 \times \dfrac{1}{10^5} = 5.6 \times 10^{-5}$

c) $0.000\,726 = \dfrac{7.26}{10\,000} = 7.26 \times \dfrac{1}{10\,000} = 7.26 \times \dfrac{1}{10^4} = 7.26 \times 10^{-4}$

> **TIP**
>
> You can write down the answer without any intermediate steps.
>
> Move the decimal point until the number is between 1 and 10.
>
> Count the number of places the point has moved: put a minus sign in front and that is the power of 10.

Discovery 7.6

Find the approximate distance, in kilometres, of each of the planets from the Sun.

Write the distances in standard form.

(You will find the distances in books such as atlases or encyclopedias or on the internet.)

EXERCISE 7.4

1 Write these numbers in standard form.

   a) 7000   **b)** 84 000   **c)** 563   **d)** 6 500 000

   e) 723 000   **f)** 27   **g)** 8 million   **h)** 39.2 million

2 Write these numbers in standard form.

   a) 0.003   **b)** 0.056   **c)** 0.000 38

   d) 0.000 006 3   **e)** 0.000 082   **f)** 0.000 000 38

3 These numbers are in standard form. Write them as ordinary numbers.

   a) 5×10^4 **b)** 3.7×10^5 **c)** 7×10^{-4} **d)** 6.9×10^6

   e) 6.1×10^{-3} **f)** 4.73×10^4 **g)** 2.79×10^7 **h)** 4.83×10^{-5}

   i) 1.03×10^{-2} **j)** 9.89×10^8 **k)** 2.61×10^{-6} **l)** 3.7×10^2

Calculating with numbers in standard form

When you need to multiply or divide numbers in standard form, you can use your knowledge of the laws of indices.

EXAMPLE 7.9

Work out these. Give your answers in standard form.

a) $(7 \times 10^3) \times (4 \times 10^4)$

b) $(7 \times 10^7) \div (2 \times 10^{-3})$

c) $(3 \times 10^8) \div (5 \times 10^3)$

Solution

a) $(7 \times 10^3) \times (4 \times 10^4) = 7 \times 4 \times 10^3 \times 10^4 = 28 \times 10^{3+4} = 28 \times 10^7 = 2.8 \times 10^8$

b) $(7 \times 10^7) \div (2 \times 10^{-3}) = \dfrac{7 \times 10^7}{2 \times 10^{-3}} = 3.5 \times 10^{7-(-3)} = 3.5 \times 10^{10}$

c) $(3 \times 10^8) \div (5 \times 10^3) = \dfrac{3 \times 10^8}{5 \times 10^3} = 0.6 \times 10^{8-3} = 0.6 \times 10^5 = 6 \times 10^4$

When you need to add or subtract numbers in standard form, it is much safer to change to ordinary numbers first.

EXAMPLE 7.10

Work out these. Give your answers in standard form.

a) $(7 \times 10^3) + (4.0 \times 10^4)$

b) $(7.2 \times 10^5) + (2.5 \times 10^4)$

c) $(5.3 \times 10^{-3}) - (4.9 \times 10^{-4})$

Solution

a)
$$\begin{array}{r} 7\,0\,0\,0 \\ +\,4\,0\,0\,0\,0 \\ \hline 4\,7\,0\,0\,0 = 4.7 \times 10^4 \end{array}$$

b)
$$\begin{array}{r} 7\,2\,0\,0\,0\,0 \\ +\quad 2\,5\,0\,0\,0 \\ \hline 7\,4\,5\,0\,0\,0 = 7.45 \times 10^5 \end{array}$$

c)
$$\begin{array}{r} 0\,.\,0\,0\,5\,3\,0 \\ -\,0\,.\,0\,0\,0\,4\,9 \\ \hline 0\,.\,0\,0\,4\,8\,1 = 4.81 \times 10^{-3} \end{array}$$

Standard form on your calculator

You can do all the calculations above on your calculator using the $\boxed{\text{EXP}}$ button.

EXAMPLE 7.11

Find $(7 \times 10^7) \div (2 \times 10^{-3})$ using your calculator.

Solution

This is part **b)** of Example 7.9. These are the keys to press on your calculator.

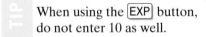

You should see 3.5×10^{10}.

> **TIP**
> When using the $\boxed{\text{EXP}}$ button, do not enter 10 as well.

On some calculators the $\boxed{(-)}$ button is marked $\boxed{+\!\!/\!\!-}$.

Sometimes your calculator will give you an ordinary number which you will have to write in standard form, if you are asked to give your answer in this format. Otherwise, your calculator will give you the answer in standard form.

Modern calculators usually give the correct version of standard form, for example 2.8×10^8.

Older calculators often give a calculator version such as 2.8^{08}. You must write this in proper standard form, 2.8×10^8, for your answer.

Some graphical calculators display standard form as, for example, 2.8 E 08.

Again you must write this in proper standard form for your answer.

Check up 7.4

Practise by checking the rest of Example 7.9 and Example 7.10 on your calculator.

EXERCISE 7.5

1 Work out these. Give your answers in standard form.

   a) $(4 \times 10^3) \times (2 \times 10^4)$ **b)** $(6 \times 10^7) \times (2 \times 10^3)$

   c) $(7 \times 10^3) \times (8 \times 10^2)$ **d)** $(4.8 \times 10^3) \div (1.2 \times 10^{-2})$

   e) $(4 \times 10^3) \times (1.3 \times 10^4)$ **f)** $(4 \times 10^9) \div (8 \times 10^4)$

   g) $(4 \times 10^3) + (6 \times 10^4)$ **h)** $(6.2 \times 10^5) - (3.7 \times 10^4)$

2 Work out these. Give your answers in standard form.

a) $(6.2 \times 10^5) \times (3.8 \times 10^7)$ 　　　b) $(6.3 \times 10^7) \div (4.2 \times 10^2)$

c) $(6.67 \times 10^8) \div (4.6 \times 10^{-3})$ 　　d) $(3.7 \times 10^{-4}) \times (2.9 \times 10^{-3})$

e) $(1.69 \times 10^8) \div (5.2 \times 10^3)$ 　　f) $(7.63 \times 10^5) + (3.89 \times 10^4)$

g) $(3.72 \times 10^6) - (2.8 \times 10^4)$ 　　h) $(5.63 \times 10^{-3}) - (4.28 \times 10^{-4})$

Challenge 7.5

Light takes approximately 3.3×10^{-9} seconds to travel 1 metre.
The distance from the Earth to the Sun is 150 000 000 km.

a) Write 150 000 000 km in metres using standard form.

b) How long does it take for light to reach Earth from the Sun?

Challenge 7.6

Paper is 0.08 mm thick.

a) Write this thickness in metres using standard form.

b) A library has 4.6×10^4 metres of shelf space.
Assuming 80% of this is filled with paper, the rest being the covers of the books,
estimate how many sheets of paper there are on the shelves.
Give your answer in standard form.

Challenge 7.7

The table shows information about the Earth.

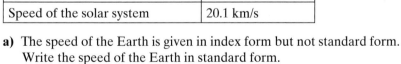

Distance from the Sun	149 503 000 km
Circumference of solar orbit	9.4×10^8 km
Speed of the Earth in solar orbit	0.106×10^6 km/h
Speed of the solar system	20.1 km/s

a) The speed of the Earth is given in index form but not standard form.
Write the speed of the Earth in standard form.

b) How far does the Earth travel at this speed in one day?
Give your answer in standard form correct to 3 significant figures.

c) How far does the solar system travel in one day?
Give your answer in standard form correct to 3 significant figures.

d) Imagine an object travelled from the Earth to the Sun and back. How far would it travel?
Give your answer in standard form correct to three significant figures.

- **The rules of indices**
  - $a^m \times a^n = a^{m+n}$
  - $a^m \div a^n = a^{m-n}$
  - $(a^m)^n = a^{m \times n}$
  - $a^0 = 1$, whatever the value of a
  - $a^{-n} = \dfrac{1}{a^n}$
  - $a^{\frac{1}{n}} = \sqrt[n]{a}$
  - $a^{\frac{m}{n}} = \sqrt[n]{a^m} = (\sqrt[n]{a})^m$

- **Standard form is used to deal with very large and very small numbers**
  - Numbers are written as $a \times 10^n$ where a is between 1 and 10 and n is a whole number
  - Large numbers such as 93 million (93 000 000) are written as 9.3×10^7
  - Small numbers such as 0.000 007 82 are written as 7.82×10^{-6}
  - You can enter numbers in standard form in your calculator using the $\boxed{\text{EXP}}$ button

MIXED EXERCISE 7

1 Work out these. Give your answers as whole numbers or fractions.

a) $9^{\frac{1}{2}}$　　b) $27^{\frac{2}{3}}$　　c) 3^{-2}　　d) $16^{-\frac{1}{2}}$　　e) $100^{\frac{5}{2}}$

2 Work out these. Give your answers as whole numbers or fractions.

a) $5^{-1} \times 4^{\frac{1}{2}}$　　　　b) $81^{\frac{1}{2}} \times 3^0 \times 144^{-\frac{1}{2}}$

c) $16^{\frac{3}{4}} + 5^0 + 1000^{\frac{2}{3}}$　　d) $8^{-\frac{1}{3}} \times 49^{\frac{1}{2}} \times 3^{-1}$

3 Work out these. Give your answers to 4 significant figures.

a) 2.6^3　　b) $4.3^{\frac{1}{2}}$　　c) 3.8^{-2}　　d) $\sqrt[5]{867}$　　e) $50^{\frac{2}{3}}$

4 Work out these. Give your answers to 4 significant figures.

a) $5.8^4 \div 2.6^3$　　　　b) $2.8^{\frac{1}{2}} \times 7.6^{-2}$

c) $(3.8 \times 1.7^4)^{\frac{1}{3}}$　　d) $5^{\frac{1}{2}} + 3^{-2} - 6^{0.7}$

5 Write these numbers as powers of prime numbers.

a) 49　　b) $\frac{1}{25}$　　c) $8^{-\frac{2}{3}}$　　d) $\sqrt[3]{121}$　　e) $\frac{1}{\sqrt{3}}$

6 Write each of these numbers as a product of powers of prime numbers. Use indices where possible.

For example, $\dfrac{8}{\sqrt{3}} = 8 \times \dfrac{1}{\sqrt{3}} = 2^3 \times 3^{-\frac{1}{2}}$

a) 12^3 **b)** $4^2 \times 18$ **c)** $\dfrac{8}{81}$ **d)** $\dfrac{125}{36}$ **e)** $\sqrt[3]{15}$

7 Write these numbers in standard form.

a) 16 500   **b)** 0.000 869

c) 53 million   **d)** 0.000 000 083

8 Write these as ordinary numbers.

a) 5.3×10^5 **b)** 6.32×10^{-3}

c) 7.26×10^8 **d)** 1.28×10^{-6}

9 Work out these. Write your answers in standard form.

a) $(2 \times 10^3) \times (4 \times 10^5)$ **b)** $(6 \times 10^8) \times (4 \times 10^{-3})$

c) $(9 \times 10^7) \div (4 \times 10^3)$ **d)** $(5 \times 10^6) \div (2 \times 10^{-3})$

e) $(4 \times 10^7) \div (5 \times 10^2)$ **f)** $(7 \times 10^{-3}) \times (3 \times 10^{-5})$

g) $(4 \times 10^4) + (6 \times 10^3)$ **h)** $(7 \times 10^{-4}) + (6 \times 10^{-5})$

i) $(6 \times 10^7) - (4 \times 10^5)$ **j)** $(8 \times 10^{-4}) - (3 \times 10^{-5})$

10 Work out these. Give your answers in standard form to 3 significant figures.

a) $(5.32 \times 10^5) \times (1.28 \times 10^3)$ **b)** $(6.23 \times 10^{-5}) \times (4.62 \times 10^{-6})$

c) $(2.8 \times 10^5) \div (3.2 \times 10^{-3})$ **d)** $(6.1 \times 10^4) \div (7.3 \times 10^8)$

e) $(6.53 \times 10^5) + (7.26 \times 10^4)$ **f)** $(2.87 \times 10^{-4}) - (8.26 \times 10^{-5})$

8 → USING A CALCULATOR

THIS CHAPTER IS ABOUT

- Using your calculator to perform more complex calculations efficiently
- Exploring examples of exponential growth and decay
- The upper and lower bounds of measurements and the effect of these on a calculation

YOU SHOULD ALREADY KNOW

- How to use the basic functions on your calculator
- How to round answers to a given number of decimal places or significant figures
- How to write a number in standard index form

The efficient use of a calculator

You can use your calculator to perform complex calculations quickly and efficiently.

> **TIP**
> Different makes and models of calculators work in different ways. Investigate how yours works and the functions that it has. You should practise using your calculator regularly so that you can use it efficiently and effectively.

EXAMPLE 8.1

Use your calculator to work out these. Give your answers to 3 significant figures.

a) $\dfrac{14.73 + 2.96}{15.25 - 7.14}$ **b)** $\sqrt{17.8^2 + 4.3^2}$

Solution

There are many ways to do these calculations. Here is a direct method for each.

a) (1 4 . 7 3 + 2 . 9 6)
÷ (1 5 . 2 5 − 7 . 1 4) = 2.181 257 ...
= 2.18 (to 3 s.f.)

b) √ (1 7 . 8 x^2 + 4 . 3 x^2) = 18.312 017 ...
= 18.3 (to 3 s.f.)

> **TIP**
> Notice the use of brackets. These make sure that calculations are performed in the correct order.

Here are some other useful function buttons that you need to be aware of.

x^{-1} $1/x$ — This is the **reciprocal** button.
This button is not essential. You could do $1 \div x$ instead.

$\wedge$ x^y y^x — This is the **power** button.

$\sqrt[x]{\ }$ $y^{1/x}$ — This is the **root** button.
Again, this button is not essential. You could use the 'power' button, with a power of $\frac{1}{x}$, instead.

$\sin$ $\cos$ $\tan$
$\sin^{-1}$ $\cos^{-1}$ $\tan^{-1}$ — These are the **trigonometry** buttons.

EXP EE — This is the **standard form** button.
Once again, this button is not essential. You could use $\times$ 1 0 $\wedge$ instead.

EXAMPLE 8.2

Use your calculator to work out these. Give your answers to 3 significant figures.

a) $\dfrac{1}{1.847}$ b) 4.2^3 c) $\sqrt[4]{15}$

d) $\cos 73°$ e) $\cos^{-1} 0.897$ f) $(3.7 \times 10^{-5}) \div (8.3 \times 10^6)$

Solution

a) 1 $\cdot$ 8 4 7 x^{-1} $=$ 0.541 (to 3 s.f.)

b) 4 $\cdot$ 2 $\wedge$ 3 $=$ 74.1 (to 3 s.f.)

c) 1 5 $\sqrt[x]{\ }$ 4 $=$ 1.97 (to 3 s.f.)

   or 4 $\sqrt[x]{\ }$ 1 5 $=$ 1.97 (to 3 s.f.)

> **TIP**
> Check you know how your calculator operates.

d) $\cos$ 7 3 $=$ 0.292 (to 3 s.f.)

e) $SHIFT$ $\cos$ 0 $\cdot$ 8 9 7 $=$ 26.2° (to 3 s.f.)

f) 3 $\cdot$ 7 EXP $(-)$ 5 $\div$ 8 $\cdot$ 3 EXP 6 $=$ 4.46×10^{-12} (to 3 s.f.)

Use your calculator to work out these. Give all your answers to 3 significant figures.

1 a) $\dfrac{1}{7.2} + \dfrac{1}{14.6}$ **b)** $\dfrac{1}{0.961} \div \dfrac{1}{0.412}$ **c)** $4.2\left(\dfrac{1}{5.5} - \dfrac{1}{7.6}\right)$

2 a) 1.562^5 **b)** 6.8^{-4} **c)** $0.32^3 + 0.51^2$

3 a) $\sqrt[4]{31.8}$ **b)** $\sqrt[3]{0.9316}$ **c)** $\sqrt[5]{8.6 \times 9.71}$

4 a) $\sin 46.2°$ **b)** $\tan 51.6°$ **c)** $\sin 12° - \cos 31°$

5 a) $\cos^{-1} 0.832$ **b)** $\sin^{-1} 0.910$ **c)** $\tan^{-1}\left(\dfrac{43.9}{16.3}\right)$

6 a) $(5.7 \times 10^4) \times (8.2 \times 10^3)$ **b)** $\dfrac{4.6 \times 10^5}{7 \times 10^{-3}}$ **c)** $(1.8 \times 10^{12}) \times (2 \times 10^{-20})$

7 a) $\dfrac{8.71 \times 3.65}{0.84}$ **b)** $\dfrac{0.074 \times 9.61}{23.1}$ **c)** $\dfrac{41.78}{0.0537 \times 264}$

8 a) $\dfrac{114}{27.6 \times 58.9}$ **b)** $\dfrac{0.432 - 0.317}{0.76}$ **c)** $\dfrac{6.51 - 0.1114}{7.24 + 1.655}$

9 a) $3\cos 14.2° - 5\sin 16.3°$ **b)** $\dfrac{3.5 \times 4.4 \times \sin 18.7°}{2}$ **c)** $\cos^{-1}\left(\dfrac{2.7^2 + 3.6^2 - 1.9^2}{2 \times 2.7 \times 3.6}\right)$

10 a) $(4.7 \times 10^5)^2$ **b)** $\sqrt{6.4 \times 10^{-3}}$ **c)** $\dfrac{(9.6 \times 10^4) \times (3.75 \times 10^7)}{8.87 \times 10^{-6}}$

Challenge 8.1

Sometimes, when the display on a calculator is turned upside down, the inverted figures 'spell' a word.

Find the words that these calculations reveal.

a) $(84 + 17) \times 5$ **b)** $566 \times 711 - 23\,617$ **c)** $\dfrac{9999 + 319}{8.47 + 2.53}$

d) $\dfrac{27 \times 2000 - 2}{0.63 \div 0.09}$ **e)** $0.008 - \dfrac{0.3^2}{10^5}$

See if you can find more calculations that result in 'calculator words'. Check them with your partner.

Using trial and improvement, or otherwise, solve each of these equations.
Give your answers correct to 2 decimal places.

a) $1.5^x = 6$ **b)** $\log x = 1.5$ **c)** $\sin 2x° = 0.9$

Find two values for x.

Exponential growth and decay

When a number is repeatedly multiplied by a fixed value **greater than one** we have **exponential growth**.

When a number is repeatedly multiplied by a fixed value **less than one** we have **exponential decay**.

The formula is the same for exponential growth and decay:

$$y = A \times b^x$$

where A is the initial value and b is the rate of growth or decay (**multiplier**).

In Chapter 6 you learned about **compound interest** and **depreciation**. These are examples of exponential growth and decay.

EXAMPLE 8.3

£2000 is invested in a bank and receives 6% per year compound interest.

a) Write down a formula for the amount of money in the bank, y, after x years.

b) How much money will there be after 5 years?

c) After how many years will the amount of money in the bank be more than doubled?

Remember that to increase by 6%, the multiplier is 1.06.

Solution

a) $y = A \times b^x$ The initial value, A, is 2000 and the rate of growth, b is 1.06.
$y = 2000 \times 1.06^x$

b) $y = 2000 \times 1.06^5$
$= £2676.45$ (to the nearest penny)

c) 12 years   Using trial and improvement, you will find that after 12 years the investment will be worth £4024.39.
Note that you do not need to do the whole calculation for each value of x: the investment will be more than double the initial value when $b^x > 2$, i.e. when $1.06^x > 2$, and you will find that $1.06^{12} = 2.012 \ldots$.

EXAMPLE 8.4

The population of a rare species of animal is falling by 2% per year.
In 2010 there were 30 000 of these animals.

a) Write down a formula for the number of animals, p, after t years.

b) How many of these animals will there be after 10 years?

c) How long will it be before the population is halved?

Solution

a) $p = 30\ 000 \times 0.98^t$

b) $p = 30\ 000 \times 0.98^{10}$
      $= 24\ 512$

c) 35 years   Using trial and improvement, we find that when $t = 35, b^t < 0.5$. The population will be 14 792.

EXERCISE 8.2

1 £5000 is invested at 3% compound interest.
  a) Write down a formula for the amount, a, the investment is worth after t years.
  b) Calculate the value of the investment after
    (i) 4 years.   **(ii)** 20 years.

2 Elaine buys a car for £9000. It depreciates in value by 12% each year.
  a) Write down a formula for the value of the car, v, after t years.
  b) Calculate the value of the car after
    (i) 3 years.   **(ii)** 8 years.
  c) Elaine wants to sell the car when its value has reduced to £5000.
     Use trial and improvement to work out how many years she will keep the car.

3 The number of a strain of bacteria decays exponentially following the formula

$$N = 1\,000\,000 \times 2^{-t}$$

where N is the number of bacteria present and t is the time in hours.

a) How many bacteria were present originally?

b) How many bacteria were present after

    (i) 5 hours?   **(ii)** 12 hours?

c) After how many hours will the bacteria no longer exist: that is, after how many hours will be the number of bacteria be less than 1?

4 A bank pays Phil compound interest on the money he invests.
Phil works out the formula that the bank will use.

$$A = 2000 \times 1.08^{n}$$

a) How much did Phil invest?

b) What was the rate of interest?

c) What does the letter n stand for in the formula?

d) How much will the investment be worth after

    (i) 5 years?   **(ii)** 15 years?

5 The population of a country is increasing at a rate of 5% per year.
In 2010 the population was 60 million.

a) Write down a formula for the population size, P, after t years.

b) What will the population be in

    (i) 2015?   **(ii)** 2105?

c) How long will it take for the population to double from its 2010 size?

6 A sample of a radioactive element has a mass of 50 g.
Its mass reduces by 10% each year.

a) Write down a formula for the mass, m, of the sample after t years.

b) Calculate the mass after

    (i) 3 years.   **(ii)** 10 years.

c) Use trial and improvement to find how long it takes for the mass to halve.
This is known as the half-life of the element.

Challenge 8.3

The curve $y = Ab^{x}$ passes through the points $(0, 5)$ and $(3, 20.48)$.

Find the values of A and b.

Bounds of measurements

Chapter 29 explains that when a value is measured to the nearest unit, the value is within half a unit of that mark.

Another way of expressing this is to say that any measurement expressed to a given unit is in **possible error** of half that unit.

EXAMPLE 8.5

A piece of wood measures 26 cm, correct to the nearest centimetre.

What are the upper and lower bounds of the length of the piece of wood?

Solution

Upper bound = 26.5 cm
Lower bound = 25.5 cm

The bounds are half a centimetre above and below the given length.
The actual length could be anywhere between 25.5 cm and 26.5 cm.

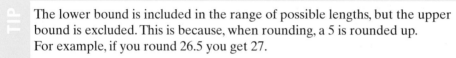

TIP

The lower bound is included in the range of possible lengths, but the upper bound is excluded. This is because, when rounding, a 5 is rounded up. For example, if you round 26.5 you get 27.

Bounds can be written as an inequality. For example, $25.5 \leqslant$ length < 26.5.

Measurements are not always given correct to the nearest unit: often they are measured correct to the nearest *part* of a unit, for example to the nearest 0.1 cm. Such measurements are in possible error of half of the part unit: in the case of a length measured to the nearest 0.1 cm, the possible error is ± 0.05 cm.

EXAMPLE 8.6

A supermarket sells chickens. Each chicken weighs 4.2 kg, correct to the nearest 0.1 kg.
a) What are the upper and lower bounds of the weight of one chicken?
b) These chickens can be bought in boxes of 10.
What are the upper and lower bounds of a box of these chickens?

Solution

a) Upper bound = 4.25 kg
Lower bound = 4.15 kg

b) Upper bound = 4.25 × 10   Multiply the bounds of one item by the number
= 42.5 kg   of items.
Lower bound = 4.15 × 10
= 41.5 kg

1 Find the upper and lower bounds for each of these measurements.
   a) The height of a tree is 4.7 m, to the nearest 0.1 m.
   b) A dog weighs 37 kg, to the nearest kilogram.
   c) A door measures 1.95 m, to the nearest centimetre.
   d) The winning time for the 200 m race was 28.45 seconds, to the nearest hundredth of a second.
   e) The volume of cola in a can is 330 ml, to the nearest millilitre.

2 A ladder is measured as 3 m to the nearest centimetre.
   Write the possible length of the ladder as an inequality.

3 A square birthday card of side 12.5 cm, to the nearest 0.1 cm, is to be put into a square envelope of side 13 cm, to the nearest centimetre.
   Is it certain that the card will fit into the envelope? Show how you decide.

4 Find the upper and lower bounds for each of these measurements.
   a) The total weight of 10 coins, each coin weighing 6 g to the nearest gram.
   b) The total length of 20 pipes, each measuring 10 m to the nearest metre.
   c) The total time to make 100 cakes when each cake takes 13.6 seconds to make, to the nearest tenth of a second.

Further calculations involving bounds of measurements

You need to be able to work out the bounds of the answer to a calculation involving more than one measurement. You do this by performing the calculation using the bounds of the measurements involved.

When adding:
• to find the upper bound, add the upper bounds.
• to find the lower bound, add the lower bounds.

When subtracting:
• to find the upper bound, subtract the lower bound from the upper bound.
• to find the lower bound, subtract the upper bound from the lower bound.

EXAMPLE 8.7

Chen and Samantha are carrying out a survey of the length of worms in a plot of land.
Chen measures a worm as 12 cm to the nearest cm and Samantha measures another worm as 10 cm to the nearest cm.
Calculate the upper and lower bounds of
a) the total length of the two worms.
b) the difference in length of the two worms.

Solution

a) Upper bound = 12.5 + 10.5   The longest possible total length occurs
              = 23 cm   with the longest possible individual lengths.

   Lower bound = 11.5 + 9.5   The shortest possible total length occurs
              = 21 cm   with the shortest possible individual lengths.

b) Upper bound = 12.5 − 9.5   The greatest possible difference occurs with
              = 3 cm   the longest possible longer length and the
              shortest possible shorter length.

   Lower bound = 11.5 − 10.5   The least possible difference occurs with
              = 1 cm   the shortest possible longer length and the
              longest possible shorter length.

When multiplying:
 - to find the upper bound, multiply the upper bounds.
 - to find the lower bound, multiply the lower bounds.

When dividing:
 - to find the upper bound, divide the upper bound by the lower bound.
 - to find the lower bound, divide the lower bound by the upper bound.

EXAMPLE 8.8

A jug has a volume of 500 cm^3, measured to the nearest 10 cm^3.
   a) Write down the least and greatest possible values of the volume of the jug.
   b) Water is poured from the jug into a tank of volume 15.5 litres measured to the
   nearest 0.1 litre.
   Explain, showing all your calculations, why it is always possible to pour water
   from 30 full jugs into the tank without overflowing.

Solution

a) Least volume of the jug = 495 cm^3
   Greatest volume of the jug = 505 cm^3

b) To show that it is always possible, we consider the worst case possible.
   This is when the jugs have greatest volume and the tank has least volume.
   Greatest volume of water from 30 jugs = 30 × 'largest jug'
                    = 30 × 505
                    = 15 150 cm^3
   Volume of the tank is 15.5 litres measured to the nearest 0.1 litre. This means
   Least volume of the tank = 15.45 litres = 15 450 cm^3
       15 150 cm^3 is less than 15 450 cm^3
   Therefore it is always possible to pour water from 30 full jugs into the tank without
   overflowing.

1   A triangle has sides of length 7 cm, 8 cm and 10 cm.
    All measurements are correct to the nearest centimetre.
    Work out the upper and lower bounds of the perimeter of the triangle.

2   Work out the largest and smallest possible areas of a rectangle measuring 27 cm by
    19 cm, where both lengths are correct to the nearest centimetre.

3   Stuffing weighing 0.5 kg is added to a chicken weighing 2.4 kg.
    Both weights are correct to the nearest 0.1 kg.
    What are the maximum and minimum possible weights of the stuffed chicken?

4   Phil and Ruby each drew a line 15 cm long, to the nearest centimetre.
    What are the greatest and least possible differences between the lengths drawn?

5   Given that $p = 5.1$ and $q = 8.6$, correct to 1 decimal place, work out the largest and
    smallest possible values of
    a) $p + q$.   **b)** $q - p$.

6   Jasbinder runs 100 m in 12.8 seconds.
    The distance is correct to the nearest metre and the time is correct to the
    nearest 0.1 second.
    Work out the upper and lower bounds of Jasbinder's speed.

7   The population of a town is 108 000, correct to the nearest 1000.
    The town covers an area of 120 square miles, to the nearest square mile.
    Work out the greatest and least possible population densities of the town.

8   Use the formula $P = \dfrac{V^2}{R}$ to work out the upper and lower bounds of P when $V = 6$
    and $R = 1$ and both values are correct to the nearest whole number.

9   Work out the upper and lower bounds of the calculation
    $$\frac{8.1 - 3.6}{11.4}$$
    if each value in the calculation is correct to 1 decimal place.

10 Concrete blocks have a mass of 15 kg measured to the nearest kg.
    a) Write down the least and greatest possible values of the mass of a concrete block.
    b) (i) Find the least and greatest possible values of the mass of 100 concrete blocks.
    (ii) Den wishes to be sure that he puts no more than 1500 kg of blocks on his lorry.
        Find the greatest number of blocks Den should put on his lorry in order to be
        sure that no more than 1500 kg is loaded.

- How to use some of the function keys on your calculator
- How to calculate exponential growth and decay
- How to find the bounds of measurements
- How to calculate the bounds in calculations involving more than one measurement

MIXED EXERCISE 8

Give your answers to questions **1** to **4** correct to 3 significant figures.

1 a) $\dfrac{1}{9.7} + \dfrac{1}{0.035}$ **b)** $9.5^3 + 3.9^5$ **c)** $\sqrt[4]{108.6}$

2 a) $3\cos 21°$ **b)** $\tan^{-1} 1.46$ **c)** $\dfrac{8.5 \times \sin 57.1°}{\sin 39.2°}$

3 a) $\dfrac{555}{10.4 + 204}$ **b)** $\dfrac{23.7 \times 0.0042}{12.4 - 1.95}$ **c)** $\dfrac{88.71 - 35.53}{26.42 + 9.76}$

4 a) $(4.2 \times 10^5) \times (3.9 \times 10^6)$ **b)** $\sqrt{4.29 \times 10^{-8}}$

5 Jack invests a sum of money at a fixed percentage yearly interest rate.
   He is told that the formula the firm will use is $A = 3000 \times 1.05^t$.
   a) How much did Jack invest?
   b) What was the rate of interest?
   c) What does the letter t stand for in the formula?
   d) How much will the investment be worth after
      (i) 4 years?   **(ii)** 12 years?

6 A car costs £12 000 when new. Its value depreciates by 13% per year.
   a) Write down a formula for the value, v, of the car after t years.
   b) Calculate the value after
      (i) 3 years.   **(ii)** 8 years.
   c) Use trial and improvement to find how long it takes for the value to halve.

7 Find the upper and lower bounds for each of these measurements.
   a) A book weighing 1.7 kg to the nearest 0.1 kg.
   b) The total weight of 10 sacks of corn, each sack weighing 25 kg to the nearest kg.

8 A rectangular room measures 4.3 m and 6.2 m. Both measurements are correct to the
   nearest 0.1 m. Work out the upper and lower bounds of the perimeter of the room.

9 A lorry driver travels 157 km in 2.5 hours. The distance is correct to the nearest kilometre
   and the time is correct to the nearest 0.1 hour. Work out the upper and lower bounds of
   the average speed of the lorry.

9 → DECIMALS AND SURDS

THIS CHAPTER IS ABOUT

- Distinguishing between terminating and recurring decimals
- Distinguishing between rational and irrational numbers
- Representing recurring decimals as fractions
- Using surds in exact calculations
- Rationalising simple fractions with a surd as the denominator

YOU SHOULD ALREADY KNOW

- How to write terminating decimals as fractions in their simplest form
- How to find the prime factors of a number
- How to work with indices
- How to expand quadratic expressions
- How to solve a quadratic equation which does not factorise

Terminating and recurring decimals

Check up 9.1

Without using a calculator, find the decimal equivalent of each of these fractions.

a) $\frac{2}{5}$ b) $\frac{2}{3}$ c) $\frac{3}{8}$ d) $\frac{4}{70}$ e) $\frac{2}{9}$ f) $\frac{7}{6}$

Some of the decimals recur.

Did you know before you began which decimals would terminate and which would recur?

You learned in Chapter 2 that if the denominator of a fraction only has factors which are factors of 10, it will give a terminating decimal. If the denominator of a fraction has factors which are not factors of 10, it will give a recurring decimal.

This is because, when a terminating decimal is written as a fraction, the denominator must be a power of 10.

The prime factors of 10 are 2 and 5. So the prime factors of a power of 10 are powers of 2 and 5.

If the fraction is cancelled down, the numerator and denominator can only be divided by powers of 2 and/or 5.

Thus the prime factors of the denominator must be powers of 2 and/or 5.

For example, $0.265 = \dfrac{265}{1000}$ 1000 written as a product of prime factors is $2^3 \times 5^3$.

$\qquad\qquad\qquad = \dfrac{53}{200}$ The highest common factor (HCF) of 265 and 1000 is 5.

The prime factors of 200 are 2^3 and 5^2.

So, to see whether a fraction can be written as a terminating decimal, write it in its simplest terms and look at the factors of its denominator. If 2 and 5 are the only prime factors, the decimal will terminate. If 2 and 5 are not the only prime factors, the decimal will recur.

EXAMPLE 9.1

State whether each of these fractions will give a terminating decimal or a recurring decimal.

a) $\dfrac{4}{25}$ **b)** $\dfrac{3}{15}$ **c)** $\dfrac{7}{40}$ **d)** $\dfrac{8}{11}$ **e)** $\dfrac{29}{30}$

Solution

a) $\dfrac{4}{25}$ terminates. $25 = 5^2$

b) $\dfrac{3}{15}$ terminates. $\dfrac{3}{15} = \dfrac{1}{5}$

c) $\dfrac{7}{40}$ terminates. $40 = 2^3 \times 5$

d) $\dfrac{8}{11}$ recurs.   11 is prime

e) $\dfrac{29}{30}$ recurs. $30 = 2 \times 3 \times 5$

Dots are used to show that a decimal recurs.

If only one digit occurs, a dot is placed over that digit the first time it occurs after the decimal point. For example, 0.166 666 ... is written as $0.1\dot{6}$ and 3.333 333 ... is written as $3.\dot{3}$.

If a group of digits recurs, a dot is placed over the first and last digits in the group. For example, 0.012 612 612 ... is written as $0.0\dot{1}2\dot{6}$.

A group of recurring digits is also known as a **period**.

EXAMPLE 9.2

Write $\dfrac{3}{11}$ as a recurring decimal.

Solution

$$\begin{array}{r} 0.2\ 7\ 2\ 7\ 2 \\ 11\overline{)3.0^8 0^3 0^8 0^3 0^8} \end{array}$$

So $\dfrac{3}{11} = 0.\dot{2}\dot{7}$

a) Write each of these fractions as a decimal.

$$\frac{1}{7} \quad \frac{2}{7} \quad \frac{3}{7} \quad \frac{4}{7} \quad \frac{5}{7} \quad \frac{6}{7}$$

These decimals recur. What patterns do you notice?

b) Write each of these fractions as a decimal.

$$\frac{1}{13} \quad \frac{2}{13} \quad \frac{3}{13} \quad \frac{4}{13} \quad \frac{5}{13}$$

Can you predict the results for $\frac{6}{13}$ to $\frac{12}{13}$?

Work them out to check your predictions.

c) Write the fractions $\frac{1}{17}$ to $\frac{4}{17}$ as decimals and predict the results for $\frac{5}{17}$ to $\frac{16}{17}$.

Decimals which recur can all be written as fractions.

EXAMPLE 9.3

Write $0.\dot{4}0\dot{2}$ as a fraction in its simplest form.

Solution

Let $a = 0.\dot{4}0\dot{2}$ (1)

$1000a = 402.\dot{4}0\dot{2}$ (2)   Multiply a by 1000 to move the recurring pattern on by a whole period.

$\begin{aligned} 1000a &= 402.\dot{4}0\dot{2} \\ a &= 0.\dot{4}0\dot{2} \end{aligned}$ Subtract a from $1000a$. This will eliminate the digits after the decimal point. Even though they go on forever, they are the same.

$999a = 402$ (2) − (1)   You now have a simple equation which can easily be solved.

$a = \dfrac{402}{999}$ Divide each side by 999.

$a = \dfrac{134}{333}$ Remember to give your answer in its simplest form.

TIP   Check your answer by dividing 134 by 333 on your calculator.

The same method can be used for all recurring decimals.

If only part of the decimal recurs, make sure you multiply by a power of 10 such that the recurring pattern is moved on by exactly one whole period.

EXAMPLE 9.4

Write $0.4\dot{2}$ as a fraction in its simplest form.

Solution

Let $a = 0.4\dot{2}$ (1)

$10a = 4.2\dot{2}$ (2)   Multiply a by 10 to move the recurring pattern on by a whole period. You do not need to move the pattern on so that it appears before the decimal point, although often it will, as in Example 9.3.

$10a = 4.2\dot{2}$ (2)   Subtract a from $10a$ to eliminate the recurring digits.

$\underline{\quad a = 0.4\dot{2} \quad}$ (1)

$9a = 3.8$ (2) − (1)   Solve the resulting equation.

$a = \dfrac{3.8}{9}$ This is not a fraction.

$a = \dfrac{38}{90}$ Multiply the numerator and denominator of the fraction by 10 to get rid of the decimal.

$a = \dfrac{19}{45}$ Remember to give your answer in its simplest form.

EXERCISE 9.1

1 Which of these fractions are equivalent to recurring decimals?

   a) $\frac{4}{15}$ **b)** $\frac{3}{20}$ **c)** $\frac{4}{35}$ **d)** $\frac{9}{125}$ **e)** $\frac{11}{16}$

2 Find the decimal equivalent of each of the fractions in question **1**.

3 When these fractions are written as decimals, which of them terminate?

   a) $\frac{2}{5}$ **b)** $\frac{17}{20}$ **c)** $\frac{38}{125}$ **d)** $\frac{7}{18}$ **e)** $\frac{3}{8}$

4 Find the decimal equivalent of each of the fractions in question **3**.

5 Find the fractional equivalent of each of these terminating decimals.
Write each fraction in its simplest form.

   a) 0.12   **b)** 0.205   **c)** 0.375   **d)** 0.3125

6 Find the fractional equivalent of each of these recurring decimals.
Write each fraction in its simplest form.

   a) $0.\dot{4}$ **b)** $0.\dot{7}$ **c)** $0.1\dot{2}$ **d)** $0.4\dot{7}$

7 Find the fractional equivalent of each of these recurring decimals.
Write each fraction in its simplest form.

   a) $0.3\dot{5}$ **b)** $0.\dot{5}\dot{4}$ **c)** $0.\dot{1}\dot{2}$ **d)** $0.\dot{1}\dot{7}$

8 Find the fractional equivalent of each of these recurring decimals.
Write each fraction in its simplest form.

   a) $0.\dot{1}0\dot{8}$ **b)** $0.\dot{1}2\dot{3}$ **c)** $0.0\dot{7}$ **d)** $0.0\dot{9}\dot{3}$

Challenge 9.2

Work in pairs.

Practise your division skills by writing some fractions as recurring decimals.

Hint: Remember that any fraction whose denominator has prime factors other than 2 or 5 will recur.

Give your recurring decimals to your partner and challenge them to turn them back into fractions.

Challenge 9.3

a) Given that $\frac{5}{33} = 0.\dot{1}\dot{5}$, write $\frac{5}{330}$ as a recurring decimal.

b) Given that $\frac{70}{333} = 0.\dot{2}1\dot{0}$, write $\frac{7}{333}$ as a recurring decimal.

Surds

A **rational** number is a number that can be written as an ordinary fraction, $\frac{m}{n}$, where m and n are integers.

An **irrational** number cannot be written in this form. Decimals which neither terminate nor recur are irrational numbers. For example, π and $\sqrt{6}$ are irrational numbers.

Surds are irrational numbers expressed in root form.
For example, $\sqrt{c}$ or $a + b\sqrt{c}$, where c is an integer which is not a perfect square and a and b are rational.

You will also meet surds in Chapter 19, when solving quadratic equations which do not factorise.

When working with surds, you deal with the rational and irrational parts of the expression separately.

EXAMPLE 9.5

Write each of these in the form $\sqrt{a}$.

a) $\sqrt{3} \times \sqrt{5}$ **b)** $(\sqrt{2})^3$

Solution

a) $\sqrt{3} \times \sqrt{5} = \sqrt{15}$ Use the result $\sqrt{a} \times \sqrt{b} = \sqrt{ab}$.

b) $(\sqrt{2})^3 = \sqrt{8}$

Some surds may be simplified by taking out perfect square factors.

EXAMPLE 9.6

Simplify these.

a) $\sqrt{18}$ **b)** $\sqrt{48}$

Solution

a) $\sqrt{18} = \sqrt{9 \times 2}$ 18 is the product of 9, which is a perfect square, and 2.

$\quad\quad\; = \sqrt{9} \times \sqrt{2}$

$\quad\quad\; = 3\sqrt{2}$

b) $\sqrt{48} = \sqrt{16 \times 3}$ 48 is the product of 16, which is a perfect square, and 3.

$\quad\quad\; = \sqrt{16} \times \sqrt{3}$

$\quad\quad\; = 4\sqrt{3}$

TIP

If you did not spot that 16 is a factor of 48, you could have simplified $\sqrt{48}$ in stages.

$\sqrt{48} = \sqrt{4 \times 12}$

$\quad\; = \sqrt{4 \times 4 \times 3}$

$\quad\; = 4\sqrt{3}$

EXAMPLE 9.7

Simplify $\sqrt{50} + \sqrt{8}$.

Solution

$\sqrt{50} + \sqrt{8} = \sqrt{25 \times 2} + \sqrt{4 \times 2}$

$\quad\quad\quad\quad\;\; = 5\sqrt{2} + 2\sqrt{2}$

$\quad\quad\quad\quad\;\; = 7\sqrt{2}$

Sometimes you will need to use your skills at expanding quadratic expressions.

EXAMPLE 9.8

$p = 3 + 5\sqrt{2}$ and $q = 4 - 7\sqrt{2}$.

Find, in the form $a + b\sqrt{2}$, the value of

a) $p + q$.   **b)** pq.   **c)** p^2.

Solution

a) $p + q = 3 + 5\sqrt{2} + 4 - 7\sqrt{2}$ $3 + 4 = 7$ and $5\sqrt{2} - 7\sqrt{2} = -2\sqrt{2}$.
$= 7 - 2\sqrt{2}$

b) $pq = (3 + 5\sqrt{2})(4 - 7\sqrt{2})$ Expand the brackets in the usual way.
$= 12 + 20\sqrt{2} - 21\sqrt{2} - 35 \times 2$ When you multiply $5\sqrt{2}$ and $-7\sqrt{2}$, use the result
$= -58 - \sqrt{2}$ $\sqrt{a} \times \sqrt{a} = a$.

c) $p^2 = (3 + 5\sqrt{2})(3 + 5\sqrt{2})$ Expand the brackets in the usual way.
$= 9 + 15\sqrt{2} + 15\sqrt{2} + 25 \times 2$ Again, use the result $\sqrt{a} \times \sqrt{a} = a$.
$= 59 + 30\sqrt{2}$

When there is a square root in the denominator of a fraction, you may be asked to simplify by **rationalising the denominator**.

This involves multiplying the numerator and denominator by a factor such that the denominator of the fraction becomes a rational number. Remember that when you multiply both numerator and denominator by the same number, the value of the fraction is unchanged.

If the denominator is of the form $a\sqrt{b}$, simply multiply the numerator and denominator by $\sqrt{b}$, and simplify.

EXAMPLE 9.9

Simplify each of these by rationalising the denominator.

a) $\dfrac{2}{\sqrt{6}}$ **b)** $\dfrac{8\sqrt{3}}{\sqrt{2}}$

Solution

a) $\dfrac{2}{\sqrt{6}} = \dfrac{2}{\sqrt{6}} \times \dfrac{\sqrt{6}}{\sqrt{6}}$ Multiply the numerator and denominator by $\sqrt{6}$.

$= \dfrac{2\sqrt{6}}{6}$ Use the result $\sqrt{a} \times \sqrt{a} = a$.

$= \dfrac{\sqrt{6}}{3}$

b) $\dfrac{8\sqrt{3}}{\sqrt{2}} = \dfrac{8\sqrt{3}}{\sqrt{2}} \times \dfrac{\sqrt{2}}{\sqrt{2}}$ Multiply the numerator and denominator by $\sqrt{2}$.

$\qquad = \dfrac{8\sqrt{6}}{2}$

$\qquad = 4\sqrt{6}$

EXERCISE 9.2

1 Simplify these.

   a) $\sqrt{3} + 5\sqrt{3}$ **b)** $12\sqrt{5} - 3\sqrt{5}$ **c)** $6\sqrt{5} - \sqrt{5}$

   d) $\sqrt{2} \times 6\sqrt{2}$ **e)** $5\sqrt{3} \times \sqrt{7}$ **f)** $3\sqrt{2} \times \sqrt{8}$

2 Write each of these expressions in the form $a\sqrt{b}$, where b is an integer which is as small as possible.

   a) $\sqrt{50}$ **b)** $2\sqrt{125}$ **c)** $6\sqrt{32}$

   d) $5\sqrt{54}$ **e)** $\sqrt{90}$ **f)** $\sqrt{2000}$

3 Simplify these.

   a) $\sqrt{12} + 5\sqrt{3}$ **b)** $8\sqrt{5} - \sqrt{45}$ **c)** $\sqrt{8} + 3\sqrt{2}$

   d) $\sqrt{60} \times 2\sqrt{3}$ **e)** $\sqrt{32} \times \sqrt{18}$ **f)** $5\sqrt{30} \times \sqrt{60}$

4 Expand and simplify these.

   a) $\sqrt{3}(2 + \sqrt{3})$ **b)** $2\sqrt{5}(\sqrt{2} + \sqrt{5})$

   c) $4\sqrt{7}(3 + \sqrt{14})$ **d)** $5\sqrt{2}(\sqrt{8} + \sqrt{6})$

5 Expand and simplify these.

   a) $(1 + 2\sqrt{3})(2 + 5\sqrt{3})$ **b)** $(3 - \sqrt{2})(5 + \sqrt{2})$

   c) $(6 - 2\sqrt{5})(3 + \sqrt{5})$ **d)** $(\sqrt{6} + 2)(\sqrt{6} + 3)$

   e) $(\sqrt{11} + 1)(\sqrt{11} - 1)$ **f)** $(10 - \sqrt{7})(3 - 2\sqrt{7})$

6 Find the value of each of these expressions, when $m = 2 + 5\sqrt{3}$ and $n = 3 - 7\sqrt{3}$. State whether your answer is rational or irrational.

   a) $4m$ **b)** $m + n$ **c)** $3m - 2n$ **d)** mn

7 Find the value of each of these expressions, when $p = 3 + 5\sqrt{2}$ and $q = 3 - 5\sqrt{2}$. State whether your answer is rational or irrational.

   a) $3q$ **b)** $\sqrt{2}p$ **c)** $p + q$

   d) pq **e)** p^2 **f)** q^2

8 Show that $(10 - 3\sqrt{7})(10 + 3\sqrt{7})$ is rational, finding its value.

9 Rationalise the denominator and simplify each of the following.

a) $\dfrac{10}{\sqrt{2}}$

b) $\dfrac{2}{\sqrt{10}}$

c) $\dfrac{4}{3\sqrt{10}}$

d) $\dfrac{14}{5\sqrt{8}}$

e) $\dfrac{2\sqrt{3}}{3\sqrt{2}}$

f) $\dfrac{12\sqrt{6}}{7\sqrt{15}}$

10 Rationalise the denominator and simplify each of the following.

a) $\dfrac{6 + 3\sqrt{2}}{\sqrt{2}}$

b) $\dfrac{15 + \sqrt{5}}{2\sqrt{5}}$

c) $\dfrac{12 + 3\sqrt{2}}{2\sqrt{3}}$

d) $\dfrac{5 + 2\sqrt{3}}{\sqrt{6}}$

Challenge 9.4

You can rationalise a denominator in the form $a \pm b\sqrt{c}$ by multiplying the numerator and denominator of the fraction by $a \mp b\sqrt{c}$.

For example, to rationalise the denominator of $\dfrac{3\sqrt{5}}{2 + \sqrt{5}}$, multiply by $\dfrac{2 - \sqrt{5}}{2 - \sqrt{5}}$.

To rationalise the denominator of $\dfrac{\sqrt{3}}{6 - 5\sqrt{3}}$, multiply by $\dfrac{6 + 5\sqrt{3}}{6 + 5\sqrt{3}}$.

Why does this method work?

WHAT YOU HAVE LEARNED

- **A fraction can be written as a terminating decimal if, in its simplest form, 2 and 5 are the only prime factors of its denominator**
- **A recurring decimal can be written as a fraction by multiplying by the correct power of 10 to move the recurring pattern on by a whole 'period', then subtracting the original decimal and dividing your resulting equation to obtain the fraction**
- **Surds are irrational numbers which can be expressed in root form**
- **Surds may be added and subtracted by adding their rational and irrational parts separately**
- **Some surds may be simplified by taking a perfect square factor out from the square root**
- **When multiplying or dividing or simplifying surds you often need to use these results:**
  $$\sqrt{a} \times \sqrt{a} = a \text{ and } \sqrt{a} \times \sqrt{b} = \sqrt{ab}$$
- **When a surd is a fraction with a denominator of the form $a\sqrt{b}$, you can rationalise the denominator by multiplying both numerator and denominator by $\sqrt{b}$**

1 Which of these fractions are equivalent to recurring decimals?

    a) $\frac{5}{6}$ **b)** $\frac{5}{8}$ **c)** $\frac{5}{18}$ **d)** $\frac{11}{25}$ **e)** $\frac{11}{15}$

2 Find the decimal equivalent of each of the fractions in question **1**.

3 Find the fractional equivalent of each of these terminating decimals.
Write each fraction in its simplest form.

    a) 0.72   **b)** 0.325   **c)** 0.625   **d)** 0.192

4 Find the fractional equivalent of each of these recurring decimals.
Write each fraction in its simplest form.

    a) $0.\dot{4}$ **b)** $0.5\dot{4}$ **c)** $0.\dot{5}\dot{4}$ **d)** $0.\dot{5}0\dot{4}$

5 Write each of these in the form $a\sqrt{b}$, where b is an integer which is as small as possible.

    a) $\sqrt{32}$ **b)** $2\sqrt{75}$ **c)** $5\sqrt{18}$
    d) $\sqrt{108}$ **e)** $\sqrt{60}$ **f)** $\sqrt{675}$

6 Simplify these.

    a) $\sqrt{18} + 4\sqrt{2}$ **b)** $6\sqrt{5} - \sqrt{20}$ **c)** $\sqrt{48} + 3\sqrt{3}$
    d) $\sqrt{90} \times 2\sqrt{10}$ **e)** $\sqrt{8} \times \sqrt{24}$ **f)** $4\sqrt{15} \times \sqrt{3}$

7 Expand and simplify these.

    a) $\sqrt{2}(5 + \sqrt{2})$ **b)** $3\sqrt{5}(\sqrt{3} + 2\sqrt{5})$
    c) $2\sqrt{11}(1 + \sqrt{22})$ **d)** $3\sqrt{3}(\sqrt{6} + \sqrt{12})$

8 Expand and simplify these.

    a) $(1 + 5\sqrt{3})(2 + \sqrt{3})$ **b)** $(4 - \sqrt{2})(1 + \sqrt{2})$ **c)** $(4 - 3\sqrt{5})(1 + \sqrt{5})$
    d) $(\sqrt{7} + 5)(\sqrt{7} + 2)$ **e)** $(\sqrt{5} + 1)(\sqrt{5} - 1)$ **f)** $(8 - \sqrt{3})(5 - 2\sqrt{3})$

9 Find the value of each of these expressions, when $p = 7 + 2\sqrt{3}$ and $q = 7 - 2\sqrt{3}$.
State whether your answer is rational or irrational.

    a) $6q$ **b)** $5\sqrt{3}p$ **c)** $p - q$
    d) pq **e)** p^2 **f)** q^2

10 Rationalise the denominator and simplify each of the following.

    a) $\dfrac{8}{\sqrt{2}}$ **b)** $\dfrac{6}{\sqrt{15}}$ **c)** $\dfrac{5}{4\sqrt{10}}$
    d) $\dfrac{2 + \sqrt{5}}{\sqrt{5}}$ **e)** $\dfrac{12 + \sqrt{3}}{\sqrt{3}}$ **f)** $\dfrac{42 + \sqrt{21}}{3\sqrt{7}}$

10 → ALGEBRAIC MANIPULATION 1

THIS CHAPTER IS ABOUT

- **Expanding or simplifying brackets in algebra**
- **Factorising expressions**
- **Index notation in algebra**

YOU SHOULD ALREADY KNOW

- **Letters can be used to stand for numbers**

Expanding brackets

Joe has a job making cheese sandwiches.
He uses two slices of bread and one slice of cheese for each sandwich.
He wants to know how much it will cost to make 25 sandwiches.

Joe uses 50 slices of bread and 25 slices of cheese to make 25 sandwiches.

You could use letters to represent the cost of the sandwich ingredients.
Let b represent the cost of each slice of bread in pence.
Let c represent the cost of each slice of cheese in pence.

You could then write one sandwich costs $b + b + c = 2b + c$.

(You write $1c$ as just c.)

25 sandwiches cost 25 times this amount. You could write $25(2b + c)$.

$2b$ and c are called **terms**. $25(2b + c)$ is called an **expression**.

To work out the cost of 25 sandwiches you could write

$$25(2b + c) = 25 \times 2b + 25 \times c = 50b + 25c.$$

This is is called **expanding the brackets**. You multiply *each* term inside the brackets by the number outside the brackets.

EXAMPLE 10.1

Work out these.

a) $10(2b + c)$ **b)** $35(2b + c)$

c) $16(2b + c)$ **d)** $63(2b + c)$

Solution

a) $10(2b + c) = 10 \times 2b + 10 \times c$
$$= 20b + 10c$$

b) $35(2b + c) = 35 \times 2b + 35 \times c$
$$= 70b + 35c$$

c) $16(2b + c) = 16 \times 2b + 16 \times c$
$$= 32b + 16c$$

d) $63(2b + c) = 63 \times 2b + 63 \times c$
$$= 126b + 63c$$

You expand brackets with other letters or other signs, with numbers or with more terms, in the same way.

EXAMPLE 10.2

Expand these.

a) $12(2b + 7g)$ **b)** $6(3m - 4n)$

c) $8(2x - 5)$ **d)** $3(4p + 2v - c)$

Solution

a) $12(2b + 7g) = 12 \times 2b + 12 \times 7g$
$$= 24b + 84g$$

b) $6(3m - 4n) = 6 \times 3m - 6 \times 4n$
$$= 18m - 24n$$

c) $8(2x - 5) = 8 \times 2x - 8 \times 5$
$$= 16x - 40$$

d) $3(4p + 2v - c) = 3 \times 4p + 3 \times 2v - 3 \times c$
$$= 12p + 6v - 3c$$

Expand these.

1 $10(2a + 3b)$ **2** $3(2c + 7d)$ **3** $5(3e - 8f)$

4 $7(4g - 3h)$ **5** $5(2u + 3v)$ **6** $6(5w + 3x)$

7 $7(3y + z)$ **8** $8(2v + 5)$ **9** $6(2 + 7w)$

10 $4(3 - 8a)$ **11** $2(4g - 3)$ **12** $5(7 - 4b)$

13 $2(3i + 4j - 5k)$ **14** $4(5m - 3n + 2p)$ **15** $6(2r - 3s - 4t)$

Challenge 10.1 ⏱

The diagram shows a rectangular room with a square rug at one end.
Write an expression for

a) the length of the uncovered section of the room.

b) the area of the uncovered section of the room.

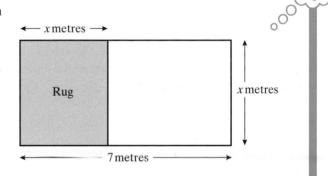

Combining terms

Terms with the same letter are called **like terms**. Like terms can be added.

One weekend Joe works on both Saturday and Sunday.
On Saturday he makes 75 cheese sandwiches and on Sunday he makes 45 cheese sandwiches.

You can calculate the total cost of the bread and the cost of the cheese he uses as follows.

Write expressions for the two days and expand the brackets.

$$75(2b + c) + 45(2b + c) = 150b + 75c + 90b + 45c$$

Simplify the expression by collecting like terms.

$$150b + 75c + 90b + 45c = 150b + 90b + 75c + 45c$$
$$= 240b + 120c$$

EXAMPLE 10.3

Expand the brackets and simplify these.

a) $10(2b + c) + 5(2b + c)$ 　　　　　　**b)** $35(b + c) + 16(2b + c)$

c) $16(2b + c) + 63(b + 2c)$ 　　　　　　**d)** $30(b + 2c) + 18(b + c)$

Solution

a)
$$10(2b + c) + 5(2b + c) = 20b + 10c + 10b + 5c$$
$$= 20b + 10b + 10c + 5c$$
$$= 30b + 15c$$

b)
$$35(b + c) + 16(2b + c) = 35b + 35c + 32b + 16c$$
$$= 35b + 32b + 35c + 16c$$
$$= 67b + 51c$$

c)
$$16(2b + c) + 63(b + 2c) = 32b + 16c + 63b + 126c$$
$$= 32b + 63b + 16c + 126c$$
$$= 95b + 142c$$

d)
$$30(b + 2c) + 18(b + c) = 30b + 60c + 18b + 18c$$
$$= 30b + 18b + 60c + 18c$$
$$= 48b + 78c$$

You expand and simplify brackets with other letters or other signs, or with numbers, in the same way. You have to take particular care when there are minus signs in the expression.

EXAMPLE 10.4

Expand the brackets and simplify these.

a) $2(3c + 4d) + 5(3c + 2d)$

b) $5(4e + f) - 3(2e - 4f)$

c) $5(4g + 3) - 2(3g + 4)$

Solution

a) Multiply all terms in the first bracket by the number in front of that bracket.

Multiply all the terms in the second bracket by the number in front of that bracket.

Then collect like terms together.

$$2(3c + 4d) + 5(3c + 2d) = 6c + 8d + 15c + 10d$$
$$= 6c + 15c + 8d + 10d$$
$$= 21c + 18d$$

b) Take care with the signs.

Think of the second part of the expression,
$-3(2e - 4f)$, as $+ (-3) \times (2e + (-4f))$.

You need to multiply both the terms in the second bracket by -3.

You learned the rules for calculating with negative numbers in Chapter 1.

$$
\begin{aligned}
5(4e + f) - 3(2e - 4f) &= 5(4e + f) + (-3) \times (2e + (-4f)) \\
&= 5 \times 4e + 5 \times f + (-3) \times 2e + (-3) \times (-4f) \\
&= 20e + 5f + (-6e) + (+12f) \\
&= 20e + (-6e) + 5f + 12f \\
&= 20e - 6e + 5f + 12f \\
&= 14e + 17f
\end{aligned}
$$

c) Again you need to take need to care with the signs.

Think of the second part of the expression,
$-2(3g + 4)$, as $+ (-2) \times (3g + 4)$.

You need to multiply both the terms in the second bracket by -2.

$$
\begin{aligned}
5(4g + 3) - 2(3g + 4) &= 5(4g + 3) + (-2) \times (3g + 4) \\
&= 20g + 15 + (-2) \times 3g + (-2) \times 4 \\
&= 20g + 15 + (-6g) + (-8) \\
&= 20g + (-6g) + 15 + (-8) \\
&= 20g - 6g + 15 - 8 \\
&= 14g + 7
\end{aligned}
$$

EXERCISE 10.2

Expand the brackets and simplify these.

1 **a)** $8(2a + 3) + 2(2a + 7)$ **b)** $5(3b + 7) + 6(2b + 3)$
    c) $2(3 + 8c) + 3(2 + 7c)$ **d)** $6(2 + 3a) + 4(5 + a)$

2 **a)** $5(2s + 3t) + 4(2s + 7t)$ **b)** $2(2v + 7w) + 5(2v + 7w)$
    c) $7(3x + 8y) + 3(2x + 7y)$ **d)** $3(2v + 5w) + 4(8v + 3w)$

3 **a)** $4(3x + 5) + 3(3x - 4)$ **b)** $2(4y + 5) + 3(2y - 3)$
    c) $5(2 + 7z) + 4(3 - 8z)$ **d)** $3(2 + 5x) + 5(6 - x)$

4 **a)** $3(2n + 7p) + 2(5n - 6p)$ **b)** $5(3q + 8r) + 3(2q - 9r)$
    c) $7(2d + 3e) + 3(3d - 5e)$ **d)** $4(2f + 7g) + 3(2f - 9g)$
    e) $3(3h - 8j) - 5(2h - 7j)$ **f)** $6(2k - 3m) - 3(2k - 7m)$

Factorising

You learned in Chapter 1 that the highest common factor of a set of numbers is the largest number that will divide into all of the numbers in the set.

Remember that in algebra you use letters to stand for numbers. For example, $2x = 2 \times x$ and $3x = 3 \times x$.

You do not know what x is but you do know that $2x$ and $3x$ both divide by x. So x is the highest common factor of $2x$ and $3x$.

When you have unlike terms, for example $2x$ and $4y$, you must assume that x and y do not have any common factors. However, you can look for common factors in the numbers. The highest common factor of 2 and 4 is 2 so the highest common factor of $2x$ and $4y$ is 2.

Factorising is the reverse of expanding brackets. You divide each of the terms in the bracket by the highest common factor and write this common factor outside the bracket.

EXAMPLE 10.5

Factorise these.

a) $(12x + 16)$ b) $(x - x^2)$ c) $(8x^2 - 12x)$

Solution

a) $(12x + 16)$ The common factor of $12x$ and 16 is 4.

    $4(\qquad)$ You write this factor outside the bracket.

You then divide each term inside the original bracket by the common factor, 4.

$12x \div 4 = 3x$ and $16 \div 4 = 4$.

$4(3x + 4)$ You write the new terms inside the bracket.

$(12x + 16) = 4(3x + 4)$

> **TIP**
> Check that your answer is correct by expanding it.
> $4(3x + 4) = 4 \times 3x + 4 \times 4 = 12x + 16$

b) $(x - x^2)$ The common factor of x and x^2 is x.
                      (Remember that x^2 is $x \times x$.)

    $x(\qquad)$ You write this factor outside the bracket.

You then divide each term inside the original bracket by the common factor, x.

$x \div x = 1$ and $x^2 \div x = x$

$(x - x^2) = x(1 - x)$

c) $(8x^2 - 12x)$ Think about the numbers and the letters separately and then combine them. The common factor of 8 and 12 is 4 and the common factor of x^2 and x is x. Therefore, the common factor of $8x^2$ and $12x$ is $4 \times x = 4x$.

$4x($ $)$ You write this factor outside the bracket.

You then divide each term inside the original bracket by the common factor, $4x$.

$8x^2 \div 4x = 2x$ and $12x \div 4x = 3$

$(8x^2 - 12x) = 4x(2x - 3)$

◉ EXERCISE 10.3

Factorise these as fully as possible.

1
   a) $(10x + 15)$ **b)** $(2x + 6)$ **c)** $(8x - 12)$ **d)** $(4x - 20)$

2
   a) $(14 + 7x)$ **b)** $(8 + 12x)$ **c)** $(15 - 10x)$ **d)** $(9 - 12x)$

3
   a) $(3x^2 + 5x)$ **b)** $(5x^2 + 20x)$ **c)** $(12x^2 - 8x)$ **d)** $(6x^2 - 8x)$

Expanding pairs of brackets

Earlier in this chapter you learned how to expand brackets of the type $25(2b + c)$. In that case you were multiplying a bracket by a single term. A term can be a number or a letter or a combination of the two, such as $3x$.

You can also expand pairs of brackets. In this case you are multiplying a bracket by another bracket. You must multiply each term inside the second bracket by each term inside the first bracket. The examples which follow show two methods for doing this.

EXAMPLE 10.6

Expand these.

a) $(a + 2)(a + 5)$

b) $(b + 4)(2b + 7)$

c) $(2m + 5)(3m - 4)$

Solution

a) Method 1

Use a grid to multiply each of the
terms in the second bracket by
each of the terms in the first.

×	a	$+2$
a	a^2	$+2a$
$+5$	$+5a$	$+10$

$= a^2 + 2a + 5a + 10$ Collect like terms together: $2a + 5a = 7a$.
$= a^2 + 7a + 10$

Method 2

Use the word FOIL to make sure you multiply each term in the second bracket
by each term in the first.

    F: first × first
    O: outer × outer
    I: inner × inner
    L: last × last

If you draw arrows to show the multiplications, you can think of a smiley face.

F   L

$(a + 2)(a + 5)$

I

O

$= a \times a + a \times 5 + 2 \times a + 2 \times 5$
$= a^2 + 5a + 2a + 10$
$= a^2 + 7a + 10$

b) Method 1

×	b	$+4$
$2b$	$2b^2$	$+8b$
$+7$	$+7b$	$+28$

$= 2b^2 + 8b + 7b + 28$
$= 2b^2 + 15b + 28$

Method 2

$(b + 4)(2b + 7)$

$= b \times 2b + b \times 7 + 4 \times 2b + 4 \times 7$
$= 2b^2 + 7b + 8b + 28$
$= 2b^2 + 15b + 28$

c) Method 1

×	$2m$	$+5$
$3m$	$6m^2$	$+15m$
-4	$-8m$	-20

$= 6m^2 + 15m - 8m - 20$
$= 6m^2 + 7m - 20$

Method 2

$(2m + 5)(3m - 4)$

$= 2m \times 3m + 2m \times -4 + 5 \times 3m + 5 \times -4$
$= 6m^2 - 8m + 15m - 20$
$= 6m^2 + 7m - 20$

> **TIP**
>
> Choose the method you prefer and
> stick to it.

Expand the brackets and simplify these. Use the method you prefer.

1 **a)** $(a + 3)(a + 7)$ **b)** $(b + 7)(b + 4)$ **c)** $(3 + c)(2 + c)$

2 **a)** $(3d + 5)(3d - 4)$ **b)** $(4e + 5)(2e - 3)$ **c)** $(2 + 7f)(3 - 8f)$

3 **a)** $(2g - 3)(2g - 7)$ **b)** $(2h - 7)(2h - 7)$ **c)** $(3j - 8)(2j - 7)$

4 **a)** $(2k + 7)(5k - 6)$ **b)** $(3 + 8m)(2 - 9m)$ **c)** $(2 + 3n)(3 - 5n)$

5 **a)** $(2 + 7p)(2 - 9p)$ **b)** $(3r - 8)(2r - 7)$ **c)** $(2s - 3)(2s - 7)$

Challenge 10.2   ?

What pairs of brackets have been expanded to give each of these expressions?

a) $m^2 + 3m + 2$ **b)** $x^2 - 5x + 6$ **c)** $2y^2 + 3y - 2$

Challenge 10.3   ?

The diagram shows a rectangular garden with a shed in one corner.

a) Write an expression for
  (i) the length of the garden.
  (ii) the width of the garden.

b) Write an expression for the area of the garden.
Simplify the expression.

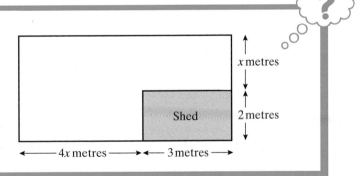

Challenge 10.4   ?

The diagram shows a rectangular picture hanging on a rectangular wall in a gallery.

The picture is placed so that it is 1 metre from the top of the wall and 1 metre from the bottom of the wall.

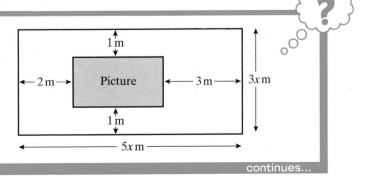

continues...

The picture is also 2 metres from the left-hand end of the wall and 3 metres from the right-hand end of the wall.

The length of the wall is $5x$ metres and its width is $3x$ metres.

a) Write an expression for
    (i) the length of the picture.
    (ii) the width of the picture.

b) Write an expression for the area of the wall occupied by the picture. Simplify the expression.

Index notation

In Chapter 1 you learned that you can write 2 to the power 4 as 2^4 and that the power, 4 in this case, is called the index. You can say that 2^4 is written in **index notation**.

You can use index notation in algebra too. You have already met x^2. This means x squared or x to the power 2.

y^5 is another example of an expression written using index notation. It means y to the power 5 or $y \times y \times y \times y \times y$. The index is 5.

EXAMPLE 10.7

Write these using index notation.
a) $5 \times 5 \times 5 \times 5 \times 5 \times 5$
b) $x \times x \times x \times x \times x \times x \times x$
c) $p \times p \times p \times p \times r \times r \times r$
d) $3w \times 4w \times 5w$

Solution

a) $5 \times 5 \times 5 \times 5 \times 5 \times 5 = 5$ to the power $6 = 5^6$

b) $x \times x \times x \times x \times x \times x \times x = x$ to the power $7 = x^7$

c) $p \times p \times p \times p \times r \times r \times r = p^4 \times r^3 = p^4 r^3$

d) $3w, 4w$ and $5w$ are like terms so you multiply the numbers together first, then the letters.
    $3w \times 4w \times 5w = (3 \times 4 \times 5) \times (w \times w \times w) = 60 \times w^3 = 60w^3$

EXERCISE 10.5

Simplify each of the following, writing your answer using index notation.

1 **a)** $3 \times 3 \times 3 \times 3$
   b) $7 \times 7 \times 7$
   c) $10 \times 10 \times 10 \times 10 \times 10$

2 **a)** $x \times x \times x \times x \times x$
   b) $y \times y \times y \times y$
   c) $z \times z \times z \times z \times z \times z \times z$

3 **a)** $m \times m \times n \times n \times n \times n$
   b) $f \times f \times f \times f \times g \times g \times g \times g \times g$
   c) $p \times p \times p \times r \times r \times r \times r$

4 **a)** $2k \times 4k \times 7k$
   b) $3y \times 5y \times 8y$
   c) $4d \times 2d \times d$

Challenge 10.5

Simplify each of the following, writing your answer using index notation.

a) $m^2 \times m^4$ **b)** $x^3 \times 5x^6$

c) $5y^4 \times 3y^3$ **d)** $2b^3 \times 3b^2 \times 4b$

WHAT YOU HAVE LEARNED

- **How to simplify brackets by collecting like terms**
- **When you expand brackets such as 25(2b + c), you multiply each of the terms inside the bracket by the number (or term) outside the bracket**
- **When you expand brackets such as (a + 2)(a + 5), you multiply each of the terms in the second bracket by each of the terms in the first bracket**
- **One way to expand brackets is to use a grid. Another way is to use the word FOIL to make sure you perform all the multiplications**
- **Factorising expressions is the reverse of expanding brackets**
- **To factorise an expression you take the common factors outside the bracket**
- **How to use index notation in algebra**

1 Expand these.

  a) $8(3a + 2b)$

  d) $9(a - 2b)$

  g) $4(5x - 3y)$

  j) $3(2j + 5k)$

  b) $5(4a + 3b)$

  e) $3(4x + 5y)$

  h) $2(4x + y)$

  k) $7(r + 2s)$

  c) $12(3a - 5b)$

  f) $6(3x - 2y)$

  i) $5(3f - 4g)$

  l) $4(3v - w)$

2 Expand the brackets and simplify these.

  a) $2(3x + 4) + 3(2x + 1)$

  c) $2(2x + 3) + 3(x + 2)$

  e) $3(3y + 5) + 2(3y - 4)$

  g) $3(2a + 4) - 3(a + 2)$

  i) $6(3p + 4) - 3(4p + 2)$

  k) $2(4j + 8) - 3(3j - 5)$

  b) $4(2x + 3) + 3(4x + 5)$

  d) $5(2y + 3) + 2(3y - 5)$

  f) $3(5y + 2) + 2(3y - 1)$

  h) $2(6m + 2) - 3(2m + 1)$

  j) $4(5t + 3) - 3(2t - 4)$

  l) $6(2w + 5) - 4(3w - 4)$

3 Factorise these as fully as possible.

  a) $(4x + 8)$

  d) $(12x - 18)$

  g) $(24 + 8x)$

  j) $(32x - 12)$

  m) $(2x - x^2)$

  b) $(6x + 12)$

  e) $(6 - 10x)$

  h) $(16x + 12)$

  k) $(20 - 16x)$

  n) $(3y - 7y^2)$

  c) $(9x - 6)$

  f) $(10 - 15x)$

  i) $(6x + 8)$

  l) $(15 + 20x)$

  o) $(5z^2 + 2z)$

4 Expand the brackets and simplify these.

  a) $(a + 5)(a + 4)$

  d) $(x - 1)(x + 8)$

  g) $(3x + 4)(x + 9)$

  j) $(2p + 4)(3p - 2)$

  b) $(a + 2)(a + 3)$

  e) $(x + 9)(x - 5)$

  h) $(y - 3)(2y + 7)$

  k) $(t - 5)(4t - 3)$

  c) $(3 + a)(4 + a)$

  f) $(x - 2)(x - 1)$

  i) $(2 - 3p)(7 - 2p)$

  l) $(2a - 3)(3a + 5)$

5 Simplify each of the following, writing your answer using index notation.

  a) $4 \times 4 \times 4 \times 4 \times 4 \times 4$

  c) $2 \times 2 \times 2 \times 2 \times 2$

  e) $j \times j \times j$

  g) $v \times v \times v \times w \times w \times w$

  i) $x \times x \times x \times y \times y \times y \times y \times y$

  b) $5 \times 5 \times 5 \times 5$

  d) $a \times a \times a \times a \times a \times a \times a$

  f) $t \times t \times t \times t \times t \times t$

  h) $d \times d \times d \times e \times e \times e \times e \times e \times e$

  j) $5p \times 4p \times 3p$

11 → ALGEBRAIC MANIPULATION 2

THIS CHAPTER IS ABOUT

- Factorising quadratic expressions
- Simplifying algebraic fractions, including those involving indices and quadratic expressions

YOU SHOULD ALREADY KNOW

- How to carry out operations on numerical fractions
- How to factorise linear expressions
- How to use index notation

Simplifying algebraic fractions

You already know how to **simplify**, or cancel, **numerical fractions**: you divide the **numerator** and **denominator** by a **common factor**.

Algebraic fractions follow the same rules as numerical fractions. However, it is important to realise that in a fraction such as $\frac{x+2}{2x+3}$, both the numerator and the denominator are expressions in their own right.

It is not possible to cancel individual terms in the numerator with individual terms in the denominator. In the fraction $\frac{x+2}{2x+3}$, there is an x in both the numerator and the denominator, but it is not a factor of either expression and so cannot be cancelled. Only complete factors can be cancelled.

In order to cancel an algebraic fraction, you must first **factorise** the numerator and denominator. For the fraction $\frac{x+2}{2x+3}$, this is not possible: it is already in its **simplest form**.

EXAMPLE 11.1

Simplify $\dfrac{x^2 + 5x}{x^2 - 2x}$.

Solution

$\dfrac{x^2 + 5x}{x^2 - 2x} = \dfrac{x(x + 5)}{x(x - 2)}$ First factorise the expressions.

$\phantom{\dfrac{x^2 + 5x}{x^2 - 2x}} = \dfrac{\cancel{x}(x + 5)}{\cancel{x}(x - 2)}$ Cancel the common factor, x.

$\phantom{\dfrac{x^2 + 5x}{x^2 - 2x}} = \dfrac{x + 5}{x - 2}$ The expression is now in its simplest form.

EXAMPLE 11.2

Simplify $\dfrac{2x + 6}{x^2 + 3x}$.

Solution

$\dfrac{2x + 6}{x^2 + 3x} = \dfrac{2(x + 3)}{x(x + 3)}$ First factorise the expressions.

$\phantom{\dfrac{2x + 6}{x^2 + 3x}} = \dfrac{2}{x}$ Cancel the common factor, $(x + 3)$.

EXERCISE 11.1

Simplify each of these algebraic fractions. If this is not possible, say so.

1 $\dfrac{3x - 9}{6x - 3}$

2 $\dfrac{2x + 6}{5x - 10}$

3 $\dfrac{8 + 2x}{6 - 4x}$

4 $\dfrac{12 - 4x}{6 - 9x}$

5 $\dfrac{3x^2 - x}{2x^2 + x}$

6 $\dfrac{2x^2 - 4x}{6x^2 - 8x}$

7 $\dfrac{9x - 6x^2}{2x^2 + 8x}$

8 $\dfrac{5x^2 + 15x}{10x - 5}$

Factorising quadratic expressions

In Chapter 10 you learned how to **expand** expressions such as $(x + 1)(x + 7)$, to give the **quadratic** expression $x^2 + 8x + 7$. You also need to be able to factorise quadratic expressions, which is the reverse of that process.

EXAMPLE 11.3

Factorise $x^2 + 7x + 12$.

Solution

$x^2 + 7x + 12 = x^2 + 4x + 3x + 12$ The 'trick' is to look for **factor pairs** of **12** (the **constant term**): $1 \times 12, 2 \times 6, 3 \times 4$.
Choose the pair that can be added or subtracted to give **7** (the **coefficient of** x).

From 1 and 12 it is possible to make 11 or 13.

From 2 and 6 it is possible to make 4 or 8.

From 3 and 4 it is possible to make 1 or **7** so the required factors are 3 and 4.

$= (x + 3)(x + 4)$

TIP

The *coefficient* is the number part of a term in an expression.

Check that you have factorised the expression correctly by expanding the brackets.

$(x + 3)(x + 4) = x^2 + 4x + 3x + 12$
$\qquad\qquad = x^2 + 7x + 12 ✓$

You do not need to write out all the steps of your solution as in Example 11.3. However, you should always expand the brackets to check your solution.

EXAMPLE 11.4

Factorise $x^2 + 3x - 10$.

Solution

$x^2 + 3x - 10 = (x + 5)(x - 2)$ The factor pairs of 10 are 1×10 and 2×5.
                                    To make 3 you need to use 5 and -2.

Check by expanding the brackets.
$(x + 5)(x - 2) = x^2 - 2x + 5x - 10$
$\qquad\qquad\quad = x^2 + 3x - 10 \checkmark$

EXAMPLE 11.5

Factorise $x^2 - 6x + 8$.

Solution

$x^2 - 6x + 8 = (x - 2)(x - 4)$ The factor pairs of 8 are 1×8 and 2×4.
                                  To make -6 you need to use -2 and -4.

Check by expanding the brackets.
$(x - 2)(x - 4) = x^2 - 4x - 2x + 8$
$\qquad\qquad\quad = x^2 - 6x + 8 \checkmark$

EXERCISE 11.2

Factorise each of these expressions.

1 a) $x^2 + 3x + 2$ **b)** $x^2 + 6x + 5$ **c)** $x^2 + 8x + 12$ **d)** $x^2 + 8x + 15$
   e) $x^2 + 10x + 9$ **f)** $x^2 + 6x + 9$ **g)** $x^2 + 9x + 20$ **h)** $x^2 + 14x + 24$

2 a) $x^2 + 5x - 6$ **b)** $x^2 + 9x - 10$ **c)** $x^2 + 2x - 8$ **d)** $x^2 + 4x - 12$
   e) $x^2 + 2x - 15$ **f)** $x^2 + 8x - 20$ **g)** $x^2 + 5x - 24$ **h)** $x^2 + 2x - 24$

3 a) $x^2 - 7x + 10$ **b)** $x^2 - 9x + 8$ **c)** $x^2 - 3x + 2$ **d)** $x^2 - 9x + 14$
   e) $x^2 - 10x + 16$ **f)** $x^2 - 14x + 24$ **g)** $x^2 - 11x + 30$ **h)** $x^2 - 5x + 6$

Work in pairs.

Write your own quadratic expression (either one that will factorise or one that will not).

Challenge your partner to say if it factorises or not, and to carry out the factorisation if it does.

If they are correct they score one point, if they are wrong they score nothing. The first to score five points wins.

Difference of two squares

Expressions such as $x^2 - 16$ look as though they cannot be factorised. However, this is a special-case quadratic expression known as the **difference of two squares** (the expression represents the difference between two squares, in this case x^2 and 16).

To factorise such an expression, find the square root of each term. Because the numerical term in the expanded expression is negative, the numerical term is positive in one of the brackets and negative in the other.

$$x^2 - a = (x + \sqrt{a})(x - \sqrt{a})$$

For example, $x^2 - 16 = (x + 4)(x - 4)$.

This can be checked by expanding the brackets.

$$(x + 4)(x - 4) = x^2 - 4x + 4x - 16 \qquad -4x + 4x = 0$$
$$= x^2 - 16$$

EXAMPLE 11.6

Factorise $x^2 - 9$.

Solution

$x^2 - 9 = (x + 3)(x - 3)$ Find the square root of each term in the expression.

Check by expanding the brackets.

$$(x + 3)(x - 3) = x^2 - 3x + 3x - 9 \qquad -3x + 3x = 0$$
$$= x^2 - 9 \checkmark$$

EXAMPLE 11.7

Factorise $8x^2 - 8$.

Solution

$$8x^2 - 8 = 8(x^2 - 1) \qquad \text{The terms are not squares but there is a common factor.}$$
$$= 8(x + 1)(x - 1) \qquad \text{Factorise the expression in brackets.}$$

Check by expanding the brackets.
$$8(x + 1)(x - 1) = 8(x^2 - x + x - 1) \qquad -x + x = 0$$
$$= 8(x^2 - 1)$$
$$= 8x^2 - 8 \checkmark$$

EXERCISE 11.3

Factorise each of these expressions.

1 **a)** $x^2 - 9$ **b)** $x^2 - 16$ **c)** $x^2 - 49$
    d) $x^2 - 81$ **e)** $x^2 - 100$ **f)** $x^2 - 144$

2 **a)** $3x^2 - 12$ **b)** $5x^2 - 45$ **c)** $3x^2 - 108$
    d) $7x^2 - 343$ **e)** $10x^2 - 4000$ **f)** $8x^2 - 200$

Challenge 11.3

Factorise each of these expressions.

a) $\dfrac{2x + 14}{x^2 - 49}$ **b)** $\dfrac{10x + 15}{4x^2 - 9}$ **c)** $\dfrac{3x^2 - 108}{2x^2 + 12x}$

d) $\dfrac{5x^3 + 5x^2 - 30x}{3x^3 - 12x}$ **e)** $\dfrac{10x^3 - 5x^2 - 30x}{12x^3 - 27x}$

Simplifying algebraic fractions involving indices

You should already be familiar with the laws of indices from Key Stage 3.

$$a^m \times a^n = a^{m+n} \text{ (add the indices)}$$
$$\text{and}$$
$$a^m \div a^n = a^{m-n} \text{ (subtract the indices)}$$

These laws can be applied when simplifying algebraic fractions.

Remember that only complete factors can be cancelled: parts of factors cannot.

EXAMPLE 11.8

Simplify $\dfrac{5x^3y^2}{2z^4} \times \dfrac{8z^2}{15xy^3}$.

Solution

$$\dfrac{5x^3y^2}{2z^4} \times \dfrac{8z^2}{15xy^3} = \dfrac{x^2}{z^2} \times \dfrac{4}{3y}$$

Look for common factors: 5 in 5 and 15
2 in 2 and 8
x in x^3 and x
y^2 in y^2 and y^3
z^2 in z^4 and z^2

$$= \dfrac{4x^2}{3yz^2}$$ Simplify.

EXAMPLE 11.9

Simplify $\dfrac{4a^5b^3}{15c^2} \div \dfrac{8a^2b^5}{9c^4}$.

Solution

$$\dfrac{4a^5b^3}{15c^2} \div \dfrac{8a^2b^5}{9c^4} = \dfrac{4a^5b^3}{15c^2} \times \dfrac{9c^4}{8a^2b^5}$$

As with division of numerical fractions, it is easier to change the operation to multiplication and to invert the fraction after the division sign.

$$= \dfrac{a^3}{5} \times \dfrac{3c^2}{2b^2}$$

Look for common factors: 4 in 4 and 8
3 in 15 and 9
a^2 in a^5 and a^2
b^3 in b^3 and b^5
c^2 in c^2 and c^4

$$- \dfrac{3a^3c^2}{10b^2}$$ Simplify.

EXERCISE 11.4

Simplify each of these expressions.

1 $\dfrac{6a^4b}{5c^2} \times \dfrac{10c^3}{9a^2b^3}$

2 $\dfrac{12x^5y^4}{7z^5} \times \dfrac{14z^3}{15x^3y^2}$

3 $\dfrac{6a^4b^5}{9c^3} \div \dfrac{10a^2b^3}{3c^6}$

4 $\dfrac{8a^7b^5}{15c^4} \div \dfrac{4ab^2}{9c}$

5 $\dfrac{6t^5v^2}{5w^2} \div \dfrac{8t^2v^5}{15w^4}$

6 $\dfrac{12e^6f^2}{5g^3} \times \dfrac{10g^4}{8e^4f^3}$

7 $\dfrac{5a^3b^2}{2c^4} \times \dfrac{8c^2d^4}{15ae^3} \times \dfrac{3e^2}{2b^3d^2}$

8 $\dfrac{8t^3v^2}{3x^5z^4} \times \dfrac{15x^2}{12v^4y^3} \div \dfrac{10t^4}{9y^2z^5}$

9 $\left(\dfrac{4a^5b^6}{5c^2d^4} \div \dfrac{12a^2b^5}{25c^4}\right) \times \dfrac{8d^2}{15ab^3}$

10 $\left(\dfrac{2e^2f^3}{3g^2} \div \dfrac{8e^4f^4}{9f^2g^4}\right) \div \dfrac{8f^3g^5}{4e^4}$

11 $\dfrac{24x^{\frac{5}{2}}}{8} \times \dfrac{x^{\frac{3}{2}}}{x}$

12 $\dfrac{36a^{-3}}{4a^2} \times \dfrac{b^4}{b^8}$

13 $30x^{\frac{3}{2}} \times \dfrac{x^{-\frac{7}{2}}}{5}$

14 $\dfrac{48a^{-6}}{4a^3}$

15 $10x^{\frac{5}{2}} \times 2x^{-\frac{3}{2}}$

Challenge 11.4

Work in pairs.

Write your own algebraic fraction problem (using either multiplication or division only).

Challenge your partner to simplify the problem to a single algebraic fraction.

Factorising quadratic expressions where the coefficient of $x^2 \neq 1$

Earlier in this chapter you learned how to factorise expressions such as $x^2 + 7x + 12$, to give $(x + 3)(x + 4)$. In all the examples, the coefficient of x^2 was 1.

It is also possible to factorise quadratic expressions where the coefficient of x^2 is not 1 in a similar way.

EXAMPLE 11.10

Factorise $3x^2 + 14x + 8$.

Solution

$3x^2 + 14x + 8 = (3x + 2)(x + 4)$

The factor pairs of 8 are 1×4 and 2×4. When forming the x term, one of the two factors will also be multiplied by 3 (the coefficient of x^2). To make 14 you need to use 2 and 4 with the 3, $(2 + 3 \times 4)$. The 4 is multiplied by the 3 so the $3x$ must not be in the same bracket as the 4.

Check by expanding the brackets.

$(3x + 2)(x + 4) = 3x^2 + 12x + 2x + 8$

$\qquad\qquad\qquad = 3x^2 + 14x + 8 \checkmark$

EXAMPLE 11.11

Factorise $2x^2 + 3x - 20$.

Solution

$2x^2 + 3x - 20 = (2x - 5)(x + 4)$

The factor pairs of 20 are $1 \times 20, 2 \times 10$ and 4×5.
To make 3 you need to use 4 and –5 with the 2,
$(2 \times 4 - 5)$.
The 4 is multiplied by the 2 so the $2x$ must not be in the same bracket as the 4.

Check by expanding the brackets.
$$(2x - 5)(x + 4) = 2x^2 + 8x - 5x - 20$$
$$= 2x^2 + 3x - 20 \checkmark$$

EXERCISE 11.5

Factorise each of these expressions.

1　**a)** $3x^2 + 17x + 20$　　　**b)** $2x^2 + 7x + 6$　　　**c)** $3x^2 + 13x + 4$

　　d) $5x^2 + 18x + 9$　　　**e)** $4x^2 + 6x + 2$　　　**f)** $3x^2 + 11x + 10$

　　g) $2x^2 + 5x + 2$　　　**h)** $4x^2 + 17x + 15$　　　**i)** $5x^2 + 8x + 3$

2　**a)** $2x^2 - x - 15$　　　**b)** $3x^2 + x - 14$　　　**c)** $5x^2 - 17x - 12$

　　d) $3x^2 - 5x - 12$　　　**e)** $4x^2 - 3x - 10$　　　**f)** $2x^2 - 7x - 15$

　　g) $4x^2 - 7x - 2$　　　**h)** $3x^2 - 16x - 12$　　　**i)** $4x^2 + 21x - 18$

3　**a)** $3x^2 - 14x + 8$　　　**b)** $5x^2 - 19x + 12$　　　**c)** $3x^2 - 26x + 35$

　　d) $2x^2 - 21x + 40$　　　**e)** $2x^2 - 11x + 12$　　　**f)** $4x^2 - 11x + 6$

　　g) $2x^2 - 21x + 40$　　　**h)** $3x^2 - 5x + 2$　　　**i)** $3x^2 - 7x + 4$

Challenge 11.5

Factorise each of these expressions.

a) $8x^2 + 10x + 3$

b) $15x^2 + 2x - 8$

c) $8x^2 - 2x - 15$

d) $6x^2 - 29x + 35$

EXAMPLE 11.12

Factorise $15x^2 + 26x + 8$.

Solution

$15x^2 + 26x + 8$

Consider $ax^2 + bx + c$. Look at the product ac:

$$ac = 15 \times 8 = 120$$

Factor pairs are $1 \times 120, 2 \times 60, 3 \times 40, \ 4 \times 30, 5 \times 24, 6 \times 20$, etc.
Notice that $6 + 20 = b$, the coefficient in the middle term in x.

For the expression $15x^2 + 26x + 8$, partition the term in x:

$$20 + 6 = 26$$

$5x(3x + 4) + 2(3x + 4)$ Factorise pairs of terms.
$(3x + 4)(5x + 2)$ Factorise the two terms.

Alternatively:

$15x^2 + 26x + 8$ must be in the form $(15x \ldots)(x \ldots)$ or $(3x \ldots)(5x \ldots)$.
Use trial and improvement for the last terms, multiplying them together to give 8.
Possible solutions could be 1×8 and 2×4.
This gives a possible solution of $(3x + 4)(5x + 2)$.
Check your answer by expanding the brackets.

EXERCISE 11.6

Factorise each of these expressions.

1	$6x^2 + 13x + 6$	**2**	$20x^2 + 19x + 3$
3	$27x^2 + 24x + 4$	**4**	$10x^2 - 11x - 6$
5	$15x^2 - 7x - 2$	**6**	$50x^2 - 35x + 6$
7	$20x^2 + 76x + 21$	**8**	$30x^2 - 42x + 12$
9	$20x^2 + 43x + 6$	**10**	$8x^2 - 19x + 6$

Simplifying fractions involving quadratic expressions

Earlier in this chapter you learned that, to cancel an algebraic fraction, you must first factorise the numerator and denominator. Sometimes the fraction will involve quadratic expressions.

EXAMPLE 11.13

Simplify $\dfrac{2x + 14}{x^2 + 6x - 7}$.

Solution

$$\frac{2x + 14}{x^2 + 6x - 7} = \frac{2(x + 7)}{(x - 1)(x + 7)}$$ First factorise the expressions.

$$= \frac{2}{x - 1}$$ Cancel the common factor, $(x + 7)$.

EXAMPLE 11.14

Simplify $\dfrac{x^2 - 25}{x^2 + 3x - 10}$.

Solution

$$\frac{x^2 - 25}{x^2 + 3x - 10} = \frac{(x + 5)(x - 5)}{(x + 5)(x - 2)}$$ First factorise the expressions.

$$= \frac{x - 5}{x - 2}$$ Cancel the common factor, $(x + 5)$.

EXAMPLE 11.15

Simplify $\dfrac{2x^2 + 4x - 16}{x^2 - 7x + 10}$.

Solution

$$\frac{2x^2 + 4x - 16}{x^2 - 7x + 10} = \frac{2(x^2 + 2x - 8)}{x^2 - 7x + 10}$$ Take the common factor, 2, out of the numerator.

$$= \frac{2(x + 4)(x - 2)}{(x - 5)(x - 2)}$$ Factorise the expressions.

$$= \frac{2(x + 4)}{x - 5}$$ Cancel the common factor, $(x - 2)$.

EXAMPLE 11.16

Simplify $\dfrac{2(x^2 + 4)^2}{3x^2 + 12}$.

Solution

$$\frac{2(x^2 + 4)^2}{3x^2 + 12} = \frac{2(x^2 + 4)(x^2 + 4)}{3(x^2 + 4)}$$

Rewrite the numerator and take the common factor, 3, out of the denominator.

$$= \frac{2(x^2 + 4)}{3}$$

Cancel the common factor, $(x^2 + 4)$.

EXERCISE 11.7

Simplify each of these algebraic fractions.

1 $\dfrac{3x + 15}{x^2 + 3x - 10}$

2 $\dfrac{6x - 18}{x^2 - x - 6}$

3 $\dfrac{x^2 - 5x + 6}{x^2 - 4x + 3}$

4 $\dfrac{x^2 - 3x - 4}{x^2 - 4x - 5}$

5 $\dfrac{x^2 - 2x - 3}{x^2 - 9}$

6 $\dfrac{3x^2 - 12}{x^2 + 2x - 8}$

7 $\dfrac{3x^2 + 5x + 2}{2x^2 - x - 3}$

8 $\dfrac{2x^2 + x - 6}{x^2 + x - 2}$

9 $\dfrac{6x^2 - 3x}{(2x - 1)^2}$

10 $\dfrac{5(x + 3)^2}{x^2 - 9}$

11 $\dfrac{(x - 3)(x + 2)^2}{x^2 - x - 6}$

12 $\dfrac{2x^2 + x - 6}{(2x - 3)^2}$

WHAT YOU HAVE LEARNED

- **When simplifying algebraic fractions only complete factors may be cancelled**
- **How to factorise quadratic expressions**
- **How to factorise the difference of two squares**

1 Simplify each of these algebraic fractions.

a) $\dfrac{5x + 10x}{15 - 5x}$

b) $\dfrac{8x + 12}{10 + 6x}$

c) $\dfrac{6x^2 - 4x}{10x^2 + 15x}$

d) $\dfrac{8x - 12x^2}{12x^2 + 8x}$

e) $\dfrac{9x + 15x^2}{5x^2 - 10x}$

f) $\dfrac{4x + 8}{3x^2 + 6x}$

g) $\dfrac{6x - 9}{4x^2 - 9}$

h) $\dfrac{5x^2 + 15x}{4x + 12}$

2 Factorise each of these expressions.

a) $x^2 + 5x + 4$

b) $x^2 + 7x + 12$

c) $x^2 + 9x + 14$

d) $x^2 + 8x + 7$

e) $x^2 + 7x + 10$

f) $x^2 + 6x + 5$

g) $x^2 + 17x + 30$

h) $x^2 + 20x + 36$

3 Factorise each of these expressions.

a) $x^2 - 81$

b) $x^2 - 64$

c) $x^2 - 169$

d) $x^2 - 225$

e) $3x^2 - 48$

f) $5x^2 - 45$

g) $7x^2 - 343$

h) $10x^2 - 1000$

4 Simplify each of these algebraic fractions.

a) $\dfrac{5a^3b^2}{2c^4} \times \dfrac{8c^2}{15ab^3}$

b) $\dfrac{6x^3}{5y^3z^4} \times \dfrac{25y^2z^2}{18x}$

c) $\dfrac{4q^5r^3}{15s^2} \div \dfrac{8q^2r^5}{9s^4}$

d) $\dfrac{6t^4v^5}{15w^3} \div \dfrac{4t^5v^3}{21w^2}$

e) $\dfrac{3x^4y^2}{5z^3} \times \dfrac{4x}{9y^5z^2} \times \dfrac{5yz^7}{8x^3}$

f) $\dfrac{2e^2n}{3t^4} \times \dfrac{3t^5}{e^4n^6} \div \dfrac{8t^2}{4en^4}$

g) $\dfrac{24x^{\frac{1}{2}} \times 2x^{-\frac{3}{2}}}{6x^{\frac{5}{2}}}$

h) $\dfrac{10a^{-2} \times 2a^{-6}}{15a^{15}}$

5 Factorise each of these expressions.

a) $5x^2 + 27x + 10$

b) $3x^2 + 16x + 21$

c) $3x^2 - 22x + 24$

d) $4x^2 - 13x + 3$

e) $2x^2 + 3x - 14$

f) $6x^2 + x - 7$

g) $3x^2 - 4x - 32$

h) $5x^2 - 12x - 9$

i) $28x^2 + 29x + 6$

j) $20x^2 + 37x - 6$

k) $30x^2 + 7x - 15$

l) $24x^2 + 25x + 6$

6 Simplify each of these algebraic fractions.

a) $\dfrac{2x + 8}{x^2 + x - 12}$

b) $\dfrac{6x - 18}{3x^2 + 6x}$

c) $\dfrac{x^2 + 3x - 4}{x^2 + 6x + 8}$

d) $\dfrac{x^2 - 2x - 3}{x^2 - 1}$

e) $\dfrac{10x^2 + 15x}{2x^2 - x - 6}$

f) $\dfrac{6x^2 + 5x - 4}{2x^2 + 5x - 3}$

g) $\dfrac{4(x - 5)^2}{2x^2 - 50}$

h) $\dfrac{2x(x + 3)^2}{x^2 + x - 6}$

12 → EQUATIONS AND INEQUALITIES 1

<table>
<tr><td>

THIS CHAPTER IS ABOUT

- **Solving simple equations**
- **Solving simple inequalities**

</td><td>

YOU SHOULD ALREADY KNOW

- **How to collect like terms**
- **How to add, subtract, multiply and divide with negative numbers**
- **The squares of whole numbers up to 10**

</td></tr>
</table>

Solving equations

Sometimes the x term in the equation is squared (x^2). If there is an x^2 term and no other x term in the equation, you can solve it by getting the x^2 term on its own and taking the square root of each side. However, you must remember that if you square a negative number, the result is positive. For example, $(-6)^2 = 36$.

When you solve an equation involving x^2, there will usually be two values that satisfy the equation.

EXAMPLE 12.1

Solve these equations.

a) $5x + 1 = 16$ **b)** $x^2 + 3 = 39$

> **TIP** Remember that you must always do each operation to the whole of both sides of the equation.

Solution

a)
$$5x + 1 = 16$$
$$5x + 1 - 1 = 16 - 1 \quad \text{First subtract 1 from each side.}$$
$$5x = 15$$
$$5x \div 5 = 15 \div 5 \quad \text{Now divide each side by 5.}$$
$$x = 3$$

b) $x^2 + 3 = 39$
$$x^2 = 36 \quad \text{First subtract 3 from each side.}$$
$$\sqrt{x^2} = \pm\sqrt{36} \quad \text{Now find the square root of each side.}$$
$$x = 6$$
$$\text{or} \quad x = -6$$

EXERCISE 12.1

Solve these equations.

1 $2x - 1 = 13$	**2** $2x - 1 = 0$	**3** $2x - 13 = 1$
4 $3x - 2 = 19$	**5** $6x + 12 = 18$	**6** $3x - 7 = 14$
7 $4x - 8 = 12$	**8** $4x + 12 = 28$	**9** $3x - 6 = 24$
10 $5x - 10 = 20$	**11** $x^2 + 3 = 28$	**12** $x^2 - 4 = 45$
13 $y^2 - 2 = 62$	**14** $m^2 + 3 = 84$	**15** $m^2 - 5 = 20$
16 $x^2 + 10 = 110$	**17** $x^2 - 4 = 60$	**18** $20 + x^2 = 36$
19 $16 - x^2 = 12$	**20** $200 - x^2 = 100$	

Solving equations with brackets

You learned how to **expand brackets** in Chapter 10.

If you are solving an equation with brackets in it, expand the brackets first.

> **TIP**
> Remember to multiply *each* term inside the brackets by the number outside the brackets.

EXAMPLE 12.2

Solve these equations.

a) $3(x + 4) = 24$ **b)** $4(p - 3) = 20$

Solution

a) $3(x + 4) = 24$

$\quad\quad 3x + 12 = 24$ Multiply each term inside the brackets by 3.

$\quad\quad\quad\quad 3x = 12$ Subtract 12 from each side.

$\quad\quad\quad\quad\quad x = 4$ Divide each side of the equation by 3.

b) $4(p - 3) = 20$

$\quad\quad 4p - 12 = 20$ Multiply each term inside the brackets by 4.

$\quad\quad\quad\quad 4p = 32$ Add 12 to each side.

$\quad\quad\quad\quad\quad p = 8$ Divide each side by 4.

Solve these equations.

1 $3(p - 4) = 36$	**2** $3(4 + x) = 21$	**3** $6(x - 6) = 6$	**4** $4(x + 3) = 16$
5 $2(x - 8) = 14$	**6** $2(x + 4) = 10$	**7** $2(x - 4) = 20$	**8** $5(x + 1) = 30$
9 $3(x + 7) = 9$	**10** $2(x - 7) = 6$	**11** $5(x - 6) = 20$	**12** $7(a + 3) = 28$
13 $3(2x + 3) = 40$	**14** $5(3x - 1) = 40$	**15** $2(5x - 3) = 14$	**16** $4(3x - 2) = 28$
17 $7(x - 4) = 28$	**18** $3(5x - 12) = 24$	**19** $2(4x + 2) = 20$	**20** $2(2x - 5) = 12$

Equations with x on both sides

Some equations, such as $3x + 4 = 2x + 5$, have x on both sides.
You should get the terms in x all together on the left-hand side of the
equation and the constant terms together on the right-hand side.

$$3x + 4 = 2x + 5$$
$$3x + 4 - 2x = 2x + 5 - 2x$$
$$x + 4 = 5$$
$$x + 4 - 4 = 5 - 4$$
$$x = 1$$

Start by subtracting $2x$ from both sides of the equation, which
will cancel the $2x$ on the right hand side and get all the x terms
together on the left hand side of the equation. Now subtract 4
from both sides which will cancel the 4 on the left-hand side of
the equation.

EXAMPLE 12.3

Solve these equations.

a) $8x - 3 = 3x + 7$

b) $18 - 5x = 4x + 9$

Solution

a)
$$8x - 3 = 3x + 7$$
$$8x - 3 - 3x = 3x + 7 - 3x$$
$$5x - 3 = 7$$
$$5x - 3 + 3 = 7 + 3$$
$$5x = 10$$
$$\frac{5x}{5} = \frac{10}{5}$$
$$x = 2$$

Start by subtracting $3x$ from both sides of the equation,
which will cancel the $3x$ on the right-hand side and get all
the x terms together on the left-hand side of the equation.
Now add 3 to both sides which will cancel the 3 on the left-
hand side of the equation.

Divide both sides by the coefficient of x, that is, by 5.

b)
$$18 - 5x = 4x + 9$$
$$18 - 5x - 4x = 4x + 9 - 4x$$
$$18 - 9x = 9$$
$$18 - 9x - 18 = 9 - 18$$
$$-9x = -9$$
$$\frac{-9x}{-9} = \frac{-9}{-9}$$
$$x = 1$$

Start by subtracting $4x$ from both sides of the equation,
which will cancel the $4x$ on the right-hand side and get all
the x terms together on the left-hand side of the equation.
Now subtract 18 from both sides which will cancel the 18
on the left-hand side of the equation.

Divide both sides by the coefficient of x, that is, by -9.

Solve these equations.

1 $7x - 4 = 3x + 8$	**2** $5x + 4 = 2x + 13$	**3** $6x - 2 = x + 8$	**4** $5x + 1 = 3x + 21$
5 $9x - 10 = 3x + 8$	**6** $5x - 12 = 2x - 6$	**7** $4x - 23 = x + 7$	**8** $8x + 8 = 3x - 2$
9 $11x - 7 = 6x + 8$	**10** $5 + 3x = x + 9$	**11** $2x - 3 = 7 - 3x$	**12** $4x - 1 = 2 + x$
13 $2x - 7 = x - 4$	**14** $3x - 2 = x + 7$	**15** $x - 5 = 2x - 9$	**16** $x + 9 = 3x - 3$
17 $3x - 4 = 2 - 3x$	**18** $5x - 6 = 16 - 6x$	**19** $3(x + 1) = 2x$	**20** $49 - 3x = x + 21$

Challenge 12.1

The length of a rectangular field is 10 metres more than its width.

The perimeter of the field is 220 metres.

What are the width and length of the field?

Hint: let x represent the width and draw a sketch of the rectangle.

Challenge 12.2

A rectangle measures $(2x + 1)$ cm by $(x + 9)$ cm.

Find the value of x for which the rectangle is a square.

Challenge 12.3

Here is an equation with boxes instead of numbers.

$\square x + \square = \square x + \square$

Jake found four numbers to put in the boxes by rolling an ordinary 6-sided dice four times. He then tried to solve the equation he had made.

a) What is the largest possible solution?

b) What is the smallest possible solution?

c) Find an equation he cannot solve.

Fractions in equations

You know that $k \div 6$ can be written as $\dfrac{k}{6}$.

You solve an equation like $\dfrac{k}{6} = 2$ by multiplying both sides of the equation by the denominator of the fraction.

Check up 12.1

Solve these equations.

a) $\dfrac{x}{3} = 10$ b) $\dfrac{m}{4} = 2$ c) $\dfrac{m}{2} = 6$ d) $\dfrac{p}{3} = 9$ e) $\dfrac{y}{7} = 4$

Some equations involving fractions take more than one step to solve.

These are solved using the same method as equations without fractions.

You can get rid of the fraction, by multiplying both sides of the equation by the denominator of the fraction, at the end.

EXAMPLE 12.4

Solve the equation $\dfrac{x}{8} + 3 = 5$.

Solution

$\dfrac{x}{8} + 3 = 5$

$\quad \dfrac{x}{8} - 2 \qquad$ Subtract 3 from each side.

$\quad x = 16 \qquad$ Multiply each side by 8.

EXERCISE 12.4

Solve these equations.

1 $\dfrac{x}{4} + 3 = 7$ **2** $\dfrac{a}{5} - 2 = 6$ **3** $\dfrac{x}{4} - 2 = 3$ **4** $\dfrac{y}{5} - 5 = 5$

5 $\dfrac{y}{6} + 3 = 8$ **6** $\dfrac{p}{7} - 4 = 1$ **7** $\dfrac{m}{3} + 4 = 12$ **8** $\dfrac{x}{8} + 8 = 16$

9 $\dfrac{x}{9} + 7 = 10$ **10** $\dfrac{y}{3} - 9 = 2$

Challenge 12.4

Try to solve this ancient puzzle.

A number plus its three-quarters, plus its half, plus its fifth, makes 49.
What is the number?

Challenge 12.5

I think of a number. I square it and add 1. The answer divided by 10 gives 17.
What is the number?

Inequalities

If you want to buy a packet of sweets costing 79p, you need at least 79p.

You may have more than that in your pocket. The amount in your pocket must be greater than or equal to 79p.

If the amount in your pocket is x, then this can be written as $x \geqslant 79$.
This is an inequality.

> The symbol $\geqslant$ means 'greater than or equal to'.
> The symbol $>$ means 'greater than'.
> The symbol $\leqslant$ means 'less than or equal to'.
> The symbol $<$ means 'less than'.

On a number line you use an open circle to represent $>$ or $<$, and a solid circle to represent $\geqslant$ or $\leqslant$.

Inequalities are solved in a similar way to equations.

EXAMPLE 12.5

Solve the inequality $2x - 1 > 8$.
Show the solution on a number line.

Solution

$2x - 1 > 8$

$\quad\quad 2x > 9$ Add 1 to each side.

$\quad\quad\quad x > 4.5$ Divide each side by 2.

$$-5 \quad -4 \quad -3 \quad -2 \quad -1 \quad 0 \quad 1 \quad 2 \quad 3 \quad 4 \quad 5$$

Negative inequalities work a bit differently.

Rules for inequalities

Inequalities behave exactly the same as equations:

- when you add or subtract the same quantity from both sides of an inequality
- when you multiply or divide both sides of an inequality by a **positive** quantity.

However, when you multiply or divide both sides of an inequality by a **negative** quantity, inequalities behave differently to equations.

Consider the inequality $$-2x \leqslant -4$$

Adding $2x$ to both sides and adding 4 to both sides gives $$-2x + 2x + 4 \leqslant -4 + 2x + 4$$

Which simplifies to $$4 \leqslant 2x$$

Dividing both sides by 2 gives us $$2 \leqslant x$$

$$\text{or} \quad x \geqslant 2$$

Therefore, if we divide both sides of the inequality $-2x \leqslant -4$ by -2, we must change $\leqslant$ into $\geqslant$ to get the correct result.

Therefore $\quad -2x \leqslant -4$ becomes on dividing both sides by -2

$$\frac{-2x}{-2} \geqslant \frac{-4}{-2}$$

giving $\quad x \geqslant 2$

This gives us the following rule for multiplying or dividing an inequality by a **negative** quantity:

Whenever you multiply or divide both sides of an inequality by a **negative** quantity, you must also reverse the inequality sign, that is change $<$ to $>$, or $\leqslant$ to $\geqslant$, and so on.

See how this works in Example 12.6.

EXAMPLE 12.6

Solve the inequality $7 - 3x \leqslant 1$.

Solution

$$7 - 3x \leqslant 1$$
$$7 - 3x - 7 \leqslant 1 - 7$$
$$-3x \leqslant -6$$
$$\frac{-3x}{-3} \geqslant \frac{-6}{-3}$$
$$x \geqslant 2$$

Subtract 7 from both sides of the equation, which will cancel the 7 on the left-hand side and get all the x terms together on the left-hand side of the equation and all the constants on the right-hand side of the equation.
Divide both sides by the coefficient of x, that is, by -3.
Remember that the $\leqslant$ sign must reverse to become $\geqslant$ because we are dividing by a negative quantity.

For each of questions **1** to **6,** solve the inequality and show the solution on a number line.

1 $x - 3 > 10$ **2** $x + 1 < 5$ **3** $5 > x - 8$

4 $2x + 1 \leqslant 9$ **5** $3x - 4 \geqslant 5$ **6** $10 \leqslant 2x - 6$

For each of questions **7** to **20,** solve the inequality.

7 $5x < x + 8$ **8** $2x \geqslant x - 5$ **9** $4 + x < -5$

10 $2(x + 1) > x + 3$ **11** $6x > 2x + 20$ **12** $3x + 5 \leqslant 2x + 14$

13 $5x + 3 \leqslant 2x + 9$ **14** $8x + 3 > 21 + 5x$ **15** $5x - 3 > 7 + 3x$

16 $6x - 1 < 2x$ **17** $5x < 7x - 4$ **18** $9x + 2 \geqslant 3x + 20$

19 $5x - 4 \leqslant 2x + 8$ **20** $5x < 2x + 12$

WHAT YOU HAVE LEARNED

- **To solve equations involving brackets, expand the brackets first**
- **To solve equations with x on both sides, get the terms in x together on the left-hand side of the equation**
- **Solve equations involving fractions in the same way as equations without fractions, and deal with the fraction at the end**
- **The symbol $\geqslant$ means 'greater than or equal to', $>$ means 'greater than', $\leqslant$ means 'less than or equal to' and $<$ means 'less than'**
- **$x \geqslant 4, x > 3, y \leqslant 6$ and $y < 7$ are inequalities**
- **Inequalities can be solved in a similar way to equations**
- **Solutions to inequalities can be represented on a number line**

Solve these equations.

1 $2(m - 4) = 10$ **2** $5(p + 6) = 40$ **3** $7(x - 2) = 42$

4 $3(4 + x) = 21$ **5** $4(p - 3) = 20$ **6** $3x^2 = 48$

7 $2x^2 = 72$ **8** $5p^2 + 1 = 81$ **9** $4x^2 - 3 = 61$

10 $2a^2 - 3 = 47$ **11** $\dfrac{x}{5} - 1 = 4$ **12** $\dfrac{x}{6} + 5 = 10$

13 $\dfrac{y}{3} + 7 = 13$ **14** $\dfrac{y}{7} - 6 = 1$ **15** $\dfrac{a}{4} - 8 = 1$

Solve each of these inequalities and show the solution on a number line.

16 $5x + 1 \leqslant 11$ **17** $10 + 3x \leqslant 5x + 4$ **18** $7x + 3 < 5x + 9$

19 $6x - 8 > 4 + 3x$ **20** $5x - 7 > 7 - 2x$

13 → EQUATIONS AND INEQUALITIES 2

Equations

You learned how to solve simple equations in Chapter 12.

Check up 13.1

Solve each of these equations.

a) $5x + 2 = 12$

b) $3x - 9 = x + 4$

c) $2(4x - 3) = 14$

d) $x - 2 = 3x + 6$

The equations that follow will bring together all the ideas you met in Chapter 12 and extend them.

Equations with brackets on both sides

In Chapter 12 you learned how to solve equations involving brackets, and equations with x on both sides. You can use the strategies needed to solve both of these types of equation to solve more complex equations.

EXAMPLE 13.1

Solve the equation $3(4x - 5) = 2(3x - 2) - 4x - 3$.

Solution

$$3(4x - 5) = 2(3x - 2) - 4x - 3$$

$$12x - 15 = 6x - 4 - 4x - 3 \qquad \text{First expand the brackets.}$$

$$12x - 6x + 4x = 15 - 4 - 3 \qquad \text{Collect all the } x \text{ terms on the left-hand side of the}$$
$$\qquad\qquad\qquad\qquad\qquad \text{equation, and the numerical terms on the right.}$$

$$10x = 8$$

$$x = \tfrac{8}{10} \qquad\qquad \text{Divide each side of the equation by the coefficient of } x, 10.$$

$$x = \tfrac{4}{5} \qquad\qquad \text{Give the answer in its simplest form.}$$

 TIP

A common error is to go from $10x = 8$ to $x = \dfrac{10}{8}$ rather than $\dfrac{8}{10}$.

Make sure you divide by the coefficient of x.

Equations involving fractions

You learned in Chapter 12 how to solve equations involving fractions and a single x term.

When an equation involves a fraction and more than one x term you need to get rid of the fraction first, by multiplying each side of the equation by the denominator of the fraction.

EXAMPLE 13.2

Solve the equation $\dfrac{x}{3} = 2x - 3$.

Solution

$$\frac{x}{3} = 2x - 3$$

$$x = 3(2x - 3) \qquad \text{First multiply each side of the equation by the denominator, 3.}$$

$$x = 6x - 9 \qquad\quad \text{Expand the brackets.}$$

$$6x - 9 = x \qquad\quad \text{Swap the sides of the equation so that the } x \text{ term with the larger}$$
$$\qquad\qquad\qquad\qquad \text{positive coefficient is on the left. You could do this at a later stage.}$$

$$5x = 9 \qquad\qquad \text{Collect all the } x \text{ terms on the left-hand side of the equation, and the}$$
$$\qquad\qquad\qquad\qquad \text{numerical terms on the right.}$$

$$x = \tfrac{9}{5} \qquad\qquad \text{Divide each side of the equation by the coefficient of } x, 5.$$

$$x = 1\tfrac{4}{5} \qquad\qquad \text{Give the answer as a mixed number.}$$

When an equation involves more than one fraction, you need to multiply each side of the equation by the lowest common multiple (LCM) of all the denominators.

EXAMPLE 13.3

Solve the equation $\dfrac{x}{4} = \dfrac{3x}{2} - \dfrac{5}{3}$.

Solution

$$\frac{x}{4} = \frac{3x}{2} - \frac{5}{3}$$

$$12 \times \frac{x}{4} = 12 \times \left(\frac{3x}{2} - \frac{5}{3} \right) \qquad \text{Multiply each side of the equation by the LCM of 4, 2 and 3, which is 12.}$$

$$3x = 18x - 20$$

$$18x - 20 = 3x \qquad \text{Swap the sides of the equation so that the } x \text{ term with the larger positive coefficient is on the left.}$$

$$15x = 20 \qquad \text{Collect all the } x \text{ terms on the left-hand side of the equation, and the numerical terms on the right.}$$

$$x = \tfrac{20}{15} \qquad \text{Divide each side of the equation by the coefficient of } x, 15.$$

$$x = \tfrac{4}{3} \qquad \text{Cancel the fraction by dividing by the common factor, 5.}$$

$$x = 1\tfrac{1}{3} \qquad \text{Give the answer as a mixed number.}$$

 TIP

A common error when multiplying through by a number is to multiply just the first term.

Use brackets to make sure.

EXAMPLE 13.4

Solve the equation $\dfrac{2x - 3}{6} + \dfrac{x + 2}{3} = \dfrac{5}{2}$.

Solution

$$\frac{2x-3}{6} + \frac{x+2}{3} = \frac{5}{2}$$

$$6 \times \left(\frac{2x-3}{6} + \frac{x+2}{3}\right) = 6 \times \frac{5}{2}$$ Multiply each side of the equation by the LCM of 6, 3 and 2, which is 6.

$$2x - 3 + 2(x + 2) = 15$$

$$2x - 3 + 2x + 4 = 15$$ Expand the brackets.

$$4x + 1 = 15$$ Simplify by collecting like terms.

$$4x = 14$$ Subtract 1 from each side of the equation.

$$x = \frac{14}{4}$$ Divide each side of the equation by the coefficient of x, 4.

$$x = \frac{7}{2}$$ Cancel the fraction by dividing by the common factor, 2.

$$x = 3\frac{1}{2}$$ Give the answer as a mixed number.

EXERCISE 13.1

Solve each of these equations.

1 $5(x - 4) = 4x$

2 $4(2x - 2) = 3(x + 4)$

3 $2(4x - 5) = 2x + 6$

4 $\frac{x}{2} = 3x - 10$

5 $\frac{x}{3} = x - 2$

6 $\frac{3x}{2} = 7 - 2x$

7 $\frac{4x}{3} = 4x - 2$

8 $\frac{2x}{3} = x - \frac{4}{3}$

9 $\frac{x}{2} = \frac{3x}{4} - 6$

10 $\frac{x}{3} = \frac{3x}{4} - \frac{1}{6}$

11 $\frac{3x}{2} = \frac{3x-2}{5} + 4$

12 $\frac{2x-1}{6} = \frac{x-3}{2} + \frac{2}{3}$

13 $\frac{x-2}{3} + \frac{2x-1}{2} = \frac{17}{6}$

14 $\frac{2x-3}{3} - \frac{2x+1}{6} + \frac{3}{2} = 0$

15 $\frac{3x-2}{2} = \frac{x-3}{3} + \frac{7}{6}$

Challenge 13.1

a) Mr Watson left his assets to be shared amongst three people.
He left one third to John, one half to Sylvia and the rest, £75 000, to Maxine.
Use algebra to find how much he left altogether.

b) At a sports club there are seven more men than women.
A quarter of the number of women is the same as one fifth of the number of men.
Use algebra to find how many women there are at the sports club.

Equations where the unknown is the denominator

In some equations the unknown will be the denominator. The first step towards solving equations of this sort is to multiply through by the denominator.

EXAMPLE 13.5

Solve the equation $\dfrac{200}{x} = 8$.

Solution

$$\frac{200}{x} = 8$$

$200 = 8x$	First multiply each side of the equation by the denominator, x.
$8x = 200$	Swap the sides of the equation so that the x term is on the left.
$x = 25$	Divide each side of the equation by the coefficient of x, 25.

If the equation involves more than one denominator, you need to multiply through by the LCM of all the denominators.

EXAMPLE 13.6

Solve the equation $\dfrac{3}{2x} = \dfrac{6}{5}$.

Solution

$$\frac{3}{2x} = \frac{6}{5}$$

$10x \times \dfrac{3}{2x} = 10x \times \dfrac{6}{5}$	Multiply each side of the equation by the LCM of $2x$ and 5, which is $10x$.
$15 = 12x$	Swap the sides of the equation so that the x term is on the left.
$12x = 15$	
$x = \frac{15}{12}$	Divide each side of the equation by the coefficient of x, 12.
$x = \frac{5}{4}$	Cancel the fraction by dividing by the common factor, 3.
$x = 1\frac{1}{4}$	Give the answer as a mixed number.

TIP

These equations are not difficult if all the steps are carried out.

Avoid the common error of going from, for example, $\dfrac{2}{x} = 8$ to $x = 4$ rather than $x = \frac{1}{4}$.

Equations involving decimals

When an equation involves decimals, the solution may not be exact.

EXAMPLE 13.7

Solve the equation $3.6x = 8.7$.

Give your answer correct to 3 significant figures.

Solution

$3.6x = 8.7$

$x = 8.7 \div 3.6$ Divide each side of the equation by 3.6.

$x = 2.416\ 666\ ...$

$x = 2.42$ (to 3 s.f.)   Round your answer to the given degree of accuracy.

EXERCISE 13.2

Solve each of these equations.

Where the answer is not exact, give your answer correct to 3 significant figures.

1 $\dfrac{20}{x} = 5$ **2** $\dfrac{4}{x} = 12$ **3** $\dfrac{75}{2x} = 3$

4 $\dfrac{16}{3x} = \dfrac{1}{6}$ **5** $\dfrac{2}{3x} = \dfrac{4}{3}$ **6** $3.5x = 9.6$

7 $5.2x = 25$ **8** $\dfrac{x}{3.4} = 2.7$ **9** $2.3(x - 1.2) = 4.6$

10 $\dfrac{3.4}{x} = 12$

Challenge 13.2

A piece of string is 84 cm long. It is cut into x pieces of equal length.

Another piece of string is 60 cm long and is cut into eight pieces, each of which is $\frac{1}{2}$ cm longer than the pieces from the 84 cm string.

Use algebra to find the value of x.

Inequalities

In Chapter 12 you also learned how to solve simple inequalities.

> ## Check up 13.2
>
> Solve each of these inequalities.
>
> **a)** $4x < 5$ **b)** $2x - 4 > 5$ **c)** $3(2x - 5) < 6$

In general, inequalities follow the same rules as equations. For example, you deal with fractions in inequalities in the same way as you do fractions in equations.

EXAMPLE 13.8

Solve the inequality $\frac{x}{3} \leqslant 2x - 3$.

Solution

$\frac{x}{3} \leqslant 2x - 3$

$x \leqslant 3(2x - 3)$ Multiply each side of the inequality by the denominator, 3.

$x \leqslant 6x - 9$ Expand the brackets.

$9 \leqslant 5x$ Collect the x terms on the side with the higher x term and the numerical terms on the other side.

$1.8 \leqslant x$ Divide each side of the inequality by the coefficient of x, 5.

$x \geqslant 1.8$ Rewrite the inequality so that x is on the left. Remember to turn the inequality sign round as well.

There is, however, one way in which the rules for inequalities are different from the rules for equations: when you multiply or divide by a negative number, you need to turn the inequality sign around.

EXAMPLE 13.9

Solve the inequality $2 - 3x > 8$.

Solution

$2 - 3x > 8$

$-3x > 6$ Subtract 2 from each side of the inequality.

$x < -2$ Divide each side of the inequality by the coefficient of x, -3. Because you are dividing by a negative number, you need to turn the inequality sign around.

Discovery 13.1

a) To see why the method used in Example 13.9 works, solve the inequality $2 - 3x > 8$ by adding $3x$ to each side.

b) Now try both methods to solve these inequalities.

    (i) $4 - x < 7$ **(ii)** $4 > 3 - 7x$

TIP

If you are not sure which way round the inequality sign should be, check by testing a number that satisfies the solution in the original inequality.

For instance, in Example 13.9, the solution is $x < -2$.
Choose a number that satisfies this solution, for example $n = -10$.
Substitute the number into the original inequality, $2 - 3x > 8$.

When $x = -10, 2 - 3 \times (-10) > 8$
$$32 > 8 \checkmark$$

EXERCISE 13.3

Solve each of these inequalities.

1 $3(2x - 6) > 12 - 4x$ **2** $12(x - 4) \leqslant 3(2x - 6)$ **3** $\dfrac{x}{2} < 3x - 10$

4 $\dfrac{2x}{3} < 3x - 14$ **5** $\dfrac{3x}{2} > \dfrac{x}{4} + 2$ **6** $\dfrac{x}{2} \geqslant \dfrac{3x}{4} - 2$

7 $\frac{1}{2}(2 - x) < 9$ **8** $7 - \dfrac{x}{3} \geqslant 3$ **9** $4 - x \leqslant 7 - 2x$

10 $3.4x - 5 \leqslant 2.1x$

Challenge 13.3

Natasha wants to organise a disco for as many of her friends as possible.

It costs £75 to hire the disco equipment and the buffet costs £3.50 for each person. She has only £150 to spend.

Use algebra to find the largest number of people that can go to the disco.

Solving inequalities with two unknowns

Discovery 13.2

For each of these inequalities, write down three or four possible pairs of values for x and y.

a) $x + y < 5$ **b)** $3x + y > 10$ **c)** $2x + 3y \leqslant 12$

There are infinitely many possible answers to Discovery 13.2!

Discovery 13.3

For each of these pairs of inequalities, write down two possible pairs of values for x and y.

The values must satisfy both the inequalities at the same time.

a) $2x + 3y \leqslant 6$ and $x > y$ **b)** $4x - 2y > 5$ and $x + y < 6$

c) $5x + 4y < 20$ and $2x + y > 4$

As the many possible answers cannot be listed, it is much better to draw them on a graph and show the possible values by shading.

Discovery 13.4

a) Draw a pair of axes and label them −4 to 4 for both x and y.
Draw and label the line $x = 2$, and label the regions $x \leqslant 2$ and $x \geqslant 2$ clearly.

b) Draw a pair of axes and label them −4 to 4 for both x and y.
Draw and label the line $y = x + 2$, and label the regions $y \leqslant x + 2$ and $y \geqslant x + 2$ clearly.

c) Draw a pair of axes and label them 0 to 6 for both x and y.
Draw and label the line $3x + 2y = 12$, and label the regions $3x + 2y \leqslant 12$ and $3x + 2y \geqslant 12$ clearly.

You can represent an inequality on a graph by drawing the graph of the line and labelling the region that satisfies the inequality.

Often the region that *does not* satisfy the inequality is shaded; the region that *does* satisfy the inequality is left unshaded.

Several inequalities can be represented on the same axes and the region where the values of x and y satisfy all of them can be found.

EXAMPLE 13.10

Draw a pair of axes and label them 0 to 8 for both x and y.
Show, by shading, the region where $x \geqslant 0, y \geqslant 0$ and $x + 2y \leqslant 8$.

Solution

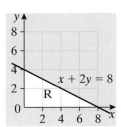

First draw the line $x + 2y = 8$.

Then shade out the regions $x \leqslant 0, y \leqslant 0$, and $x + 2y \geqslant 8$. These are the regions which are *not required* for the solution; that is, where the values of x and y *do not* satisfy the inequalities.

The required region, where the values of x and y satisfy all three inequalities, is labelled R.

> **TIP**
>
> When you have drawn a line, test which side is the required region: choose a simple point on one side of the line and see whether or not it satisfies the inequality.

When an inequality involves the signs $<$ or $>$, the points on the line are *not* included in the solution. For example, if $x < 2$, the points $(2, -2)$, $(2, -1), (2, 0), (2, 1), (2, 2)$ and so on do not satisfy the inequality. Inequalities involving the signs $<$ or $>$ can be represented on a graph using a broken line.

EXAMPLE 13.11

Draw a pair of axes and label them 0 to 5 for both x and y.
Show, by shading, the region where $x \geqslant 0, y \geqslant 0, 5x + 4y < 20$ and $2x + y > 4$.

Solution

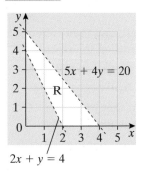

Draw the two lines $5x + 4y = 20$ and $2x + y = 4$.

Because they represent the inequalities $5x + 4y < 20$ and $2x + y > 4$ the lines should be broken, not solid.

Label them clearly.

Shade out the regions $x \leqslant 0, y \leqslant 0, 5x + 4y > 20$ and $2x + y < 4$.

The required region is labelled R.

1   Draw a pair of axes and label them 0 to 6 for both x and y.
    Show, by shading, the region where $x \geq 0$, $y \geq 0$ and $x + 2y \leq 6$.

2   Draw a pair of axes and label them -3 to 4 for both x and y.
    Show, by shading, the region where $x \geq 0$, $y \leq 3$ and $y \geq 2x - 3$.

3   Draw a pair of axes and label them -4 to 4 for both x and y.
    Show, by shading, the region where $x > -2$, $y < 3$ and $y > 2x$.

4   Draw a pair of axes and label them 0 to 6 for both x and y.
    Show, by shading, the region where $x < 4$, $y < 3$ and $3x + 4y > 12$.

5   Draw a pair of axes and label them -1 to 5 for both x and y.
    Show, by shading, the region where $y \geq 0$, $y \leq x + 1$ and $3x + 5y < 15$.

Challenge 13.4

At James' and Nicola's wedding, 56 people need to be taken to the reception.

Harry's Hire Cars has nine four-seater cars costing £25 each and five eight-seater cars costing £35 each.

James and Nicola hire x four-seater and y eight-seater cars.

a) One inequality is $x \leq 9$.
    Write down two other inequalities that must be satisfied.

b) Draw these three inequalities on a graph and shade out the regions not required.

c) Find the combination of cars that will cost the least.
    State how many of each type of car James and Nicola should hire, and the total cost.

WHAT YOU HAVE LEARNED

- **To solve equations with brackets on both sides, first expand the brackets, and then collect all the x terms on the left-hand side of the equation and the numerical terms on the right.**
- **To solve equations involving fractions, first multiply every term in the equation by the lowest common multiple of the denominators, expand any brackets and then rearrange as usual**
- **Linear inequalities are solved using the same rules as equations, except that when you multiply or divide by a negative number the inequality sign must be changed around**
- **To illustrate a number of inequalities on a graph, draw the lines and shade out the regions not required. Inequalities involving the signs $<$ or $>$ are represented using a broken line**

For questions **1** to **14**, solve the equation or inequality.

Where the answer is not exact, give your answer correct to 3 significant figures.

1 $4(2x - 5) = 3(4 - x) + 1$

2 $2(x - 3) = x - 1$

3 $\dfrac{3x}{2} = 2 + x$

4 $\dfrac{x}{2} = 3x + 5$

5 $\dfrac{x}{2} = \dfrac{3x}{4} - \dfrac{1}{2}$

6 $\dfrac{3x}{2} = \dfrac{3x - 2}{5} + 1$

7 $\dfrac{2x - 1}{6} = \dfrac{2(x - 3)}{3} + 1$

8 $\dfrac{4}{x} = 24$

9 $\dfrac{75}{2x} = 5$

10 $7.3x = 18.2$

11 $3.6x - 2.4 = 7.6$

12 $\dfrac{7x}{2} < 3x + 2$

13 $2(x - 4) \leqslant 3(2x - 3)$

14 $2.5x - 7.3 > 4.2$

15 Draw a pair of axes and label them 0 to 6 for x and -2 to 10 for y.
Show, by shading, the region where $x > 0$, $y > 3x - 2$ and $4x + 3y < 24$.

14 → GRAPHS 1

THIS CHAPTER IS ABOUT

- Drawing straight-line graphs from equations given in explicit or implicit form
- Distance–time graphs
- Drawing and interpreting graphs of real-life situations
- Drawing graphs of quadratic functions
- Solving equations using quadratic graphs

YOU SHOULD ALREADY KNOW

- How to plot and read points in all four quadrants
- How to substitute numbers into equations
- How to draw graphs with equations of the form $y = 2$, $x = 3$, etc
- How to rearrange equations
- How to add, subtract and multiply negative numbers
- How to plot and interpret simple straight-line graphs involving conversions, distance–time and other real-life situations
- How to use the relationship between distance, speed and time

Drawing straight-line graphs

The most common straight-line graphs have equations of the form $y = 3x + 2$, $y = 2x - 3$, etc.

This can be written in a general form as

$$y = mx + c$$

To draw a straight-line graph, work out three pairs of coordinates by substituting values of x into the formula to find y.

You can draw a straight line with only two points, but you should always work out a third point as a check.

EXAMPLE 14.1

Draw the graph of $y = -2x + 1$ for values of x from -4 to 2.

Solution

Find the values of y when $x = -4, 0$ and 2.

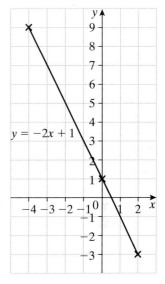

$y = -2x + 1$

When $x = -4$
$$y = -2 \times -4 + 1$$
$$y = 9$$

When $x = 0$
$$y = -2 \times 0 + 1$$
$$y = 1$$

When $x = 2$
$$y = -2 \times 2 + 1$$
$$y = -3$$

The y values needed are -3 to 9.

Draw the axes and plot the points $(-4, 9)$, $(0, 1)$ and $(2, -3)$.

Join them with a straight line. Label it $y = -2x + 1$.

TIP

Always use a ruler to draw a straight-line graph.

TIP

If axes have been drawn for you, check the scale before plotting points or reading values.

EXERCISE 14.1

1   Draw the graph of $y = 4x$ for values of x from -3 to 3.

2   Draw the graph of $y = x + 3$ for values of x from -3 to 3.

3   Draw the graph of $y = 3x - 4$ for values of x from -2 to 4.

4   Draw the graph of $y = 4x - 2$ for values of x from -2 to 3.

5   Draw the graph of $y = -3x - 4$ for values of x from -4 to 2.

Harder straight-line graphs

Sometimes you will be asked to draw graphs with equations of a different form.

For equations such as $2y = 3x + 1$, work out three points as before, remembering to divide by 2 to find the y value.

EXAMPLE 14.2

Draw the graph of $2y = 3x + 1$ for values of x from -3 to 3.

Solution

Find the values of y when $x = -3, 0$ and 3.

$2y = 3x + 1$

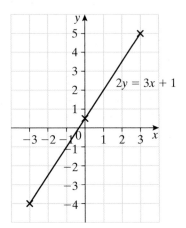

When $x = -3$

$$2y = 3 \times -3 + 1$$
$$2y = -8$$
$$y = -4$$

When $x = 0$

$$2y = 3 \times 0 + 1$$
$$2y = 1$$
$$y = \tfrac{1}{2}$$

When $x = 3$

$$2y = 3 \times 3 + 1$$
$$2y = 10$$
$$y = 5$$

The values of y needed are -4 to 5.

Draw the axes and plot the points $(-3, -4)$, $(0, \tfrac{1}{2})$ and $(3, 5)$.
Join them with a straight line. Label it $2y = 3x + 1$.

For equations such as $4x + 3y = 12$, work out y when $x = 0$, and x when $y = 0$. These are easy to work out. Find a third point as a check after you have drawn the line.

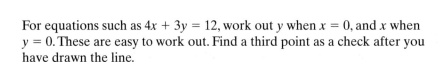

EXAMPLE 14.3

Draw the graph of $4x + 3y = 12$.

Solution

Find the value of y when $x = 0$ and the value of x when $y = 0$.

$4x + 3y = 12$

When $x = 0$
$\quad 3y = 12 \qquad 4 \times 0 = 0$ so the x term 'disappears'.
$\quad\ \ y = 4$

When $y = 0$
$\quad 4x = 12 \qquad 3 \times 0 = 0$ so the y term 'disappears'.
$\quad\ \ x = 3$

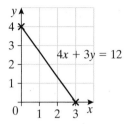

The values of x needed are 0 to 3. The values of y needed are 0 to 4.

Draw axes and plot the points $(0, 4)$ and $(3, 0)$.
Join them with a straight line. Label it $4x + 3y = 12$.

Choose a point on the line you have drawn and check it by substituting the x and y values into the equation.

> **TIP**
> Take care when plotting the points. Do not put $(0, 4)$ at $(4, 0)$ by mistake.

For example, the line passes through $(1\frac{1}{2}, 2)$.

$4x + 3y = 12$
$4 \times 1\frac{1}{2} + 3 \times 2 = 6 + 6 = 12$ ✓

EXERCISE 14.2

1 Draw the graph of $2y = 3x - 2$ for $x = -2$ to 4.
2 Draw the graph of $2x + 5y = 15$.
3 Draw the graph of $7x + 2y = 14$.
4 Draw the graph of $2y = 5x + 3$ for $x = -3$ to 3.
5 Draw the graph of $2x + y = 7$.

Challenge 14.1

The entry fee at Radium hot springs is \$6.50 each. It costs \$2 to hire towels.
A coach party of 40 people went in the hot springs and n of them hired towels.

a) Write down an equation for the total amount they spent, S, in terms of n.

b) Draw a graph of S against n, for values of n up to 40.

c) Use your graph to find how many towels were hired if the total spent was \$310.

Distance–time graphs

Check up 14.1

James walked to the bus stop and waited for the bus.

When the bus arrived he got on the bus and it took him to school without stopping.

Which of these distance–time graphs best shows James's journey to school?

Explain your answer.

a)
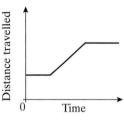

b)
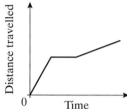

c)
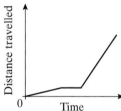

d)
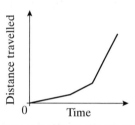

Discovery 14.1

James walked to the bus stop at 4 km/h.
This took him 15 minutes.

He waited 5 minutes at the bus stop.

The bus journey was 12 km and took
20 minutes.

The bus went at a constant speed.

a) Copy these axes and draw an
accurate graph of James's journey.

b) What was the speed of the bus
in km/h?

c) After 30 minutes, how far was
James away from home?

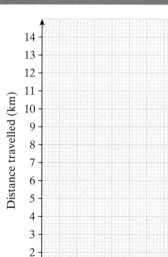

When a graph illustrates real quantities, how steeply it goes up or down is called the **rate of change**.

When the graph shows distance (vertical) against time (horizontal), the rate of change is equal to the **speed**.

Real-life graphs

When you are asked to answer questions about a given graph, you should:

- look carefully at the labels on the axes to see what the graph represents.
- check what the units are on each of the axes.
- look to see whether the lines are straight or curved.

If the graph is straight, the rate of change is constant. The steeper the line, the higher the rate of change.

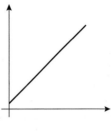

A horizontal line represents a part of the graph where there is no change in the quantity on the vertical axis.

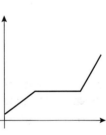

If the graph is a convex curve (viewed from below), the rate of change is increasing.

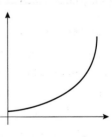

If the graph is a concave curve (viewed from below), the rate of change is decreasing.

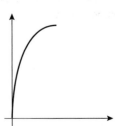

EXAMPLE 14.4

The graph shows the cost of printing tickets.

a) Find the total cost of printing 250 tickets.

b) The cost consists of a fixed charge and an additional charge for each ticket printed.

    (i)   What is the fixed charge?

    (ii) Find the additional charge for each ticket printed.

    (iii) Find the total cost of printing 800 tickets.

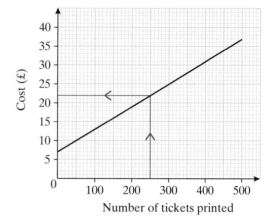

Solution

a) £22   Draw a line from 250 on the Number of tickets printed axis, to meet the straight line. Then draw a horizontal line and read off the value where it meets the Cost axis.

b) (i)   £7   Read from the graph the cost of zero tickets (where the graph cuts the Cost axis).

   (ii) 250 tickets cost £22.
        Fixed charge is £7.
        So the additional charge for 250 tickets is $22 - 7 = £15$.
        The additional charge per ticket is $\frac{15}{250} = £0.06$ or 6p

   (iii) Cost in pounds $= 7 +$ number of tickets $\times 0.06$
         Cost of 800 tickets $= 7 + 800 \times 0.06$
                          $= 7 + 48$
                          $= £55$

> **TIP**
> Work in either pounds or pence. If you work in pounds, you won't have to convert your final answer back from pence.

EXERCISE 14.3

1   Jane and Halima live in the same block of flats and go to the same school.
    The graphs represent their journeys home from school.

    a) Describe Halima's journey home.

    b) After how many minutes did Halima overtake Jane?

    c) Calculate Jane's speed in
       (i) kilometres per minute.
       (ii) kilometres per hour.

    d) Calculate Halima's fastest speed in
       (i) kilometres per minute.
       (ii) kilometres per hour.

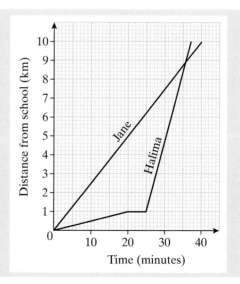

2 Anne, Britney and Catherine run a 10 km race.

Their progress is shown by the lines A, B and C respectively on the graph.

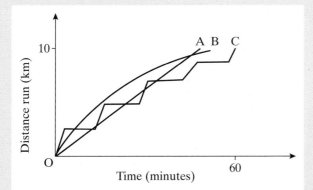

Imagine you are a commentator and give a description of the race.

3 A taxi driver charges according to the following rates.

A fixed charge of £*a*

+

x pence per kilometre for the first 20 km

+

40 pence for each kilometre over 20 km

The graph shows the charges for the first 20 km.

a) What is the fixed charge, £*a*?

b) Calculate *x*, the charge per kilometre for the first 20 km.

c) Copy the graph and add a line segment to show the charges for distances from 20 km to 50 km.

d) What is the total charge for a journey of 35 km?

e) What is the average cost per kilometre for a journey of 35 km?

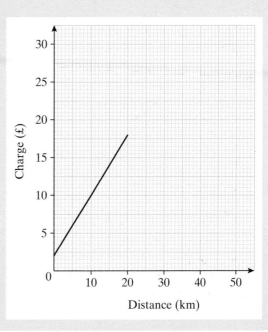

4 Water is poured into each of these glasses at a constant rate until they are full.

a) **b)**  **c)** **d)**

These graphs show depth of water (*d*) against time (*t*).
Choose the most suitable graph for each glass

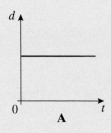

A

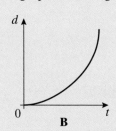

B

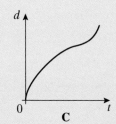

C

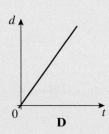

D

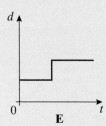

E

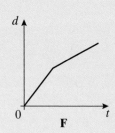

F

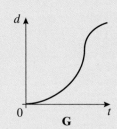

G

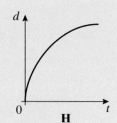

H

5 An office supplies firm advertises the following price structure for boxes of computer paper.

Number of boxes	1 to 4	5 to 9	10 or more
Price per box	£6.65	£5.50	£4.65

a) How much do 9 boxes cost?

b) How much do 10 boxes cost?

c) Draw a graph to show the total cost of orders for 1 to 12 boxes.
Use a scale of 1 cm to 1 box on the horizontal axis and 2 cm to £1 on the vertical axis.

6 A water company makes the following charges for customers with a water meter.

Basic charge	£20.00
Charge per cubic metre for the first 100 cubic metres used	£1.10
Charge per cubic metre for water used over 100 cubic metres	£0.80

   a) Draw a graph to show the charge for up to 150 cubic metres.
   Use a scale of 1 cm to 10 cubic metres on the horizontal axis and 1 cm to £10 on
   the vertical axis.

   b) Customers can choose instead to pay a fixed amount of £120.
   For what amounts of water is it cheaper to have a water meter?

Challenge 14.2

The graph shows the speed (v m/s) of a train at time t seconds.

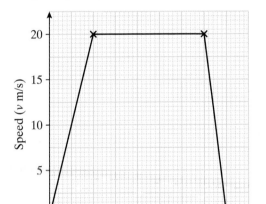

a) What is happening between the times $t = 100$ and $t = 350$?

b) (i) What is the rate of change between $t = 0$ and $t = 100$?
   (ii) What quantity does the rate of change represent?
   (iii) What are the units of the rate of change?

c) (i) What is the rate of change between $t = 350$ and $t = 400$?
   (ii) What quantity does the rate of change represent?

Quadratic graphs

A **quadratic function** is a function where the highest power of x is 2.

So the function will have an x^2 term.
It may also have an x term and a numerical term.
It will *not* have a term with any other power of x.

The function $y = x^2 + 2x - 3$ is a typical quadratic function.

Check up 14.2

State whether or not each of these functions is quadratic.

a) $y = x^2$

b) $y = x^2 + 5x - 4$

c) $y = \dfrac{5}{x}$

d) $y = x^2 - 3x$

e) $y = x^2 - 3$

f) $y = x^3 + 5x^2 - 2$

g) $y = x(x - 2)$

As with all graphs of functions of the form '$y =$ ', to plot the graph you must first choose some values of x and complete a table of values.

The simplest quadratic function is $y = x^2$.

x	-3	-2	-1	0	1	2	3
$y = x^2$	9	4	1	0	1	4	9

> **TIP** Remember that the square of a negative number is positive.

The points can then be plotted and joined up with a smooth curve.

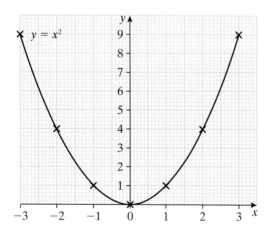

> **TIP** The y scale does not have to be the same as the x scale.

> **TIP** Turn your paper round and draw the curve from the inside. The sweep of your hand will give a smoother curve.
>
> Draw the curve without taking your pencil away from the paper.
>
> Look ahead to the next point as you draw the curve.

You can read off from your graph the value of y for any value of x or the value of x for any value of y.

For some quadratic graphs you may need extra rows in your table to get the final y values.

EXAMPLE 14.5

a) Complete the table of values for $y = x^2 - 2x$.

b) Plot the graph of $y = x^2 - 2x$.

c) Use your graph to

    (i) find the value of y when $x = 2.6$.

    (ii) solve $x^2 - 2x = 5$.

Solution

a)

x	-2	-1	0	1	2	3	4
x^2	4	1	0	1	4	9	16
$-2x$	4	2	0	-2	-4	-6	-8
$y = x^2 - 2x$	8	3	0	-1	0	3	8

> **TIP**
>
> The second and third rows are included in the table only to make the calculation of the y values easier: for this graph, add the numbers in the second and third rows to find the y values.
>
> The values you plot are the x values (first row) and y values (last row).

b) $y = x^2 - 2x$

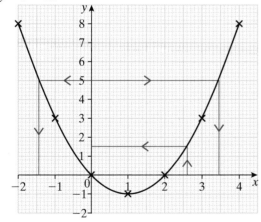

c) **(i)** $y = 1.5$ Read up from $x = 2.6$.

    (ii) $x = -1.45$ or $x = 3.45$ $x^2 - 2x = 5$ means that $y = 5$.

                                               Reading across from 5, you will see that there are two possible answers.

EXAMPLE 14.6

a) Complete the table of values for $y = x^2 + 3x - 2$.

b) Plot the graph of $y = x^2 + 3x - 2$.

c) Use your graph to
   (i) find the value of y when $x = -4.3$.
   (ii) solve $x^2 + 3x - 2 = 0$.

Solution

a)

x		-5	-4	-3	-2	-1	0	1	2
x^2		25	16	9	4	1	0	1	4
$3x$		-15	-12	-9	-6	-3	0	3	6
-2		-2	-2	-2	-2	-2	-2	-2	-2
$y = x^2 + 3x - 2$		8	2	-2	-4	-4	-2	2	8

b)

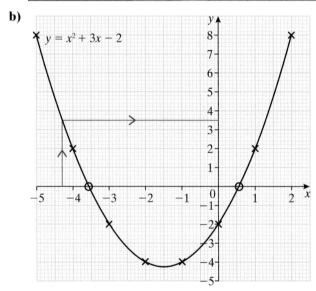

> **TIP**
>
> The lowest y values in the table are both -4, but the actual curve goes below -4. In such situations, it is often useful to work out the coordinates of the lowest (or highest) point of the curve.
>
> Because the curve is symmetrical, the lowest point of $y = x^2 + 3x - 2$ must lie halfway between $x = -2$ and $x = -1$, that is, at $x = -1.5$.
>
> When $x = -1.5$, $y = (-1.5)^2 + 3 \times -1.5 - 2 = 2.25 - 4.5 - 2 = -4.25$.

c) (i) $y = 3.5$ Read up from $x = -4.3$.
    (ii) $x = -3.6$ or $x = 0.6$ $x^2 + 3x - 2 = 0$ means that $y = 0$.
                                         Reading off the graph when $y = 0$, you will see that there are two possible answers.

All quadratic graphs are the same basic shape. This shape is called a **parabola**.

The three you have seen so far were ∪-shaped. In these graphs, the x^2 term was positive.

If the x^2 term is negative, the parabola is the other way up (∩).

If your graph is not shaped like a parabola, go back and check your table.

EXAMPLE 14.7

a) Complete the table of values for $y = 5 - x^2$.

b) Plot the graph of $y = 5 - x^2$.

c) Use your graph to solve
   (i) $5 - x^2 = 0$.   (ii) $5 - x^2 = 3$.

Solution

a)

x	-3	-2	-1	0	1	2	3
5	5	5	5	5	5	5	5
$-x^2$	-9	-4	-1	0	-1	-4	-9
$y = 5 - x^2$	-4	1	4	5	4	1	-4

b)

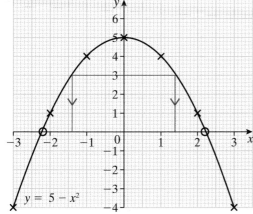

c) (i) $x = -2.25$ or $x = 2.25$ Read off at $y = 0$.
   (ii) $x = -1.4$ or $x = 1.4$ Read off at $y = 3$.

1 a) Copy and complete the table of values for $y = x^2 - 2$.

x	-3	-2	-1	0	1	2	3
x^2	9					4	
-2	-2					-2	
$y = x^2 - 2$	7					2	

b) Plot the graph of $y = x^2 - 2$.
Use a scale of 2 cm to 1 unit on the x-axis and 1 cm to 1 unit on the y-axis.

c) Use your graph to
  (i) find the value of y when $x = 2.3$.   **(ii)** solve $x^2 - 2 = 4$.

2 a) Copy and complete the table of values for $y = x^2 - 4x$.

x	-1	0	1	2	3	4	5
x^2					9		
$-4x$					-12		
$y = x^2 - 4x$					-3		

b) Plot the graph of $y = x^2 - 4x$.
Use a scale of 2 cm to 1 unit on the x-axis and 1 cm to 1 unit on the y-axis.

c) Use your graph to
  (i) find the value of y when $x = 4.2$.   **(ii)** solve $x^2 - 4x = -2$.

3 a) Copy and complete the table of values for $y = x^2 + x - 3$.

x	-4	-3	-2	-1	0	1	2	3
x^2			4					
x			-2					
-3			-3					
$y = x^2 + x - 3$			-1					

b) Plot the graph of $y = x^2 + x - 3$.
Use a scale of 2 cm to 1 unit on the x-axis and 1 cm to 1 unit on the y-axis.

c) Use your graph to
  (i) find the value of y when $x = 0.7$.   **(ii)** solve $x^2 + x - 3 = 0$.

4 a) Make a table of values for $y = x^2 - 3x + 4$. Choose values of x from -2 to 5.

b) Plot the graph of $y = x^2 - 3x + 4$.
Use a scale of 2 cm to 1 unit on the x-axis and 1 cm to 1 unit on the y-axis.

c) Use your graph to
    (i) find the minimum value of y.   **(ii)** solve $x^2 - 3x + 4 = 10$.

5 a) Copy and complete the table of values for $y = 3x - x^2$.

x	-2	-1	0	1	2	3	4	5
$3x$				3			12	
$-x^2$				-1			-16	
$y = 3x - x^2$				2			-4	

b) Plot the graph of $y = 3x - x^2$.
Use a scale of 2 cm to 1 unit on the x-axis and 1 cm to 1 unit on the y-axis.

c) Use your graph to
    (i) find the maximum value of y.   **(ii)** solve $3x - x^2 = -2$.

6 a) Make a table of values for $y = x^2 - x - 5$. Choose values of x from -3 to 4.

b) Plot the graph of $y = x^2 - x - 5$.
Use a scale of 2 cm to 1 unit on the x-axis and 1 cm to 1 unit on the y-axis.

c) Use your graph to solve
    (i) $x^2 - x - 5 = 0$.   **(ii)** $x^2 - x - 5 = 3$.

7 a) Make a table of values for $y = 2x^2 - 5$. Choose values of x from -3 to 3.

b) Plot the graph of $y = 2x^2 - 5$.
Use a scale of 2 cm to 1 unit on the x-axis and 1 cm to 1 unit on the y-axis.

c) Use your graph to solve
    (i) $2x^2 - 5 = 0$.   **(ii)** $2x^2 - 5 = 10$.

8 The total surface area (A cm^2) of this cube is given by $A = 6x^2$.

a) Make a table of values for $A = 6x^2$. Choose values of x from 0 to 5.

b) Plot the graph of $A = 6x^2$.
Use a scale of 2 cm to 1 unit on the x-axis and 1 cm to 10 units on the A-axis.

c) Use your graph to find the side of a cube with surface area
    (i) 20 cm^2.   **(ii)** 80 cm^2.

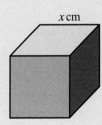
x cm

The diagram shows a sheep pen. Three sides are made of fencing. A wall is used for the fourth side.

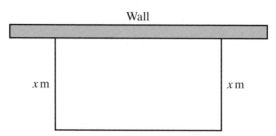

Wall

x m x m

The sides of the pen are x metres in length.
A total of 50 metres of fencing is used.

a) Explain why the area of the pen is given by $A = x(50 - 2x)$

b) Make a table of values for A using $0, 5, 10, 15, 20, 25$ as the values of x.

c) Plot a graph with x on the horizontal axis and A on the vertical axis.

d) Use your graph to find
    (i) the area of the pen when $x = 8$.
    (ii) the values of x when the area is 150 m².
    (iii) the maximum area of the pen.

WHAT YOU HAVE LEARNED

- You need only two points to draw a straight-line graph, but should always check with a third point
- When you are asked to answer questions about a given graph, look carefully at the labels and units on the axes and see whether the line is straight or curved
- A straight line represents a constant rate of change, and the steeper the line, the greater the rate of change
- A horizontal line means that there is no change in the quantity on the y-axis
- A convex curve viewed from below) represents an increasing rate of change
- A concave curve viewed from below) represents a decreasing rate of change
- The rate of change on a distance–time graph is the speed
- On a cost graph, the value where the graph cuts the cost axis is the fixed charge
- A quadratic function has x^2 as the highest power of x. It may also have an x term and a numerical term. It will not have a term with any other power of x
- The shape of all quadratic graphs is a parabola. If the x^2 term is positive the curve is U-shaped. If the x^2 term is negative the curve is ∩-shaped

1 Draw the graph of $y = 2x - 1$ for values of x from -1 to 4.

2 Draw the graph of $2x + y - 8 = 0$ for values of x from 0 to 4.

3 The graph shows an energy supplier's quarterly charges for up to 500 kWh of electricity.

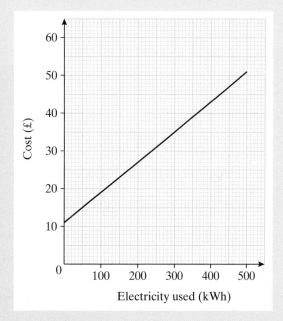

a) What is the cost if 350 kWh are used?

The cost is made up of a fixed charge plus an amount per kWh of electricity used.

b) (i) What is the fixed charge?

(ii) Calculate the cost per kWh in pence.

4 The same energy supplier makes a fixed charge for gas of £15 per quarter. In addition to this there is a charge of 2p per kWh.

a) Draw a graph to show the quarterly bill for up to 1500 kWh of gas used. Use a scale of 1 cm to 100 kWh on the horizontal axis and 2 cm to £10 on the vertical axis.

b) Look at this graph and the graph in question **3**. Is it cheaper to buy 400 kWh of electricity or 400 kWh of gas? By how much?

5 Water is poured into these vases at a constant rate. Sketch the graphs of depth of water (vertical) against time (horizontal).

a)

b)

6   The graph shows Kerry's Saturday morning shopping trip.

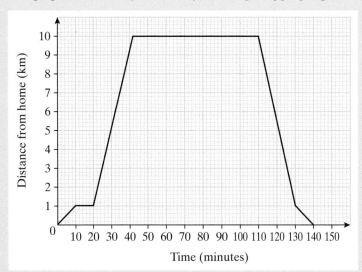

a) What happened between 10 and 20 minutes after Kerry left home?

b) How long did she spend at the shops?

c) She caught a bus home. What was the speed of the bus?

d) How far from Kerry's home is
  (i) the bus stop?   **(ii)** the shopping centre?

7   Which of these functions are quadratic?
For each of the functions that is quadratic, state whether the graph is ∪-shaped or ∩-shaped.

a) $y = x^2 + 3x$ **b)** $y = x^3 + 5x^2 + 3$ **c)** $y = 5 + 3x - x^2$

d) $y = (x + 1)(x - 3)$ **e)** $y = \dfrac{4}{x^2}$ **f)** $y = x^2(x + 1)$

g) $y = x(5 - 2x)$

8   **a)** Copy and complete the table of values for $y = x^2 + 3x$.

x	−5	−4	−3	−2	−1	0	1	2
x^2	25			4				4
$3x$	−15			−6				6
$y = x^2 + 3x$	10			−2				10

b) Plot the graph of $y = x^2 + 3x$.
Use a scale of 2 cm to 1 unit on the x-axis and 1 cm to 1 unit on the y-axis.

c) Use your graph to
  (i)   find the minimum value of y.
  (ii)   solve $x^2 + 3x = 3$.

9 a) Copy and complete the table of values for $y = (x + 3)(x - 2)$.

x	−4	−3	−2	−1	0	1	2	3
$(x + 3)$			1		3			6
$(x - 2)$			−4		−2			1
$y = (x + 3)(x - 2)$			−4		−6			6

  b) Plot the graph of $y = (x + 3)(x - 2)$.
  Use a scale of 2 cm to 1 unit on the x-axis and 1 cm to 1 unit on the y-axis.

  c) Use your graph to
  (i) find the minimum value of y.
  (ii) solve $(x + 3)(x - 2) = -2$.

10 a) Make a table of values for $y = x^2 - 2x - 1$. Choose values of x from −2 to 4.
  b) Plot the graph of $y = x^2 - 2x - 1$.
  Use a scale of 2 cm to 1 unit on the x-axis and 1 cm to 1 unit on the y-axis.

  c) Use your graph to solve
  (i) $x^2 - 2x - 1 = 0$.
  (ii) $x^2 - 2x - 1 = 4$.

11 a) Make a table of values for $y = 5x - x^2$. Choose values of x from −1 to 6.
  b) Plot the graph of $y = 5x - x^2$.
  Use a scale of 2 cm to 1 unit on the x-axis and 1 cm to 1 unit on the y-axis.

  c) Use your graph to
  (i) solve $5x - x^2 = 3$.
  (ii) find the maximum value of y.

15 → GRAPHS 2

THIS CHAPTER IS ABOUT

- Finding the gradient of a straight-line graph
- Finding the equation of a line in the form $y = mx + c$, given the gradient and intercept
- Solving simultaneous equations graphically
- Solving simultaneous equations algebraically

YOU SHOULD ALREADY KNOW

- How to plot and read points in all four quadrants
- How to draw straight-line graphs from equations given in explicit or implicit form
- How to substitute numbers into equations
- How to rearrange equations
- How to add, subtract and multiply negative numbers

Finding the gradient of a straight-line graph

The **gradient** of a graph is the mathematical way of measuring its steepness or rate of change.

$$\text{Gradient} = \frac{\text{increase in } y}{\text{increase in } x}$$

To find the gradient of a straight-line graph, mark two points, then draw a horizontal line and a vertical line to form a right-angled triangle.

For this line, gradient $= \dfrac{4}{3}$

$= 1.33$ (correct to 2 decimal places)

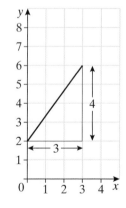

 Choose two points which are as far apart as possible. Make sure the increase in x is a whole number. This will make the arithmetic easier.

For this line, *y decreases* by 8 between the two points chosen. So the *increase* is −8.

$$\text{gradient} = \frac{-8}{2}$$

$$= -4$$

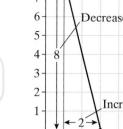

You can find the gradient of a line without drawing a diagram.

EXAMPLE 15.1

Find the gradient of the line joining each of these pairs of points.
a) (3, 5) and (8, 7)
b) (2, 7) and (6, 1)

Solution

a) (3, 5) and (8, 7)

Increase in $y = 7 − 5$
$$= 2$$

Increase in $x = 8 − 3$
$$= 5$$

Gradient $= \dfrac{2}{5}$
$$= 0.4$$

b) (2, 7) and (6, 1)

Increase in $y = 1 − 7$ Take care with the signs.
$$= -6$$

Increase in $x = 6 − 2$
$$= 4$$

Gradient $= \dfrac{-6}{4}$
$$= -1.5$$

You may prefer, however, to draw the diagram first.

EXAMPLE 15.2

Find the gradient of this distance–time graph.

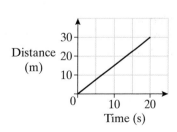

Solution

Gradient $= \dfrac{30}{20}$

$= 1.5$

For a distance–time graph, $\dfrac{\text{increase in } y}{\text{increase in } x} = \dfrac{\text{distance travelled}}{\text{time passed}}$.

You know that velocity (speed) $= \dfrac{\text{distance}}{\text{time}}$.

So the gradient of a distance–time graph gives the velocity.

In this example the units are metres and seconds.
So the velocity $= 1.5$ m/s.

◎ EXERCISE 15.1

1 Find the gradient of each of these lines.

a)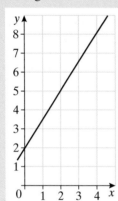

b)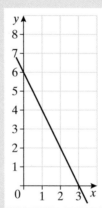

c)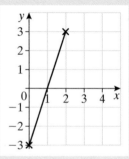

2 Find the gradient of the line joining each of these pairs of points.

a) $(3, 2)$ and $(4, 8)$ **b)** $(5, 3)$ and $(7, 7)$ **c)** $(0, 4)$ and $(2, -6)$

d) $(3, -1)$ and $(-1, -5)$ **e)** $(1, 1)$ and $(6, 1)$

3 Find the gradient of each side of the triangle ABC.

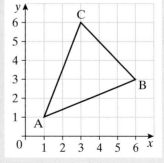

4 Find the gradient of each of these lines.

a)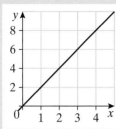

b)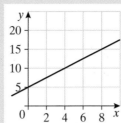

c)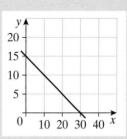

5 Draw the graph of each of these straight lines and find its gradient.

a) $y = 2x + 3$ **b)** $y = 5x - 2$ **c)** $y = -2x + 1$

d) $y = -x$ **e)** $3x + 2y = 12$

6 Find the velocity for each of these distance–time graphs.

a)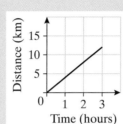

b)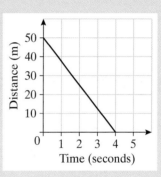

Finding the equation of a straight-line graph

Look again at the answers to question **5** in Exercise 15.1. Can you see a connection between the equation of a line and its gradient?

Discovery 15.1

a) Draw a pair of axes and label them -10 to 10 for both x and y. Draw each of these lines on the same axes.

$y = 2x$ $y = 2x + 1$ $y = 2x + 2$ $y = 2x + 4$ $y = 2x - 2$ $y = 2x - 4$

(i) What do you notice about the gradient of the lines?

(ii) What do you notice about where the lines cut the y-axis? The point where a straight line cuts the y-axis is called the **y-intercept**.

(iii) Check your ideas with these lines.

$y = x$ $y = x + 1$ $y = x + 2$ $y = x - 3$ $y = x - 4$ $y = x - 1$

b) Write down the gradient and y-intercept of each of these lines without drawing a diagram.

(i) $y = 4x + 2$ (ii) $y = 5x - 3$ (iii) $y = 2x$

(iv) $y = -3x + 2$ (v) $y = x + 4$ (vi) $y = -4x + 6$

Parallel lines have the same gradient.

An equation of the form $y = mx + c$ has gradient m and y-intercept c.

You can use this fact to find the equation of a line from its graph.

EXAMPLE 15.3

Find the equation of this line.

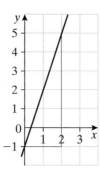

Solution

Gradient, $m = \dfrac{6}{2} = 3$

y-intercept, $c = -1$ The line crosses the y-axis at $(0, -1)$.

So the equation is $y = 3x - 1$.

You can also find the gradient and y-intercept of a line from its equation.
Sometimes you will have to rearrange the equation first.

EXAMPLE 15.4

The equation of a straight line is $5x + 2y = 10$.
a) Find the gradient of the line.
b) Find the y-intercept of the line.

Solution

First you need to rewrite the equation in the form $y = mx + c$.

$5x + 2y = 10$
$\qquad 2y = -5x + 10$ Subtract $5x$ from each side so that the y term is on its own.
$\qquad\ \ y = -2.5x + 5$ Divide each side by 2.

a) Gradient, $m = -2.5$
b) y-intercept, $c = 5$

1 Write down the equations of straight lines with these gradients and y-intercepts.

a) Gradient 3, y-intercept 2   **b)** Gradient -1, y-intercept 4

c) Gradient 5, y-intercept 0

2 Find the equation of each of these lines.

a)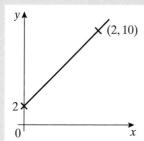

b)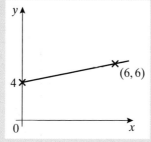

c)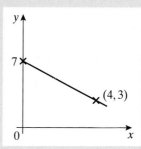

3 Find the gradient and y-intercept of each of these lines.

a) $y = 3x - 2$ **b)** $y = 2x + 4$ **c)** $y = -x + 3$

d) $y = -3x + 2$ **e)** $y = 2.5x - 6$

4 Find the gradient and y-intercept of each of these lines.

a) $y + 2x = 5$ **b)** $4x + 2y = 9$ **c)** $6x + 5y = 12$

d) $3x - 2y = 6$ **e)** $3x + 4y = 12$

5 Find the equation of each of these lines.

a)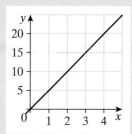

b)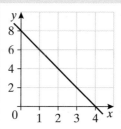

c)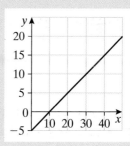

Solving simultaneous equations graphically

Discovery 15.2

a) (i) Draw the graph of $y = 2x - 4$ for values of x from 0 to 6.

    (ii) On the same axes draw the graph of $2x + 3y = 12$.

b) Write down the coordinates of the point where the two lines cross.

c) What can you say about the values of x and y that you have found?

The coordinates of the point where two lines cross satisfy the equations of both lines. You can check this by substituting the values into the equations.

For example, when $x = 3$ and $y = 2$

$y = 2x - 4$	$2x + 3y = 12$
$y = 2 \times 3 - 4$	$2 \times 3 + 3 \times 2 = 12$
$y = 2$ ✓	$12 = 12$ ✓

Only the point with these coordinates lies on both the line $y = 2x - 4$ and the line $2x + 3y = 12$. Only the values $x = 3$ and $y = 2$ satisfy both equations at the same time. The values $x = 3$ and $y = 2$ are the solution of the **simultaneous equations** $y = 2x - 4$ and $2x + 3y = 12$. *Simultaneous* means 'at the same time'.

EXAMPLE 15.5

Solve graphically the simultaneous equations $y = 2x$ and $y = 3x - 4$.
Use values of x from 0 to 5.

Solution

Work out and plot three points on each of the lines.
Three points on $y = 2x$ are $(0, 0), (2, 4)$ and $(5, 10)$.
Three points on $y = 3x - 4$ are $(0, -4), (2, 2)$, and $(5, 11)$.

Find the point where the lines cross and write down its coordinates.
The lines cross at $(4, 8)$.

So the solution to the simultaneous equations $y = 2x$ and $y = 3x - 4$ is $x = 4, y = 8$.

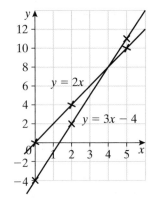

EXERCISE 15.3

Solve graphically each of these pairs of simultaneous equations.

1 $y = x + 1$ and $y = 4x - 5$. Use values of x from 0 to 5.

2 $y = 2x$ and $y = 8 - 2x$. Use values of x from -2 to 4.

3 $y = 3x + 5$ and $y = x + 3$. Use values of x from -3 to 2.

4 $y = 2x - 7$ and $y = 5 - x$. Use values of x from -1 to 5.

5 $2y = 2x + 1$ and $x + 2y = 7$. Use values of x from 0 to 7.

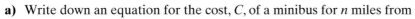

Challenge 15.1

Delaney's Cabs charge £20 plus £2 per mile to hire a minibus.
Tracey's Cars charge £50 plus £1.50 per mile to hire a minibus.

a) Write down an equation for the cost, C, of a minibus for n miles from

    (i) Delaney's Cabs.   **(ii)** Tracey's Cars.

b) On one graph draw the two lines for the equations connecting C and n, for values of n up to 100.

c) Use your graph to find the number of miles for which the two firms charge the same amount.

Solving simultaneous equations algebraically

You can also solve simultaneous equations using algebra. Further coverage of algebraic methods of solving simultaneous equations can be found in Chapter 20.

By adding or subtracting the two equations, you can eliminate either the x term or the y term from one of them. You can then go on to find x and y. This is known as solving simultaneous equations by the **method of elimination**.

EXAMPLE 15.6

Use algebra to solve the simultaneous equations $x + y = 4$ and $2x - y = 5$.

Solution

$$x + y = 4 \quad (1)$$
$$2x - y = 5 \quad (2)$$

Set out the two equations one under the other and label them.

Look to see whether either of the unknowns (x or y) has the same **coefficient** in both equations. The coefficient of x is the number in front of the x. The coefficient of y is the number in front of the y.

In this case there is y in (1) and $-y$ in (2).
Because the signs are different, the two y terms will be eliminated (cancel each other out) if the two equations are added.

$$
\begin{array}{ll}
x + y = 4 & (1) \\
2x - y = 5 & (2) \\
\hline
3x \quad\;\; = 9 & (1) + (2) \\
x \quad\;\;\; = 3 & \text{Divide by 3 to find the value of } x.
\end{array}
$$

Now that you know the value of x, you can find the value of y by substituting $x = 3$ into one of the equations. In this case, it is easiest to use equation (1).

$$x + y = 4$$
$$3 + y = 4$$
$$y = 1$$

So the solution to the simultaneous equations $x + y = 4$ and $2x - y = 5$ is $x = 3$, $y = 1$.

To check your answer, substitute the solution into the second equation.

In this case, substitute $x = 3$ and $y = 1$ into equation (2).

$2x - y = 2 \times 3 - 1 = 5$ ✓

EXAMPLE 15.7

Use algebra to solve the simultaneous equations $2x + 5y = 9$ and $2x - y = 3$.

Solution

$2x + 5y = 9$ (1)   Set out and label the equations.
$2x - y = 3$ (2)

This time $2x$ occurs in both equations. Because the signs are the same, the two x terms will be eliminated if the two equations are subtracted.

$2x + 5y = 9$ (1)
$\underline{2x - y = 3}$ (2)
      $6y = 6$ (1) − (2)   Take care with the signs: $5y - (-y) = 5y + y$.
       $y = 1$ Divide by 6 to find the value of y.

Substitute $y = 1$ into equation (1).
$2x + 5y = 9$
$2x + 5 = 9$
   $2x = 4$
    $x = 2$

So the solution to the simultaneous equations $2x + 5y = 9$ and $2x - y = 3$ is $x = 2, y = 1$.

Check your answer in equation (2).
$2x - y = 2 \times 2 - 1 = 3$ ✓

> **TIP**
> Make sure you read the question. If it says 'use algebra', you must use algebra!

EXERCISE 15.4

Use algebra to solve each of these pairs of simultaneous equations.

1 $x + y = 5$
     $2x - y = 7$

2 $3x + y = 9$
     $2x + y = 7$

3 $2x + 3y = 11$
     $2x + y = 5$

4 $2x + y = 7$
     $4x - y = 5$

5 $x + 3y = 8$
     $x - y = 4$

6 $3x + y = 7$
     $3x + 2y = 8$

7 $2x - 3y = 0$
     $4x + 3y = 18$

8 $2x + 3y = 17$
     $2x - 3y = -1$

9 $x + y = 4$
     $4x - y = 11$

10 $2x - y = 7$
      $3x + y = 8$

Challenge 15.2

a) Two numbers, x and y, have a sum of 57, and twice the first number minus the second is equal to 24.

    (i) Write down two equations in x and y.

    (ii) Solve them to find x and y.

b) At Corner Café, John bought two teas and one coffee and paid £4.70. Erica bought two teas and three coffees and paid £6.40.

    (i) Let x be the cost of a tea and y the cost of a coffee. Write down two equations involving x and y.

    (ii) Solve the equations and find the cost of a tea.

Solving harder simultaneous equations algebraically

Sometimes neither of the unknowns (x or y) has the same coefficient in both equations. In such cases you first need to multiply one of the equations.

EXAMPLE 15.8

Use algebra to solve the simultaneous equations $x + 3y = 10$ and $3x + 2y = 16$.

Solution

$$x + 3y = 10 \quad (1)$$
$$3x + 2y = 16 \quad (2)$$

In this example the coefficients of x and y are different in the two equations.

Multiply equation (1) by 3 to make the coefficient of x the same as in equation (2).

$$
\begin{array}{llll}
x & + & 3y & = & 10 & \quad (1) \\
3 \times x + 3 \times 3y & = & 3 \times 10 & \quad (1) \times 3 & \text{Remember to multiply } each \text{ term in the equation by 3.} \\
3x & + & 9y & = & 30 & \quad (3) & \text{Label the new equation.}
\end{array}
$$

You can now subtract equation (2) from equation (3) to eliminate the x term.

$$
\begin{array}{ll}
3x + 9y = 30 & \quad (3) \\
\underline{3x + 2y = 16} & \quad (2) \\
\quad\quad 7y = 14 & \quad (3) - (2) \\
\quad\quad\ y = 2 & \text{Divide by 7 to find the value of } y.
\end{array}
$$

> **TIP**
> When subtracting, you can do either $(3) - (2)$ or $(2) - (3)$. It is better to do the one that leaves the remaining term positive.

Substitute $y = 2$ into equation (1).

$$x + 3y = 10$$
$$x + 3 \times 2 = 10$$
$$x + 6 = 10$$
$$x = 4$$

So the solution to the simultaneous equations $x + 3y = 10$ and $3x + 2y = 16$ is $x = 4$, $y = 2$.

Check your answer in equation (2).

$$3x + 2y = 3 \times 4 + 2 \times 2$$
$$= 12 + 4$$
$$= 16 \checkmark$$

You must always write down clearly what you are doing, but you do not need to write as much detail as in Example 15.8. The next example shows what is required.

EXAMPLE 15.9

Use algebra to solve the simultaneous equations $4x - y = 10$ and $3x + 2y = 13$.

Solution

$$4x - y = 10 \qquad (1)$$
$$3x + 2y = 13 \qquad (2)$$

$$8x - 2y = 20 \qquad (1) \times 2 = (3)$$
$$\underline{3x + 2y = 13} \qquad (2)$$
$$11x \qquad = 33 \qquad (3) + (2)$$
$$x \qquad = 3$$

Substitute $x = 3$ in (1).

$$12 - y = 10$$
$$-y = -2$$
$$y = 2$$

Solution is $x = 3, y = 2$.

Check in (2).

$$3x + 2y = 9 + 4$$
$$= 13 \checkmark$$

Use algebra to solve each of these pairs of simultaneous equations.

1 $x + 3y = 5$
$2x - y = 3$

2 $3x + y = 9$
$x + 2y = 8$

3 $x + 3y = 9$
$2x - y = 4$

4 $2x + y = 7$
$x - 2y = -4$

5 $2x + 3y = 13$
$x - y = 4$

6 $3x + y = 7$
$2x + 3y = 7$

7 $2x - 3y = 2$
$3x + y = 14$

8 $2x + 3y = 5$
$x + 2y = 4$

9 $x + y = 4$
$4x - 2y = 7$

10 $4x - 2y = 10$
$3x + y = 5$

Challenge 15.3

Solve each of these pairs of simultaneous equations.

a) $3x + 4y = 10$
$5x - 3y = 7$

b) $2x + 3y = 3$
$3x + 4y = 5$

c) $3x + 5y = 15$
$5x - 3y = 8$

WHAT YOU HAVE LEARNED

- Gradient of a straight line $= \dfrac{\text{increase in } y}{\text{increase in } x}$
- Lines with positive gradient slope up from left to right (╱)
- Lines with negative gradient slope down from left to right (╲)
- Lines with the same gradient are parallel
- The equation of a line can be written in the form $y = mx + c$, where m is the gradient of the line and c is the y-intercept
- To solve simultaneous equations graphically, draw the lines and find the point where they cross
- To solve simultaneous equations algebraically, make the coefficient of one of the letters the same in both equations, if necessary. If the signs are the same, subtract the equations. If they are different, add the equations

1 Find the gradient of the line joining each of these pairs of points.

    a) $(1, 3)$ and $(3, 6)$ **b)** $(2, 1)$ and $(6, -3)$

2 Find the gradient of each of these lines.

    a) 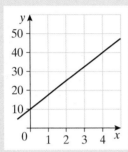   **b)**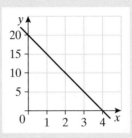

3 Find the gradient and y-intercept of each of these lines.

    a) $y = 3x - 2$ **b)** $2y = 5x - 4$

    c) $3x + 2y = 8$ **d)** $2x - y = 7$

4 Find the equation of each of these lines.

    a) 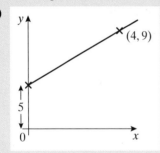   **b)**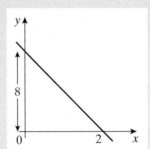

5 Solve each of these pairs of simultaneous equations graphically.

    a) $y = 3x - 1$ and $y = 4x - 3$. Use values of x from -1 to 4.

    b) $2x + 3y = 12$ and $y = 3x - 7$. Use values of x from 0 to 6.

6 Solve each of these pairs of simultaneous equations algebraically.

    a) $x + y = 6$ **b)** $2x + 5y = 13$ **c)** $3x + y = 14$ **d)** $x + 4y = 11$

          $2x + y = 10$ $2x + 3y = 11$ $x - y = 2$ $x + y = 2$

    e) $x + y = 5$ **f)** $x + y = 1$ **g)** $x + 3y = 5$ **h)** $4x - 3y = 9$

          $2x + 3y = 12$ $3x - 2y = 8$ $7x + 2y = -3$ $2x + y = 7$

16 → GRAPHS 3

THIS CHAPTER IS ABOUT

- **Velocity–time graphs and distance–time graphs**
- **Solving simultaneous equations and quadratic equations graphically**
- **Drawing the graphs of simple functions**

YOU SHOULD ALREADY KNOW

- **How to use your calculator efficiently**
- **How to draw graphs of straight lines**
- **How to interpret simple real-life graphs, including rate of change**

Velocity–time graphs

In Chapter 14 you looked at some real-life graphs, including distance–time graphs. You also saw how the rate of change was represented on the graph. In the case of distance–time graphs, the rate of change is the speed or, more correctly, **velocity**. Velocity has a direction whereas speed does not. The reverse direction has a negative gradient, as you saw in Chapter 15.

EXAMPLE 16.1

This is a distance–time graph for a small particle moving in a straight line.

Describe the motion of the particle.

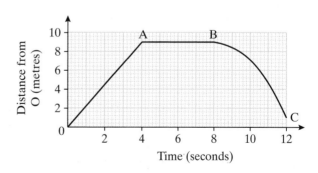

Solution

From O to A the particle moves 9 metres in 4 seconds at a constant velocity.

(Constant velocity means the velocity is the same at all times.)

From A to B the particle is stationary.

From B to C the particle moves back towards the start so it has a negative velocity. The velocity is not constant. The distance decreases slowly at first and then more rapidly.

In Example 16.1, the rate of change of distance is the velocity but the velocity from B to C is not constant.

Discovery 16.1

a) Work out the average velocity of the particle in Example 16.1 between the times
   (i)   0 and 2 seconds.   (ii) 2 and 4 seconds.
   What do you notice?

b) Now work out the average velocity of the particle between the times
   (i)   8 and 9 seconds.   (ii) 9 and 10 seconds.
   (iii) 10 and 11 seconds.   (iv) 11 and 12 seconds.
   What do you notice now?

When velocity is not constant, there is an acceleration (sometimes called deceleration when it is negative). The rate of change of velocity is called **acceleration**.

EXAMPLE 16.2

This graph is a velocity–time graph for the movement of the particle.
Describe the movement.

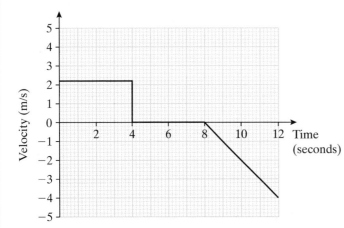

When $t = 4$ the velocity instantly becomes zero. This could not actually happen, of course, but it is usual to show rapid changes in velocity in this way for simplicity.

Solution

From $t = 0$ to $t = 4$, the velocity is constant at 2.2 m/s. There is no acceleration.
From $t = 4$ to $t = 8$, the velocity is zero.
From $t = 8$ to $t = 12$, the velocity decreases (increases backwards) with constant acceleration.

The acceleration is the gradient of the line $= \dfrac{-4}{4} = -1$ m/s².

> **TIP**
> The units of velocity are metres per second, written m/s or ms⁻¹.
> The units of acceleration are metres per second per second, written m/s² or ms⁻².

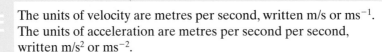

Gradient of a curve

The gradient of a curve at a point can be found by considering the limit of the gradients of a series of chords from the point on the curve. However, for now a 'by eye' method is commonly used. This method involves the drawing of a tangent to the curve at the required point. The gradient of the tangent is then equal to the gradient of the curve at the point. However this can be inaccurate due to the tangent being drawn 'by eye'.

The gradient of the tangent at the point (x, y) on the curve is equal to the gradient of the curve at the point (x, y).

$$\text{Gradient} = \frac{y_2 - y_1}{x_2 - x_1}$$

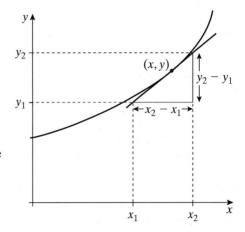

EXAMPLE 16.3

The graph shows the curve $y = 4x^2 + 10$.
Find the gradient of the curve at the point $(2, 26)$.

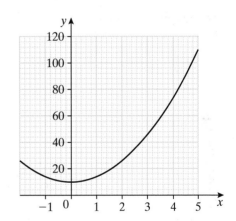

Solution

An accurate graph of $y = 4x^2 + 10$ must be drawn on graph paper. Because the question asks for the gradient at the point $x = 2$, only a section of the graph around this point actually needs to be drawn, say from $x = 0$ to $x = 4$.

The tangent at the point $x = 2$ is drawn.
The gradient of the tangent, as a straight line, is the difference in the y-values divided by the difference in the x-values.
Gradient of the curve at the point $x = 2$ is

$$\frac{42 - 10}{3 - 1} = 16$$

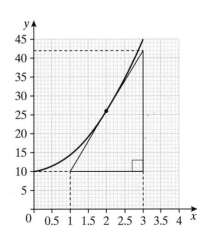

Tangents to estimate rates of change

The velocity at any given time can be found from a distance–time graph by drawing the tangent on the curve at that given time and calculating the gradient.

The gradient of a distance–time curve gives the velocity, the rate of change of distance with time.

The acceleration at any given time can be found from a velocity–time graph by drawing the tangent on the curve at that given time and calculating the gradient.

The gradient of a velocity–time curve gives the acceleration, the rate of change of velocity with time.

EXAMPLE 16.4

The distance–time graph shows a journey between 1 p.m. and 7 p.m. Find an estimate for the velocity in km/h at 2 p.m.

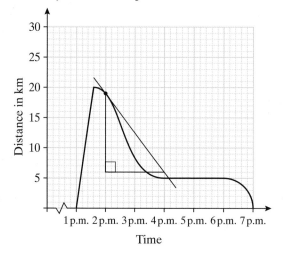

Solution

Notice that there is a negative gradient at 2 p.m. This indicates the return journey. Draw the tangent to the curve at 2 p.m. to calculate the gradient.

The units of the answer required are important. The difference in distance divided by the difference in time should give km/h so the values should be read in, or converted to be in, the correct units.

$$\text{Velocity} \approx -\frac{13}{2}$$

$$= -6.5 \text{ km/h}$$

Area under a curve

The approximate value for the area under a curve can be found by drawing ordinates to split the area into shapes which approximate to trapeziums. The sum of the areas of the trapeziums gives an approximate area under the curve.

EXAMPLE 16.5

Use three strips to estimate the area of the region enclosed by the curve $y = \dfrac{x}{10}$ and the x-axis between $x = 1$ and $x = 4$.

Solution

Draw the ordinates $x = 1, x = 2, x = 3$ and $x = 4$ on the graph to form three trapeziums.

Calculate the y-values of the curve $y = \dfrac{x}{10}$ at the points $x = 1, x = 2, x = 3$ and $x = 4$.

When $x = 1, y = \dfrac{10}{1} = 10$

$\quad\quad x = 2, y = \dfrac{10}{2} = 5$

$\quad\quad x = 3, y = \dfrac{10}{3} = 3\tfrac{1}{3}$

$\quad\quad x = 4, y = \dfrac{10}{4} = 2.5$ etc.

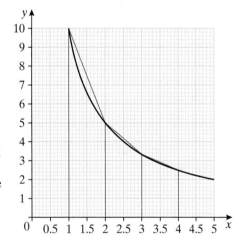

Now calculate the area of each of the trapeziums using the formula $\tfrac{1}{2}h(a + b)$, where a and b are the lengths of the parallel sides and h is the perpendicular distance between them.

The approximate area under the curve is equal to the sum of the areas of the trapeziums.

Approximate total area $\approx \tfrac{1}{2} \times 1 \times (10 + 5) + \tfrac{1}{2} \times 1 \times \left(5 + 3\tfrac{1}{3}\right) + \tfrac{1}{2} \times 1 \times \left(3\tfrac{1}{3} + 2.5\right)$

$\quad\quad\quad\quad\quad\quad\quad = 14.6$ square units

Challenge 16.1

The Trapezium Rule

Estimate the area under a curve using strips of equal width h with the ordinates $x = x_0, x = x_1, x = x_2, x = x_3$ and $x = x_4$ given that $(x_0, y_0), (x_1, y_1), (x_2, y_2), (x_3, y_3)$ and (x_4, y_4) are points on the curve.

Can you write down a formula for estimating the area under a curve using any number of equal width strips? This is known as the Trapezium Rule.

Area under a velocity–time graph

The area under a velocity–time graph represents the distance travelled.

EXAMPLE 16.6

The velocity–time graph of a particle is shown below. Find the approximate distance travelled by the particle in the first 40 seconds, using ordinates every 10 seconds.

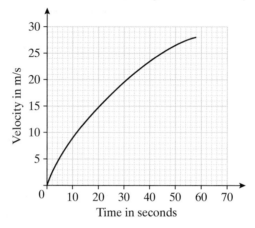

Solution

Split the area into strips of width 10 seconds, from time 0 seconds to time 40 seconds. The approximate distance travelled in the first 40 seconds is the sum of the areas of the triangle and three trapeziums.

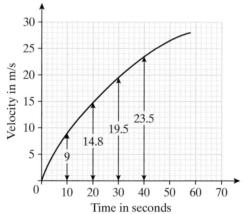

By the Trapezium Rule:

Approximate distance travelled $\approx \frac{1}{2} \times 10 \times \{0 + 2 \times (9 + 14.8 + 19.5) + 23.5\}$

$$= 550.5 \text{ km}$$

Or, alternatively: as the sum of the areas:

$\frac{1}{2} \times 10 \times 9 + \frac{1}{2} \times 10 \times (9 + 14.8) + \frac{1}{2} \times 10 \times (14.8 + 19.5) + \frac{1}{2} \times 10 \times (19.5 + 23.5)$

$= 45 + 119 + 171.5 + 215$

$= 550.5 \text{ km}$

1 a) Draw a velocity–time graph with the Time axis (t) from 0 to 10 seconds and the Velocity axis (v) from 0 to 20 m/s.
Show the velocity increasing from 0 m/s at $t = 0$ to 18 m/s when $t = 10$.

b) Calculate the acceleration.

2 a) Draw a velocity–time graph with the Time axis (t) from 0 to 10 seconds and the Velocity axis (v) from -10 to 10 m/s.
Show a constant velocity of 6 m/s from $t = 0$ to $t = 3$ and a constant acceleration of -1.5 m/s² from $t = 3$ to $t = 10$.

b) What is the velocity when $t = 10$?

3 Describe what is happening in this velocity–time graph.
Take appropriate readings and show your calculations.

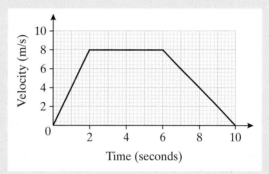

4 Here is a distance–time graph.
Draw the equivalent velocity–time graph.

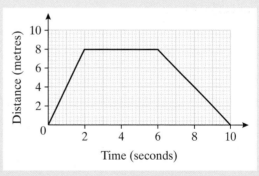

5 a) Draw the graph of $y = x^2 - 5x$, for values of x from $x = -1$ to $x = 6$.

b) Use your graph to find the gradient of the tangent to the curve at the point $x = 2$.

c) Use your graph to find the value of x where the gradient of the curve is zero.

6 a) Draw the graph of $y = x^2 + 3x + 4$ between $x = 0$ and $x = 4$.

b) Find an approximate value for the area under the curve $y = x^2 + 3x + 4$ between $x = 0$ and $x = 4$ by considering the ordinates $x = 0, x = 1, x = 2, x = 3$ and $x = 4$ to divide the area into strips of equal width.

c) State, giving a reason for your answer, whether your approximate value for the area under the curve $y = x^2 + 3x + 4$ between $x = 0$ and $x = 4$ is greater or less than the actual value for the area.

7 The velocity of a particle, from time $t = 0$ seconds to $t = 100$ seconds, is shown on the velocity–time graph.

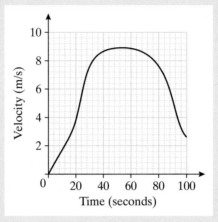

a) Calculate the acceleration at time $t = 70$ seconds.

b) Calculate the approximate total distance travelled in the first 30 seconds by splitting the area into three strips of equal width.

8 The distance–time graph shows a journey between 1 p.m. and 4 p.m. Find an estimate for the velocity in km/h at 2.30 p.m.

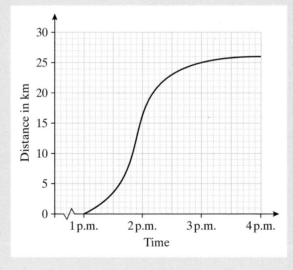

Solving simultaneous equations graphically

In Chapter 15 you saw that you could solve a pair of linear equations graphically by finding where the lines intersected (crossed). You can solve a pair of simultaneous equations where one is linear and one is quadratic by drawing the graphs and finding where the line crosses the curve. You will see that, in most cases, the line crosses the curve twice.

EXAMPLE 16.7

Solve the simultaneous equations $y = x^2 + 3x - 7$ and $y = x - 3$ graphically.
Take values of x from -5 to 2.

Solution

First make tables of values for the two equations.

x	-5	-4	-3	-2	-1	0	1	2
x^2	25	16	9	4	1	0	1	4
$+3x$	-15	-12	-9	-6	-3	0	3	6
-7	-7	-7	-7	-7	-7	-7	-7	-7
$y = x^2 + 3x - 7$	3	-3	-7	-9	-9	-7	-3	3

x	-5	0	2
$y = x - 3$	-8	-3	-1

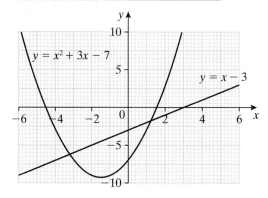

You can see from the graph that the line and the curve cross at $(-3.2, -6.2)$
and $(1.2, -1.8)$.

The solutions are, approximately, $x = -3.2, y = -6.2$ and $x = 1.2, y = -1.8$.

EXERCISE 16.2

1 a) Draw the graph of $y = x^2 - 5x + 5$ for values of x from -2 to 5.
   b) On the same axes, draw the line $y = 9 - 2x$.
   c) Write down the coordinates of the points where the line and the curve intersect.
   d) Write down the coordinates of the point on the curve where the gradient is zero.

2 a) Draw the graph of $y = x^2 - 3x - 1$ for values of x from -4 to 3.
   b) On the same axes, draw the line $y + 4x = 5$.
   c) Write down the coordinates of the points where the line and the curve intersect.
   d) Write down the coordinates of the point on the curve where the gradient is zero.

3 a) Draw the graph of $y = x^2 + 3$ for values of x from -2 to 5.
   b) On the same axes, draw the line $y = 3x + 7$.
   c) Write down the coordinates of the points where the line and the curve intersect.

4 a) Draw the graph of $y = x^2 - 2x + 3$ for values of x from 0 to 6.
   b) On the same axes, draw the line $y = 4x + 1$.
   c) Write down the coordinates of the points where the line and the curve intersect.

Using graphs to solve quadratic equations

Sometimes you may have drawn a quadratic graph and then need to solve
an equation that is different from the graph you have drawn. Rather than
drawing another graph, it may be possible to rearrange the equation to
obtain the one you have drawn.

EXAMPLE 16.8

a) Draw the graph of $y = 2x^2 - x - 3$ for values of x between -3 and 4.
b) Use this graph to solve these equations.
   (i) $2x^2 - x - 3 = 6$ **(ii)** $2x^2 - x = x + 5$

Solution

a)

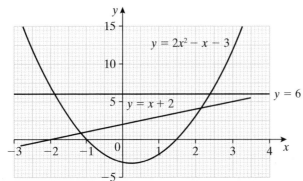

b) (i) To solve $2x^2 - x - 3 = 6$ you draw the line $y = 6$ on the same axes.
At the points of intersection of the line and the curve, $y = 6$ and
$2x^2 - x - 3$, and therefore $2x^2 - x - 3 = 6$.
The curve and line cross at approximately $x = -1.9$ and $x = 2.4$.

(ii) To solve $2x^2 - x = x + 5$ you must rearrange the equation so that
you have $2x^2 - x - 3$ on the left-hand side.

You do this by subtracting 3 from both sides.
$$2x^2 - x = x + 5$$
$$2x^2 - x - 3 = x + 5 - 3$$
$$2x^2 - x - 3 = x + 2$$
Now draw the line $y = x + 2$.

The points where the line and the curve intersect are the solution
to $2x^2 - x = x + 5$.
The curve and line cross at approximately $x = -1.2$ and $x = 2.2$.

EXERCISE 16.3

1 a) Draw the graph of $y = x^2 - 2x - 3$ for values of x from -2 to 4.

b) Use your graph to solve these equations.
- **(i)** $x^2 - 2x - 3 = 0$
- **(ii)** $x^2 - 2x - 3 = -2$
- **(iii)** $x^2 - 2x - 3 = x$
- **(iv)** $x^2 - 2x - 5 = 0$

2 a) Draw the graph of $y = x^2 - 2x + 2$ for values of x from -2 to 4.

b) Use your graph to solve these equations.
- **(i)** $x^2 - 2x + 2 = 8$
- **(ii)** $x^2 - 2x + 2 = 5 - x$
- **(iii)** $x^2 - 2x - 5 = 0$

3 a) Draw the graph of $y = 2x^2 + 3x - 9$ for values of x from -3 to 2.

b) Use your graph to solve these equations.
- **(i)** $2x^2 + 3x - 9 = -1$
- **(ii)** $2x^2 + 3x - 4 = -5$

4 a) Draw the graph of $y = x^2 - 5x + 3$ for values of x from -2 to 8.

b) Use your graph to solve these equations.
- **(i)** $x^2 - 5x + 3 = 0$
- **(ii)** $x^2 - 5x + 3 = 5$
- **(iii)** $x^2 - 7x + 3 = 0$

For questions **5** and **6**, do not draw the graphs.

5 The graph of $y = x^2 - 8x + 2$ has been drawn.
What other line needs to be drawn to solve the equation $x^2 - 8x + 6 = 0$?

6 The graph of $y = x^3 - 2x^2$ has been drawn.
What other curve needs to be drawn to solve the equation $x^3 - x^2 - 4x + 3 = 0$?

a) Draw the graph of $y = x^3 - x$.

b) What are the roots (or solutions) of $x^3 - x = 0$?

c) On the same axes, draw these lines.

    (i) $y = \frac{1}{2}$ **(ii)** $y = 5$ **(iii)** $y = 2x$

d) Write down the equations whose roots are given by the various intersections.

e) How many roots does each equation have?

f) What can you say about the number of roots of the equations $x^3 - x = kx$ and $x^3 - x = k$?

Drawing and recognising other curves

There are different types of functions and each type has its own characteristic shape. So far you have met linear functions, which give straight-line graphs, and quadratic functions, which, as you saw in Chapter 14, give a parabola when drawn. You will meet the graphs of trigonometrical functions in Chapter 33.

In this section you will meet the graphs of cubic, exponential and reciprocal functions. Cubic functions are functions which have x cubed as the highest power of x.

EXAMPLE 16.9

Draw the graph of $y = x^3$.

Solution

First draw up a table of values.

x	-3	-2	-1	0	1	2	3
y	-27	-8	-1	0	1	8	27

Then plot the points and join them with a smooth curve.

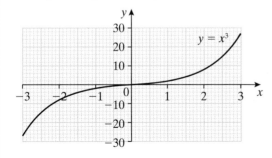

The next two examples are of exponential functions. You met these in Chapter 8.

EXAMPLE 16.10

The number of bacteria in a solution doubles each hour.
The number present at the start of timing is 300.
Draw a graph to show the number of bacteria in the solution over a 4-hour period.

Solution

First make a table of values for the number of bacteria in the solution.

Number of hours (h)	0	1	2	3	4
Number of bacteria (n)	300	600	1200	2400	4800

Then plot the points and join them with a smooth curve.

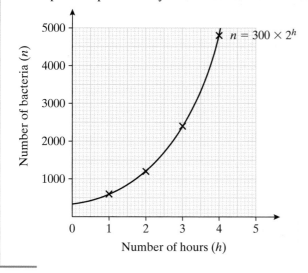

The graph in Example 16.10 is an exponential curve.

The equation is $n = 300 \times 2^h$.

The 300 is in the equation because this is the initial value.

The 2 is because the number of bacteria doubles each hour.

$n = 300 \times 2^h$ is a typical exponential equation.

EXAMPLE 16.11

a) Draw the graph of $y = 3^x$ for values of x from -2 to 3.

b) Use your graph to estimate the value of y when $x = 2.4$.

c) Use your graph to estimate the solution to the equation $3^x = 20$.

Solution

a) First make a table of values.

x	-2	-1	0	1	2	3
y	0.111	0.333	1	3	9	27

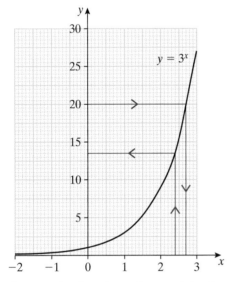

b) You read values from the graph in the usual way.
When $x = 2.4$, $y = 13.5$.

c) The solution to the equation $3^x = 20$ is the value of x when $y = 20$.
From the graph you can see that this is $x = 2.7$.

Another function you have met before is the reciprocal function. The graph of a reciprocal function has two branches, as shown in the next example.

EXAMPLE 16.12

Draw the graph of $y = \dfrac{2}{x}$ for values of x from -4 to 4.

Solution

The table of values is shown below.

x	−4	−3	−2	−1	0	1	2	3	4
y	−0.5	−0.7	−1	−2	–	2	1	0.7	0.5

Notice that the value of y cannot be found when x is zero − the value is at infinity.

It is difficult to judge where to draw the line between the values -1 and $+1$ so it is helpful to calculate the values of y when x is -0.5 and $+0.5$.

When $x = -0.5$, $y = -4$.
When $x = 0.5$, $y = 4$.

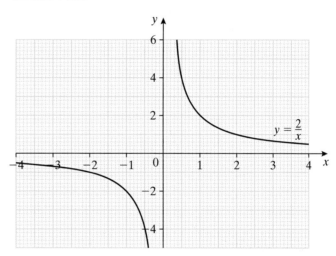

EXAMPLE 16.13

The data in the table was recorded during an experiment. Results were recorded for the two variables x and y.

x	1	2	3	4	5
y	30.9	52.2	86.8	136.3	199.0

It is known that y is approximately equal to $ax^2 + b$.

Draw a graph to estimate the values of a and b, and hence write down the relationship between x and y.

Solution

x	1	2	3	4	5
x^2	1	4	9	16	25
y	30.9	52.2	86.8	136.3	199.0

Plotting the graph $y = ax^2 + b$ shows the approximate relationship.

By comparing $y = ax^2 + b$ with the equation of a straight line $y = mx + c$ we can deduce that

a is the gradient, and

b is the point of intersection with the y-axis.

So $a = \dfrac{\text{difference in } y \text{ values}}{\text{difference in } x \text{ values}} = 7$ and $b = 24$

Hence $y = 7x^2 + 24$.

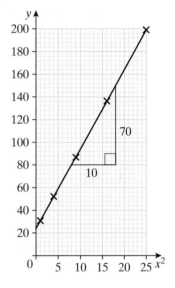

EXAMPLE 16.14

Given that the graph represents the relationship $y = pq^x$, find values for p and q and hence write down the relationship in terms of x and y.

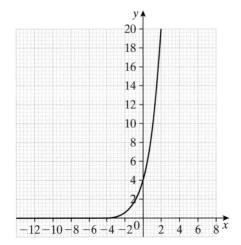

Solution

From the graph when $x = 0, y = 4$, so $4 = pq^0$

As $q^0 = 1, 4 = p \times 1$, giving $p = 4$.

So we know that $y = 4q^x$.

Selecting another value for x we read the y value from the graph. For example, when $x = 1$, $y = 12$ so by substitution into $y = 4q^x$ we find that $12 = 4 \times q^1$.

This gives $q = 3$.

The relationship is $y = 4 \times 3^x$

EXERCISE 16.4

1 **a)** Draw the graph of $y = x^3$ for values of x between -4 and 4.

   b) Use your graph to solve the equation $x^3 - 4x = 0$.

   c) Draw the graph of $y = x^3 - 4x$.

2 The population of a species of bird is dropping by 20% every 10 years.

   a) Copy and complete this table of values.

Year (y)	1970	1980	1990	2000	2010
Number of birds (n)	50 000		32 000		

   b) Draw a graph to show this relationship.

   c) Write down the equation for this relationship.

3 **a)** Draw a graph of $y = 2^{-x}$ for values of x from -4 to 2.

   b) Use your graph to estimate

   (i) the value of y when $x = 0.5$.

   (ii) the solution to the equation $2^{-x} = 10$.

4 The number of a species of animal in a park is shown in the table.

Year (y)	2005	2006	2007	2008	2009	2010
Number of animals (n)	2	6	18	54		

   a) Find the formula for n in terms of y.

   b) Copy and complete the table.

5 **a)** Plot the graph of $y = 2^x$ for values of x from -2 to 5.

   b) Use your graph to estimate

   (i) the value of y when $x = 3.2$.

   (ii) the solution of the equation $2^x = 10$.

6 a) Draw the graph of $y = \dfrac{3}{x}$.

   b) Use your graph to estimate the value of y when $x = 1.8$.

7 The data in the table was recorded during an experiment.
   Results are shown by the two variables x and y.

x	1	2	3	4
y	29.1	25.9	21.0	14.2

   a) On graph paper plot the values of y against the values of x^2.

   b) Before starting the experiment it was already known that y is approximately equal to $ax^2 + b$. Use your graph to estimate a and b.

Challenge 16.3

a) Radon is a naturally occurring gas which seeps out of some rocks.

The gas is radioactive with a half-life of 4 seconds.

This means that if there are 500 atoms of radon in a sample of gas, 4 seconds later there will be half that number so there will be 250 atoms of radon in the sample.

The formula for n, the number of atoms left after time t seconds, is $n = 1000 \times 2^{-\frac{1}{4}t}$.
   (i) Copy and complete this table of values.

t	0	4	8	12	16	20
n						

   (ii) Draw the graph of n against t.
   (iii) Use your graph to estimate the time, in seconds, when the number of atoms is 300.

b) Another radioactive element now has a mass of 50 grams.
   Its decay rate reduces its mass, m grams, by 20% each year.
   (i) Write down a formula for the mass after t years.
   (ii) By drawing the graph of m against t, estimate the time when the mass was 70 g and the time when it will be 35 g.
   Give your answers correct to the nearest tenth of a year.

WHAT YOU HAVE LEARNED

- **How to interpret velocity–time graphs and distance–time graphs**
- **To solve simultaneous equations graphically, plot the graphs and find the points of intersection**
- **When using graphs to solve equations, rearrange the equation to be solved so that the equation already plotted is on the left-hand side**
- **The shape of cubic, exponential and reciprocal graphs**
- **How to find mathematical functions from data sets**

1 Describe what is happening in this velocity–time graph.

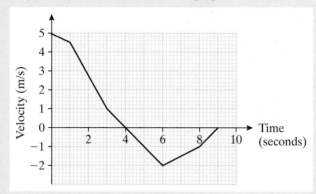

2 **a)** Draw the graph of $y = x^2 - 5x + 3$ for values of x from -2 to 4.

  b) On the same axes draw the line $7x + 2y = 11$.

  c) Write down the coordinates of the points where the line and the curve intersect.

3 **a)** Draw the graph of $y = x^2 - 2x$ for values of x from -1 to 4.

  b) Use your graph to solve the equation $x^2 - 2x = x + 1$.

4 **a)** Draw the graph of $y = x^2 - 7x$ for values of x from 0 to 7.

  b) Use your graph to solve these equations.
     (i) $x^2 - 7x + 9 = 0$ **(ii)** $x^2 - 5x + 1 = 0$

5 **a)** Draw the graph of $y = x^2 - 4x + 3$ for values of x from -2 to 8.

  b) Use your graph to solve the equation $x^2 - 4x + 1 = 0$.

6 **a)** Draw the graph of $y = x^3 - 1$ for values of x from -3 to 3.

  b) Use your graph to solve these equations.
     (i) $x^3 - 2 = 0$ **(ii)** $x^3 - x - 1 = 0$

7 **a)** Copy and complete the table of values for $y = 4^{-x}$.

x	−2.5	−2	−1.5	−1	−0.5	0	0.5	1
y		16		4			0.5	

  b) Plot the graph of $y = 4^{-x}$ for values of x from -2.5 to 1.

  c) Use your graph to estimate
     (i) the value of y when $x = -1.8$.
     (ii) the solution of the equation $4^{-x} = 25$.

8 Draw the graph of $y = 3x^3 + 2x - 3$ accurately for values of x from -2 to 3.

From your graph estimate the gradient of the curve at the points $x = -1$ and $x = 2$.

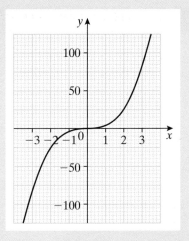

9 The graph of $y = x^3 - 4x^2 - 7x + 10$ is shown.

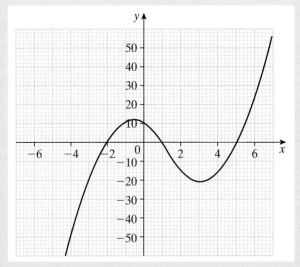

Use the trapezium rule with four strips to estimate the area of the region enclosed by the curve and the x-axis between $x = 1$ and $x = 5$.

10 The distance–time graph shows a journey between 2 p.m. and 5 p.m.

Find the velocity at 3 p.m.

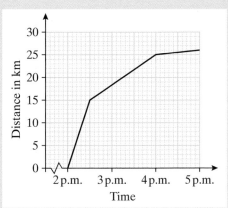

11 From the velocity–time graph shown, find the approximate distance travelled in the first 20 seconds using ordinates every 5 seconds.

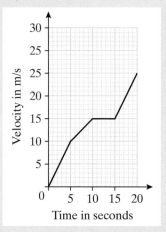

12 The data in the table was recorded during an experiment.
Results were recorded for the two variables x and y.

x	1	2	3	4	5
y	3.98	34.12	83.87	155.01	243.96

a) On graph paper plot the values of y against the values of x^2.

b) Before starting the experiment it was already known that y is approximately equal to $ax^2 + b$.
Use your graph to find the approximate relationship between y and x^2.

13 Given that the graph represents the relationship $y = pq^x$, find the values for p and q and hence write down the relationship in terms of x and y.

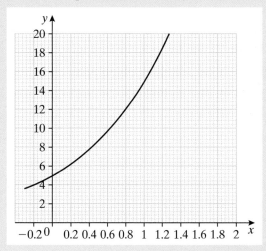

17 → PROPORTION AND VARIATION

THIS CHAPTER IS ABOUT

- **Solving problems using simple proportion**
- **Solving problems using more complex proportion (variation)**

YOU SHOULD ALREADY KNOW

- **How to use ratio**
- **How to use fractions as multipliers**
- **How to manipulate simple algebraic expressions**

Direct proportion

You met **proportion** in Chapter 3, where you learned how to solve problems using ratios.

In this sort of problem, both quantities increase at the same rate: for example, if you need 200 g of oats to make a small batch of 10 flapjacks, you need 400 g of oats to make a large batch of 20 flapjacks.

Amount of oats needed for the large batch = amount of oats needed for the small batch × 2

Number of flapjacks in the large batch = number of flapjacks in the small batch × 2

Both quantities are multiplied by the same number, called the **multiplier**.

To find the multiplier, you divide either pair of quantities. For example, $\frac{400}{200} = 2$ or $\frac{20}{10} = 2$.

So a quicker way to solve this type of problem is to find the multiplier and use it to find the unknown quantity. This second method is quicker than the first method because it combines two steps in one.

EXAMPLE 17.1

A car uses 20 litres of petrol when making a journey of 160 miles. How many litres of petrol would be used when making a similar journey of 360 miles?

Solution

Method 1

This is the method you learned in Chapter 3.

$160 : 360 = 4 : 9$ First write down the ratio linking the distances travelled in the two journeys and divide by 40 so that it is in its lowest terms.

$20 \div 4 = 5$ Divide the amount of petrol needed for the first journey by the first part of the ratio, 4.

$5 \times 9 = 45$ Multiply the second part of the ratio by 5.

45 litres of petrol would be used making a journey of 360 miles.

Method 2

$\dfrac{360}{160} = \dfrac{9}{4}$ Write the distances travelled in the two journeys as a fraction.

You want to find the petrol needed for a 360 mile journey, so make 360 the numerator and 160 the denominator. Cancel the fraction so that it is in its simplest form.

$20 \times \dfrac{9}{4} = 45$ Multiply the amount of petrol needed for the first journey by the multiplier.

45 litres of petrol would be used making a journey of 360 miles.

Notice that in Method 1 you divided by 4 and then multiplied by 9.

In Method 2 you multiplied by $\dfrac{9}{4}$, combining the two steps into one.

Questions of this type are examples of **direct proportion**. As one quantity increases (miles travelled, x) so must the other (petrol used, y). Simply, the more miles you travel, the more petrol you use.

This relationship can be written as $y \propto x$, which is expressed in words as 'y is proportional to x' or 'y varies as x'.

This graph shows direct proportion.

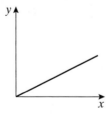

The gradient of the line could be any positive value (dependent on the multiplier) but the line will always pass through the origin.

The multiplier is sometimes called the **constant of proportionality** and given the letter k.

EXAMPLE 17.2

An excavator can dig a trench that is 560 metres long in 21 days.
How long would it take the excavator to dig a trench that is 240 metres long?

Solution

$\dfrac{240}{560} = \dfrac{3}{7}$ First find the multiplier and cancel so that it is in its simplest form.

You want to find the time required to dig a 240 m trench, so make 240 the numerator of the fraction and 560 the denominator.

$21 \times \dfrac{3}{7} = 9$ days   Multiply the known time by the multiplier to find the unknown time.

EXERCISE 17.1

For each of these questions

a) write down the multiplier.

b) calculate the required quantity.

1 An express train travels 165 metres in 3 seconds.
How far would it travel in 8 seconds?

2 An aircraft travels 216 miles in 27 minutes.
How far did it travel in 12 minutes?

3 £50 is worth $90.
How much is £175 worth?

4 A ladder that is 7 metres long has 28 rungs.
How many rungs would there be in a ladder that is 5 metres long?

5 A piece of string 27 metres long has a mass of 351 grams.
What is the mass of 15 metres of the string?

6 A rabbit can dig a tunnel that is 4 metres long in a total of 26 hours.
How long would it take in total to dig a tunnel 7 metres long?

7 A landscape gardener can paint 15 fence panels in 6 hours.
How many hours would it take him to paint 40 fence panels?

8 The cost of 12 printer cartridges is £90.00.
What is the cost of five of these cartridges?

9   A carpet with an area of 18 m^2 costs £441.
What is the cost of 14 m^2 of the same carpet?

10   A piece of balsa wood with a volume of 2.5 m^3 has a mass of 495 kilograms.
What is the mass of a piece of balsa wood with a volume of 0.9 m^3?

Inverse proportion

It is not always the case with proportion that, as one quantity increases, so does the other.

Sometimes, as one quantity increases, the other decreases.

In such cases, you need to divide by the multiplier, rather than multiply.

EXAMPLE 17.3

If three excavators can dig a hole in 8 hours, how long would it take four excavators to dig the hole?

Solution

Clearly, as more excavators are to be used, the digging will take less time.

The multiplier is $\frac{4}{3}$.

$8 \div \frac{4}{3} = 6$ Divide the known time by the multiplier to find the unknown time.

Questions of this type are examples of **inverse proportion**. As one quantity increases (number of diggers, x) the other quantity decreases (hours taken, y). Simply, the more excavators you have, the less time you need to dig the hole.

This relationship can be written as $y \propto \frac{1}{x}$, which is expressed in words as

'y is inversely proportional to x' or 'y varies inversely as x'.

This graph shows inverse proportion.

For each of these questions

a) write down the multiplier.

b) calculate the required quantity.

1 A journey takes 18 minutes at a constant speed of 32 kilometres per hour.
How long would the journey take at a constant speed of 48 kilometres per hour?

2 It takes a team of 8 men six weeks to paint a bridge.
How long would the painting take if there were 12 men?

3 A pool is normally filled using four inlet valves in a period of 18 hours.
One of the inlet valves is out of action.
How long will it take to fill the pool using just three inlet valves?

4 A journey can be completed in 44 minutes at an average speed of 50 miles per hour.
How long would the same journey take at an average speed of 40 miles per hour?

5 A supply of hay is enough to feed 12 horses for 15 days.
How long would the same supply feed 20 horses?

6 It takes three harvesters 6 hours to harvest a crop of wheat.
How long would it take to harvest the wheat if only two harvesters were available?

7 On the outward leg of a journey a cyclist travels at an average speed of 12 kilometres per hour for a period of 4 hours.
The return journey took 3 hours.
What was the average speed for the return journey?

8 It takes a team of 18 men 21 weeks to dig a canal.
How long would it take to dig the canal if there were 14 men?

9 A tank can be emptied using six pumps in a period of 18 hours.
How long will it take to empty the tank using eight pumps?

10 A gang of 9 bricklayers can build a wall in 20 days.
How long would a gang of 15 bricklayers take to build the wall?

Challenge 17.1

The time taken to complete a race is inversely proportional to the speed.

What happens to the time taken if the speed is

a) doubled? **b)** halved? **c)** increased by 20%?

Finding formulae

In Example 17.2, y is proportional to x ($y \propto x$).

The variables are metres and days, and can be written in a table.

x (metres)	560	240
y (days)	21	9

The first pair of x and y values are 560 and 21. The ratio $\dfrac{y}{x}$ is $\dfrac{21}{560}$, or $\dfrac{3}{80}$ in its simplest form.

The second pair of x and y values are 240 and 9. The ratio $\dfrac{y}{x}$ is $\dfrac{9}{240}$, or $\dfrac{3}{80}$ in its simplest form.

The same constant ratio (k) would apply to all pairs of x and y values.

So the formula for this relationship is $y = \dfrac{3}{80}x$.

The formula for any relationship of direct proportion is $y = kx$.

In Example 17.3, y is inversely proportional to x $\left(y \propto \dfrac{1}{x}\right)$.

The variables are excavators (x) and hours (y), and can be written in a table.

x (excavators)	3	4
y (hours)	8	6

The first pair of x and y values are 3 and 8. The value xy is 24.

The second pair of x and y values are 4 and 6. The value xy is 24.

The same constant (k) would apply to all pairs of x and y values.

So the formula for this relationship is $xy = 24$ or $y = \dfrac{24}{x}$.

The formula for any relationship of inverse proportion is $xy = k$ or $y = \dfrac{k}{x}$.

EXAMPLE 17.4

For each of these relationships
(i) state the type of proportion.
(ii) find the formula.
(iii) find the missing y value in the table.

a)

x	6	10	22
y	15	25	

b)

x	20	15	12
y	6	8	

Solution

a) (i) Direct proportion   As x increases, so does y.

$\dfrac{15}{6} = \dfrac{5}{2}, \dfrac{25}{10} = \dfrac{5}{2}$

(ii) $y = \dfrac{5}{2}x$

(iii) $y = \dfrac{5}{2} \times 22$

$= 55$

b) (i) Inverse proportion   As x decreases, y increases.

$20 \times 6 = 120, 15 \times 8 = 120$

(ii) $y = \dfrac{120}{x}$

(iii) $y = \dfrac{120}{12}$

$= 10$

EXERCISE 17.3

For each of these relationships
a) state the type of proportion.
b) find the formula.
c) where appropriate, find the missing y value in the table.

1

x	40	200
y	3	15

2

x	4	7	10
y	28	49	

3

x	12	18
y	16	24

4

x	20	90	150
y	8	36	

5

x	18	15
y	66	55

6

x	40	24	15
y	6	10	

7

x	30	18.75
y	5	8

8

x	60	80	200
y	24	18	

9

x	10	24
y	36	15

10

x	72	96	160
y	25	18.75	

Challenge 17.2

The variables A and B are such that $B = 50$ when $A = 20$.

a) (i) Write down a formula where A is directly proportional to B.
   (ii) Write down a formula where A is inversely proportional to B.

b) Calculate B when $A = 25$ for each of the two formulae you have found.

Other types of proportion

You have seen that, if $y \propto x$, then as x increases so does y.

For the following relationship, as the value of x increases so does the value of y.

x	4	12
y	8	72

However, there is no constant linking each pair of x and y values.

For the first pair of values the multiplier is 2.
For the second pair of values the multiplier is 6.
Therefore this is not an example of $y \propto x$.

This relationship is an example of $y \propto x^2$.

This graph shows $y \propto x^2$

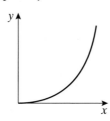

The gradient of the curve will be positive and the curve will always pass through the origin.

EXAMPLE 17.5

$y \propto x^2$

Find the formula for

x	4	12
y	8	72

Solution

The formula for this relationship is $y = kx^2$, where k is a constant (the constant of proportionality).

$y = kx^2$ To find the value of k, substitute the x value of 4 and
$8 = k \times 4^2$ the y value of 8 into the equation.
$8 = k \times 16$ You could use $x = 12$ and $y = 72$ instead.
$k = 0.5$

The formula is $y = \frac{1}{2}x^2$.

Check the formula by substituting the second pair of values into the equation.

$y = \frac{1}{2}x^2$
$72 = \frac{1}{2} \times 12^2$
$72 = 72 \checkmark$

You have seen that, if $y \propto \frac{1}{x}$, then as x increases, y decreases.

For the following relationship, as the value of x increases, the value of y decreases.

x	5	10
y	8	2

However, for the first pair of values $xy = 40$.
For the second pair of values $xy = 20$

Therefore this is not an example of $y \propto \frac{1}{x}$.

This relationship is an example of $y \propto \frac{1}{x^2}$.

This graph shows $y \propto \frac{1}{x^2}$.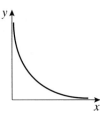

The gradient of the curve is negative.

EXAMPLE 17.6

$y \propto \dfrac{1}{x^2}$

Find the formula for

x	5	10
y	8	2

Solution

The formula for this relationship is $x^2y = k$, where k is a constant.

$x^2y = k$ To find the value of k, substitute the x value of 5
$5^2 \times 8 = k$ and the y value of 8 into the equation.
$25 \times 8 = k$ You could use $x = 10$ and $y = 2$ instead.
$k = 200$

The formula is $x^2y = 200$ or $y = \dfrac{200}{x^2}$.

Check the formula by substituting the second pair of values into the equation.

$x^2y = 200$
$10^2 \times 2 = 200$
$200 = 200$ ✓

EXERCISE 17.4

For each of these relationships
a) state the type of proportion.
b) find the formula.
c) where appropriate, find the missing y value in the table.
Hint: This exercise includes all four types of proportion that you have met in this chapter.

1

x	4	16
y	16	36

2

x	2	5	8
y	8	50	

3

x	6	9
y	12	27

4

x	5	25	35
y	10	250	

5

x	3	6
y	10.8	43.2

6

x	3	6	15
y	25	6.25	

7

x	5	10
y	50	12.5

8

x	2	5	10
y	5	0.8	

9

x	10	20
y	8	2

10

x	8	10	16
y	4	2.56	

Challenge 17.3

There are yet more types of proportion in addition to $y \propto x$, $y \propto \dfrac{1}{x}$, $y \propto x^2$ and $y \propto \dfrac{1}{x^2}$.

For each of the relationships in parts **a)** to **d)**
(i) state the type of proportion.
(ii) find the formula.

a)

x	2	5
y	24	375

b)

x	4	25
y	5	12.5

c)

x	2	4
y	256	32

d)

x	4	25
y	5	2

e) A truck is tested in a wind tunnel.
The wind resistance is proportional to the square of the speed of the truck.
What happens to the wind resistance if the speed of the truck is
(i) halved?
(ii) increased by 50%?

f) A marble is rolled down a slope.
After travelling d metres down the slope, the speed of the marble is v metres per second, where v is proportional to the square root of d.
Given that $v = 3.5$ when $d = 3$, calculate v when $d = 5$.

The tables below show examples of the following variations.

$$y \propto x^2, \quad y \propto \frac{1}{x^2}, \quad y \propto \sqrt{x} \quad \text{and} \quad y \propto x^3$$

For each of these relationships
(i) state the type of proportion.
(ii) find the formula.

a)

x	16	25
y	12	15

b)

x	2	4
y	12	48

c)

x	2	3
y	40	135

d)

x	3	6
y	4	1

e)

x	3	6
y	9	72

f)

x	4	16
y	8	16

g)

x	3	10
y	5	0.45

h)

x	10	8
y	150	96

a) It takes a team of 8 men six days to dig a trench 60 metres long.
How long would it take 10 men to dig a similar trench 50 metres long?

b) A 2250 m³ pool is normally filled using four inlet valves in a period of 18 hours.
How long will it take to fill a 1500 m³ pool using three similar inlet valves?

c) A total of 3200 washers can be made in 42 minutes using five punch machines.
How long would it take three similar machines to make 4800 washers?

d) A total of 90 bales of hay is enough to feed 12 horses for 15 days.
How many bales of hay would be needed to feed 10 horses for 8 days?

- Direct proportion is a linear relationship where, as one quantity increases, so does the other y is directly proportional to x, $y \propto x$, the equation is $y = kx$ where k is a constant.

- Inverse proportion is a non-linear relationship where, as one quantity increases, the other decreases.

  y is inversely proportional to x, $y \propto \dfrac{1}{x}$, the equation is $y = \dfrac{k}{x}$ where k is a constant

- Other forms of variation include $y \propto x^2$ and $y \propto \dfrac{1}{x^2}$

MIXED EXERCISE 17

1 a) A bicycle wheel rotates 115 times in a journey of 300 metres.
How many revolutions will the wheel make during a journey of 420 metres?

b) An express train travels 260 metres in 5 seconds.
How far would it travel at the same speed in 19 seconds?

c) A car travelling at a constant speed covers 38 miles in 57 minutes.
How far would it travel in 24 minutes?

d) A piece of card with an area of 640 square centimetres has a mass of 16 grams.
What is the mass of 1000 square centimetres of the card?

e) Painting a wall with an area of 18 m² uses 6.3 litres of paint.
How much paint is needed to paint a wall with an area of 28 m²?

2 a) It takes a team of five men 8 hours to lay a path.
How long would it take to lay the path if there were four men?

b) A tank can be emptied using five pumps in a period of 21 hours.
How long will it take to empty the tank using six pumps?

c) A stack of hay can feed 33 horses for 12 days.
For how long would the same stack of hay feed 44 horses?

d) A supply of oil is enough to run six generators for 12 days.
For how long would the same supply run eight generators?

e) A gang of 15 bricklayers can build a wall in 8 days.
How long would a gang of 12 bricklayers take to build the wall?

3 For each of these relationships
   (i) state the type of proportion.
   (ii) find the formula.
   (iii) where appropriate, find the missing y value in the table.

a)

x	2	10
y	8	40

b)

x	4	20	32
y	15	75	

c)

x	6	15
y	4	10

d)

x	30	125	275
y	6	25	

e)

x	3	6
y	4	2

f)

x	4	6	10
y	6	4	

4 A pebble is thrown vertically upwards with a speed of s metres per second. The pebble reaches a maximum height of h metres, before falling vertically downwards. It is known that h is directly proportional to the square of s.

   a) Given that a pebble thrown with a speed of 10 metres per second reaches a maximum height of 5 metres, find the expression for h in terms of s.

   b) Calculate the maximum height reached when a pebble is thrown with a speed of 3.5 metres per second.

   c) The pebble reaches a maximum height of 4.5 metres. Calculate the speed at which the pebble was thrown.

5 y is inversely proportional to x, and $y = 6$ when $x = 4$.

   a) Find the expression for y in terms of x.

   b) Calculate y when $x = \frac{1}{2}$.

6 A packaging machine is faulty. It is noticed that the distance, d metres, the conveyor belt moves before the warning bell sounds is inversely proportional to the square of its speed, s metres per second. The warning bell sounded when the speed was 5 m/s and the distance moved by the conveyor belt was 10 m.

   a) Find the expression for d in terms of s.

   b) Calculate
      (i) d when $s = 10$ m/s   **(ii)** s when $d = \frac{1}{4}$ m.

Finding the equation of a straight line in the form $y = mx + c$

Check up 18.1

Find the gradient of the line joining each of these pairs of points.

a) $(0, 2)$ and $(2, 8)$ **b)** $(2, 3)$ and $(3, 7)$ **c)** $(0, 2)$ and $(2, -2)$

d) $(-3, -1)$ and $(-1, -5)$ **e)** $(-1, 1)$ and $(-5, -1)$

Check up 18.2

Find the gradient and y-intercept of each of these lines.

a) $y = 2x - 5$ **b)** $2y = 4x - 9$ **c)** $6x + 2y = 5$

d) $3x - 4y = 6$ **e)** $2x + 4y = 5$

You learned how to find the equation of a straight line in Chapter 15.

EXAMPLE 18.1

Find the equation of this straight line.

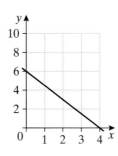

Solution

Gradient, $m = \dfrac{\text{increase in } y}{\text{increase in } x}$ First find the gradient of the line.

$\qquad\qquad = \dfrac{-6}{4} = -\dfrac{3}{2}$

y-intercept, $c = 6$ The line crosses the y-axis at $(0, 6)$.

$\qquad\qquad y = mx + c$

$\qquad\qquad y = -\frac{3}{2}x + 6$ Substitute m and c into the equation for a straight line.

$\qquad\qquad 2y = -3x + 12$ Multiply each side of the equation by 2 to get rid of the fraction.

$\qquad 3x + 2y = 12$ Add $3x$ to both sides of the equation to get rid of the negative term.

EXAMPLE 18.2

Find the equation of the line with gradient $-\frac{2}{3}$ that passes through the point $(4, 0)$.

Solution

The equation of a straight line is $y = mx + c$.

The equation is $y = -\frac{2}{3}x + c$.   You are given the gradient, m.

$y = -\frac{2}{3}x + c$

$0 = -\frac{2}{3} \times 4 + c$ To find the y-intercept, c, substitute the coordinates of the given point ($x = 4$ and $y = 0$) into the equation.

$0 = -\frac{8}{3} + c$

$c = \frac{8}{3}$

So $\qquad y = -\frac{2}{3}x + \frac{8}{3}$

$\qquad\quad 3y = -2x + 8$ Multiply each side of the equation by 3 to get rid of the fraction.

$\qquad 2x + 3y = 8$ Add $2x$ to both sides to get rid of the negative term.

EXAMPLE 18.3

Find the equation of the line that passes through $(4, 6)$ and $(6, 2)$.

Solution

$$\text{Gradient, } m = \frac{\text{increase in } y}{\text{increase in } x}$$ First find the gradient of the line.

$$= \frac{2 - 6}{6 - 4}$$

$$= \frac{-4}{2}$$

$$= -2$$

The equation is $y = -2x + c$

$$y = -2x + c$$
$$6 = -2 \times 4 + c$$ To find the y-intercept, c, substitute the coordinates of either
$$6 = -8 + c$$ of the given points (here, $x = 4$ and $y = 6$) into the equation.
$$6 + 8 = c$$
$$c = 14$$

So $y = -2x + 14$
$$2x + y = 14$$ Add $2x$ to each side of the equation to get rid of the negative term.

You could also have found the equation of the line in Example 18.3 by drawing the line through the two points and extending it to where it cuts the y-axis, then finding the gradient and y-intercept from the graph.

> **TIP**
>
> You should be able to find the equation of a line without drawing it, but it is useful to do a sketch as a check.

EXERCISE 18.1

Find the equation of each of these straight lines.

1
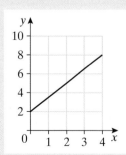

2
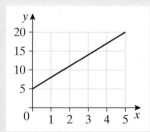

3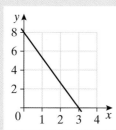

4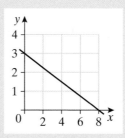

5 A line with gradient $\frac{2}{3}$, passing through the point $(2, 3)$.

6 A line with gradient $-\frac{3}{4}$, passing through the point $(3, 0)$.

7 A line passing through $(1, 4)$ and $(4, 7)$.

8 A line passing through $(2, 3)$ and $(5, 9)$.

9 A line passing through $(-1, 5)$ and $(3, -7)$.

10 A line passing through $(3, 1)$ and $(6, -1)$.

Challenge 18.1

a) Look at these graphs.

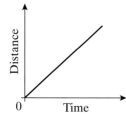

What does the gradient of each of these graphs represent?

b) A particle is accelerating at 40 m/s^2 (metres per second squared).
Its initial velocity was 25 m/s.
Find the equation connecting the velocity, v, and the time, t.

c) A ball is thrown into the air.
Its velocity after 2 seconds is 15 m/s.
The deceleration due to gravity is 10 m/s^2. (To decelerate means to slow down.)
Find the equation connecting the velocity, v, and the time, t.

Exploring gradients

Discovery 18.1

a) Draw each of the following pairs of lines on a separate pair of axes. Use values of x and y from -6 to 6.

(i) $y = x$ and $y = -x$ (ii) $y = 2x$ and $y = -\frac{1}{2}x$

(iii) $y = 5x$ and $y = -\frac{1}{5}x$ (iv) $y = 4x$ and $y = -0.25x$

(v) $y = 2x$ and $x + 2y = 6$

b) What do you notice about each pair you have drawn?
Can you spot a connection between them?
Find more pairs that give you the same result.

You learned in Chapter 15 that **parallel lines** have the same gradient.

There is also a connection between the gradients of **perpendicular lines**:

If a line has gradient m, then the perpendicular line has gradient $-\dfrac{1}{m}$.

The product of the two gradients is -1.

Proof 18.1

Use the properties of triangles to prove that if a line has gradient m, then the perpendicular line has gradient $-\dfrac{1}{m}$.

Solution

Two lines, AB and CD, cross at right angles at P.

Let the gradient of AB be $m = \dfrac{BC}{AC}$ and the gradient of CD

be $n = -\dfrac{AD}{AC}$.

In triangles ABC and ADC:

Angle ACB = angle DAC	Both are 90°.
Angle BAC = angle ADC	Both are 90° − angle PAD.
So triangles ABC and ADC are similar	They are equiangular.

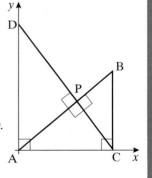

continues ...

Proof 18.1 continued

Therefore $\dfrac{BC}{AC} = \dfrac{AC}{AD}$ Ratio of corresponding sides.

So $m = -\dfrac{1}{n}$ or $mn = -1$ $-\dfrac{1}{n} = -\dfrac{1}{-\frac{AD}{AC}} = \dfrac{AC}{AD}.$

Check up 18.3

On a pair of axes, draw two lines that are perpendicular. Check that their gradients satisfy the rule above.

Repeat this for a number of different pairs of lines.

TIP
When drawing perpendicular lines, make sure the scale is the same on both axes.

If you know the equation of a line, you can find the equation of any line that is either parallel or perpendicular to it.

EXAMPLE 18.4

Find the equation of the line that is parallel to the line $2y = 3x + 4$ and passes through the point $(3, 2)$.

Solution

The gradient of the line $2y = 3x + 4$ is 1.5.
So the equation of the parallel line is $y = 1.5x + c$.
(The parallel line has the same gradient.)

$y = 1.5x + c$

$2 = 1.5 \times 3 + c$ To find the y-intercept, c, substitute the

$2 = 4.5 + c$ coordinates of the given point ($x = 3$ and $y = 2$)

$c = 2 - 4.5$ into the equation.

$c = -2.5$

So $y = 1.5x - 2.5$

      $2y = 3x - 5$ Multiply each side of the equation by 2 to make all
                    the coefficients whole numbers.

EXAMPLE 18.5

Find the equation of the line that crosses the line $y = 2x - 5$ at right angles at the point $(3, 1)$.

Solution

The gradient of the line $y = 2x - 5$ is 2.

The gradient of the perpendicular line is $-\frac{1}{2}$. $\left(\text{Since the gradient of the perpendicular line is } -\frac{1}{m}.\right)$

So the equation of the perpendicular line is $y = -\frac{1}{2}x + c$.

$$y = -\frac{1}{2}x + c$$
$$1 = -\frac{1}{2} \times 3 + c$$ To find the y-intercept, c, substitute the coordinates of the given point $(x = 3, y = 1)$ into the equation.
$$2 = -3 + 2c$$
$$2c = 5$$
$$c = 2.5$$

So $y = -\frac{1}{2}x + 2.5$
$$x + 2y = 5$$ Multiply each side of the equation by 2 to make all the coefficients whole numbers.

A line parallel to $y = mx + c$ will be of the form $y = mx + d$, where c and d are constants.

A line perpendicular to $y = mx + c$ will be of the form $y = -\dfrac{1}{m}x + e$, where c and e are constants.

EXAMPLE 18.6

Draw a pair of axes and label them 0 to 4 for x and -1 to 7 for y. Draw the line $y = 2x - 1$. Draw

a) a line parallel to $y = 2x - 1$.

b) a line perpendicular to $y = 2x - 1$.

Find the equations of the two lines.

Solution

There are many solutions to this question.

The lines drawn on this diagram are

a) $y = 2x + 1$ Parallel to $y = 2x - 1$: same gradient, y-intercept is 1.

b) $y = -\frac{1}{2}x + 4$ Perpendicular to $y = 2x - 1$: gradient is $-\frac{1}{2}$, y-intercept is 4.

$x + 2y = 8$

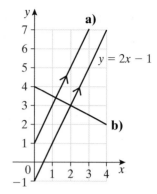

◎ EXERCISE 18.2

1 Find the gradient of a line perpendicular to the line joining each of these pairs of points.

     a) $(1, 2)$ and $(5, -6)$ **b)** $(2, 4)$ and $(3, 8)$ **c)** $(-2, 1)$ and $(2, -4)$

2 Find the equation of the line that passes through $(1, 5)$ and is parallel to $y = 3x - 3$.

3 Find the equation of the line that passes through $(0, 3)$ and is parallel to $3x + 2y = 7$.

4 Draw the line $y = 2x$.
On the same graph, draw a line parallel to it and a line perpendicular to it.
Find the equation of each of these two lines.

5 **a)** Draw the line $x + 3y = 6$. State its gradient.

     b) Draw a line perpendicular to this. State its gradient.

6 Find the equation of the line that passes through $(1, 5)$ and is perpendicular to $y = 3x - 1$.

7 Find the equation of the line that passes through $(0, 3)$ and is perpendicular to $2y + 3x = 7$.

8 Which of these lines are

     a) parallel?   **b)** perpendicular?

     $y = 4x + 3$ $2y - 3x = 5$ $6y + 4x = 1$ $4x - y = 5$

9 Two lines cross at right angles at the point $(5, 3)$. One passes through $(6, 0)$.
What is the equation of the other line?

10 In the diagram AB and BC are two sides of a square, ABCD.

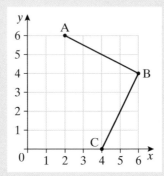

Work out

a) the equation of the line AD.

b) the equation of the line DC.

c) the coordinates of D.

Challenge 18.2

a) What happens when you try to solve these simultaneous equations?

$$2x - y = 2$$
$$2x - y = -1$$

b) Explain the result in terms of the graphs of the lines.

WHAT YOU HAVE LEARNED

- **How to find the equation of any straight line in the form $y = mx + c$**
- **That if perpendicular lines have gradients m and n, then $m = -\dfrac{1}{n}$**

MIXED EXERCISE 18

1 Draw each of these lines on a separate pair of axes. Use values of x and y as stated.

   a) $2x + 3y = 9$ x: 0 to 5 y: 0 to 5

   b) $5x + 2y + 10 = 0$ x: −2 to 0 y: −5 to 0

2 Find the equation of each of these lines.

   a)   **b)**   **c)**

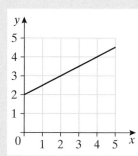
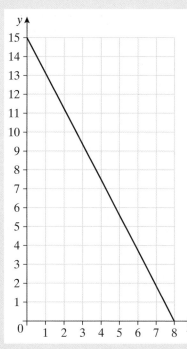
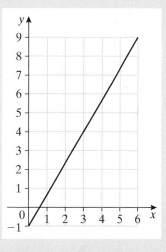

3 Find the equation of each of these lines.

   a) A line with gradient 3, passing through $(0, 2)$.

   b) A line with gradient −2, passing through $(1, 4)$.

   c) A line with gradient $\frac{1}{2}$, passing through $(2, 6)$.

4 Find the gradient of each of these lines.

   a) $y = 4x - 1$ **b)** $5y = 2x + 3$ **c)** $2x + 3y = 18$ **d)** $6x + y = 4$

5 Find the equation of the line that joins each of these pairs of points.

   a) $(4, 0)$ and $(6, 5)$ **b)** $(1, 2)$ and $(6, 4)$ **c)** $(2, 3)$ and $(5, -6)$

6 Find the equation of each of the three sides of the triangle ABC.

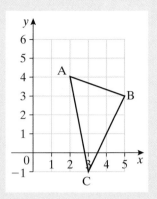

7 Find the equation of each of these lines.
   a) A line parallel to $y = 7x + 4$, passing through $(5, 1)$.
   b) A line parallel to $2y = 5x + 4$, passing through $(4, 0)$.
   c) A line parallel to $5x + 4y = 3$, passing through $(2, -1)$.

8 Find the equation of each of these lines.
   a) A line perpendicular to $y = 2x + 4$, passing through $(2, 1)$.
   b) A line perpendicular to $3y = 2x + 7$, passing through $(0, 1)$.
   c) A line perpendicular to $2x + 5y = 4$, passing through $(1, -3)$.

9 Draw axes and label them 0 to 8 for both x and y.
   a) **(i)** Draw the line passing through $(0, 8)$ and $(6, 0)$.
      (ii) What is the equation of this line?
   b) **(i)** Draw the parallel line that passes through $(0, 6)$.
      (ii) Where does this line cross the x-axis?
   c) **(i)** Draw the line perpendicular to both these lines that passes through the origin.
      (ii) What is the equation of this line?

10 Three corners of a parallelogram are A$(1, 6)$, B$(3, 4)$ and C$(5, 8)$.
   a) Find the equation of AD.
   b) Without drawing the graph, find the coordinates of D.

19 → QUADRATIC EQUATIONS

Solving quadratic equations

A **quadratic** function is a function where the highest power of x is 2.
In Chapter 11 you learned how to **factorise** a quadratic expression.

You also need to be able to solve quadratic equations.

Discovery 19.1

Find two numbers which multiply together to give zero.

If two numbers, A and B, are such that $A \times B = 0$, either $A = 0$ or $B = 0$.

We can use this fact to solve equations where the left-hand side is factorised and the right-hand side is zero.

EXAMPLE 19.1

Solve each of these quadratic equations.
a) $x(x - 3) = 0$
b) $(x + 2)(x - 3) = 0$
c) $(2x - 5)(x + 1) = 0$

Solution

a) $x(x - 3) = 0$

Either $x = 0$ or $x - 3 = 0$ The two factors are x and $x - 3$.
$$\text{If } x - 3 = 0$$
$$x = 3$$

There are two possible solutions to the equation: $x = 0$ or $x = 3$.
Both these answers are valid solutions to the equation.

b) $(x + 2)(x - 3) = 0$

Either $x + 2 = 0$ or $x - 3 = 0$
If $x + 2 = 0$ If $x - 3 = 0$
$x = -2$ $x = 3$
Solutions: $x = -2$ or $x = 3$

c) $(2x - 5)(x + 1) = 0$

Either $2x - 5 = 0$ or $x + 1 = 0$
If $2x - 5 = 0$ If $x + 1 = 0$
$2x = 5$ $x = -1$
$x = 2\frac{1}{2}$
Solutions: $x = 2\frac{1}{2}$ or $x = -1$

> **TIP**
> A quadratic equation always has two solutions, although they may be the same.

> **TIP**
> The sign of the answer is opposite to the sign in the bracket.

◎ EXERCISE 19.1

Solve each of these quadratic equations.

1 $x(x + 2) = 0$

2 $(x - 5)(x - 1) = 0$

3 $(x - 4)(x + 3) = 0$

4 $(x + 2)(x + 7) = 0$

5 $2x(x - 3) = 0$

6 $(x + 4)(x - 4) = 0$

7 $(x + 10)(x - 6) = 0$

8 $(x - 7)(2x - 3) = 0$

9 $(x + 4)(2x - 1) = 0$

10 $(2x - 5)(3x + 1) = 0$

11 $(3x + 7)(2x - 9) = 0$

12 $5x(4x + 1) = 0$

Solving quadratic equations by factorising

If the left-hand side of the quadratic equation is not already factorised,
factorise it first.

EXAMPLE 19.2

Solve each of these quadratic equations.

a) $x^2 + 5x = 0$ **b)** $x^2 - 25 = 0$ **c)** $x^2 - 5x + 6 = 0$

d) $2x^2 - 7x - 15 = 0$ **e)** $x^2 - 6x + 9 = 0$

Solution

a) $x^2 + 5x = 0$

$x(x + 5) = 0$ Take out the common factor, x.

$x = 0$ or $x + 5 = 0$

                       $x = -5$

Solutions: $x = 0$ or $x = -5$

b) $x^2 - 25 = 0$

     $(x - 5)(x + 5) = 0$ This is the difference of two squares.

     $x - 5 = 0$ or $x + 5 = 0$

        $x = 5$ $x = -5$

Solutions: $x = 5$ or $x = -5$

 These solutions can be written as $x = \pm 5$.

c) $x^2 - 5x + 6 = 0$

    $(x - 2)(x - 3) = 0$ Factorise using two brackets.

    $x - 2 = 0$ or $x - 3 = 0$

       $x = 2$ $x = 3$

Solutions: $x = 2$ or $x = 3$

d) $2x^2 - 7x - 15 = 0$

    $(2x + 3)(x - 5) = 0$ Factorise using two brackets.

          $2x + 3 = 0$ or $x - 5 = 0$

              $2x = -3$ $x = 5$

               $x = -1\frac{1}{2}$

Solutions: $x = -1\frac{1}{2}$ or $x = 5$

e)
$$x^2 - 6x + 9 = 0$$
$$(x - 3)(x - 3) = 0$$
$$x - 3 = 0 \quad \text{or} \quad x - 3 = 0$$
$$x = 3 \qquad\qquad x = 3$$

Factorise using two brackets.

Solution: $x = 3$

The two solutions are the same when the quadratic expression factorises into a 'perfect square', in this case $(x - 3)^2$. This is sometimes called a 'repeated solution'.

The difference of two squares can be solved by another method.

For example, $\quad x^2 - 25 = 0$
$$x^2 = 25 \qquad \text{Add 25 to each side.}$$
$$x = \pm 5 \qquad \text{Take the square root of each side.}$$

If you use this method, make sure you don't forget about the negative solution.

EXERCISE 19.2

Solve each of these quadratic equations.

1 $x^2 + 3x + 2 = 0$

2 $x^2 - 6x + 8 = 0$

3 $x^2 + 3x - 4 = 0$

4 $x^2 - 4x - 5 = 0$

5 $x^2 + x - 12 = 0$

6 $x^2 + 7x = 0$

7 $x^2 - 8x + 15 = 0$

8 $x^2 - 4x - 21 = 0$

9 $2x^2 - 8x = 0$

10 $x^2 - 36 = 0$

11 $x^2 + 10x + 16 = 0$

12 $x^2 - 5x - 24 = 0$

13 $x^2 + 5x - 6 = 0$

14 $x^2 - 11x + 18 = 0$

15 $4x^2 - 25 = 0$

16 $x^2 + 8x - 20 = 0$

17 $3x^2 - x = 0$

18 $x^2 - 7x - 30 = 0$

19 $3x^2 + 15x = 0$

20 $x^2 - 9x + 20 = 0$

21 $3x^2 + 4x + 1 = 0$

22 $2x^2 - 7x + 3 = 0$

23 $2x^2 + 3x - 5 = 0$

24 $3x^2 + 7x + 2 = 0$

25 $2x^2 - x - 6 = 0$

26 $3x^2 - 11x + 6 = 0$

27 $5x^2 - 24x - 5 = 0$

28 $6x^2 + x - 2 = 0$

29 $x^2 - 10x + 25 = 0$

30 $15x^2 - 4x - 3 = 0$

31 $2x^2 - 200 = 0$

32 $81 - 4x^2 = 0$

33 $6x^2 - 5x - 6 = 0$

Equations that need rearranging first

If the equation is not in the form where the right-hand side is zero, rearrange it before factorising.

Solve each of these quadratic equations.

a) $x^2 + 4x = 5$ **b)** $x^2 = 5x - 6$ **c)** $3x - x^2 - 2 = 0$

Solution

a)
$$x^2 + 4x = 5$$
$$x^2 + 4x - 5 = 0 \qquad \text{Subtract 5 from each side.}$$
$$(x + 5)(x - 1) = 0 \qquad \text{Factorise using two brackets.}$$

$$x + 5 = 0 \quad \text{or} \quad x - 1 = 0$$
$$x = -5 \qquad\qquad x = 1$$

Solutions: $x = -5$ or $x = 1$

b)
$$x^2 = 5x - 6$$
$$x^2 - 5x = -6 \qquad \text{Subtract } 5x \text{ from each side.}$$
$$x^2 - 5x + 6 = 0 \qquad \text{Add 6 to each side.}$$
$$(x - 2)(x - 3) = 0$$

$$x - 2 = 0 \quad \text{or} \quad x - 3 = 0$$
$$x = 2 \qquad\qquad x = 3$$

Solutions: $x = 2$ or $x = 3$

c)
$$3x - x^2 - 2 = 0$$

Although the right-hand side of this equation is zero, the order of the terms and the fact that the x^2 term is negative are problems.

$$3x - 2 = x^2 \qquad \text{Add } x^2 \text{ to each side.}$$
$$-2 = x^2 - 3x \qquad \text{Subtract } 3x \text{ from each side.}$$
$$0 = x^2 - 3x + 2 \qquad \text{Add 2 to each side.}$$
$$x^2 - 3x + 2 = 0 \qquad \text{Swap the sides so that the zero is on the right.}$$
$$(x - 2)(x - 1) = 0 \qquad \text{Factorise using two brackets.}$$

$$x - 2 = 0 \quad \text{or} \quad x - 1 = 0$$
$$x = 2 \qquad\qquad x = 1$$

Solutions: $x = 2$ or $x = 1$

> **TIP**
>
> Instead of rearranging the equation in Example 19.3 part **c)**, you could have multiplied each side by -1, giving $-3x + x^2 + 2 = 0$.

Solve each of these quadratic equations.

1 $x^2 - x = 12$ **2** $x^2 = 3x + 10$ **3** $x^2 = 5x + 14$ **4** $x^2 = 7x - 6$

5 $x^2 = 6x$ **6** $x^2 = 5 - 4x$ **7** $x^2 = 8 + 2x$ **8** $2x^2 = 2 - 3x$

9 $15 + 2x - x^2 = 0$ **10** $4 - 3x - x^2 = 0$

Challenge 19.1

I think of a number. I add 6 to it. I multiply the result by the original number and add 3. The answer is 58.

Write down an equation and solve it to find the original number.

Challenge 19.2

The diagram shows a rectangular pen made of fencing on three sides and a wall on the fourth side.

a) If 50 m of fencing is used, find the length of the pen in terms of x.

b) If the area of the pen is 272 m², write down an equation in x and show that it simplifies to

$$x^2 - 25x + 136 = 0.$$

c) Solve your equation to find the length and width of the pen.

Challenge 19.3

The area of the triangle is 6.5 cm².

a) Write down an equation in x and simplify it.

b) Hence solve the equation to find the lengths of AB and BC.

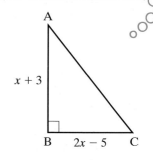

Completing the square

a) Expand and simplify each of these expressions.
  (i) $(x + m)^2$ **(ii)** $(x - m)^2$

b) Copy and complete each of these expansions.
  Use your answers to part **a)**.
  (i) $(x + 1)^2 = x^2 + \text{.......} + 1$ **(ii)** $(x + 3)^2 = x^2 + \text{.......} + 9$
  (iii) $(x - 2)^2 = x^2 - \text{.......} + 4$ **(iv)** $(x - 5)^2 = x^2 - \text{.......} + 25$
  (v) $(x + 4)^2 = x^2 + \text{.......} + \text{.......}$ **(vi)** $(x - 10)^2 = x^2 - \text{.......} + \text{.......}$
  (vii) $(x + \frac{1}{2})^2 = x^2 + \text{.......} + \text{.......}$

c) Look at your answers to part **b)**.
  (i) If $(x + m)^2 = x^2 + nx + m^2$, what is the relationship between m and n?
  (ii) If $(x - m)^2 = x^2 - nx + m^2$, what is the relationship between m and n?

d) Copy and compete each of these factorisations.
  Use your answers to part **c)**.
  (i) $x^2 + 4x + 4 = (x + \text{.......})^2$ **(ii)** $x^2 - 6x + 9 = (x - \text{.......})^2$
  (iii) $x^2 + 10x + 25 = (x + \text{.......})^2$ **(iv)** $x^2 - 12x + 36 = (x - \text{.......})^2$
  (v) $x^2 - 8x + 16 = (x - \text{.......})^2$ **(vi)** $x^2 + 14x + 49 = (x + \text{.......})^2$

e) Copy and fill in the gaps in each of these equations.
  Use your answers to part **d)**.
  (i) $x^2 + 4x = (x + \text{.......})^2 - \text{.......}$ **(ii)** $x^2 - 6x = (x - \text{.......})^2 - \text{.......}$
  (iii) $x^2 + 10x = (x + \text{.......})^2 - \text{.......}$ **(iv)** $x^2 - 12x = (x - \text{.......})^2 - \text{.......}$
  (v) $x^2 - 8x = (x - \text{.......})^2 - \text{.......}$ **(vi)** $x^2 + 14x = (x + \text{.......})^2 - \text{.......}$

You can solve a quadratic equation that does not factorise by **completing the square**.

This involves writing the left-hand side of the equation in the form $(x \pm m)^2 \pm n$, as you did in part **e)** of Discovery 19.2.

EXAMPLE 19.4

a) Write each of these quadratic expressions in the form $(x \pm m)^2 \pm n$.
  (i) $x^2 + 6x$ **(ii)** $x^2 - 2x$ **(iii)** $x^2 + 3x$

b) Write each of these quadratic expressions in the form $(x \pm m)^2 \pm n$.
  Use your answers to part **a)**.
  (i) $x^2 + 6x + 10$ **(ii)** $x^2 - 2x - 3$ **(iii)** $x^2 + 3x - 2$

Solution

a) (i) $m = 6 \div 2$ m is always half the coefficient of x.
$\qquad = 3$
$\qquad n = 3^2$ n is always equal to m^2.
$\qquad = 9$

So $x^2 + 6x = (x + 3)^2 - 9$ The sign in the bracket is always the same as the one in front of the x term.

(ii) $m = 2 \div 2$
$\qquad = 1$
$\qquad n = 1^2$
$\qquad = 1$

So $x^2 - 2x = (x - 1)^2 - 1$

(iii) $m = 3 \div 2$
$\qquad = 1.5$
$\qquad n = 1.5^2$
$\qquad = 2.25$

So $x^2 + 3x = (x + 1.5)^2 - 2.25$

b) (i) Since $\qquad x^2 + 6x = (x + 3)^2 - 9$
then $\quad x^2 + 6x + 10 = (x + 3)^2 - 9 + 10$ As 10 has been added to the left-hand
$\qquad x^2 + 6x + 10 = (x + 3)^2 + 1$ side, you need to add 10 to the right-hand side, to maintain equality.

(ii) Since $\qquad x^2 - 2x = (x - 1)^2 - 1$
then $\quad x^2 - 2x - 3 = (x - 1)^2 - 1 - 3$ This time you need to subtract 3 from
$\qquad = (x - 1)^2 - 4$ the right-hand side.

(iii) Since $\qquad x^2 + 3x = (x + 1.5)^2 - 2.25$
then $\quad x^2 + 3x - 4 = (x + 1.5)^2 - 2.25 - 4$ Subtract 4 from the right-hand side.
$\qquad = (x + 1.5)^2 - 6.25$

TIP

You may be asked to write a quadratic expression in the form $(x + m)^2 + n$. Remember that m and n could be positive or negative.

Alternatively, you may simply be asked to 'complete the square'. This means 'write the quadratic expression in the form $(x + m)^2 + n$'.

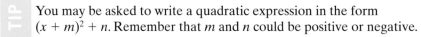 EXERCISE 19.4

1 a) Write each of these quadratic expressions in the form $(x + m)^2 + n$.
     (i) $x^2 + 8x$ **(ii)** $x^2 - 10x$ **(iii)** $x^2 + 12x$ **(iv)** $x^2 + x$

   b) Write each of these quadratic expressions in the form $(x + m)^2 + n$.
     Use your answers to part **a)**.
     (i) $x^2 + 8x - 3$ **(ii)** $x^2 - 10x + 31$ **(iii)** $x^2 + 12x - 5$ **(iv)** $x^2 + x + 2$

2 For each of these quadratic expressions, complete the square.

a) $x^2 + 2x - 3$ b) $x^2 + 4x - 1$ c) $x^2 - 6x + 12$

d) $x^2 + 10x - 6$ e) $x^2 - 20x - 50$ f) $x^2 + 12x - 1$

g) $x^2 + 8x + 19$ h) $x^2 - 3x - 3$ i) $x^2 + 5x + 10$

Challenge 19.4

a) Complete the square to express $4x^2 + 12x - 7$ in the form $(kx + m)^2 - n$, where k, m and n are positive integers.

b) Use your answer to part **a)** to solve $4x^2 + 12x - 7 = 0$.

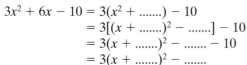

Challenge 19.5

a) Complete the following to write $3x^2 + 6x - 10$ in the form $k(x + m)^2 - n$, where k, m and n are positive integers.

$$3x^2 + 6x - 10 = 3(x^2 + \text{.......}) - 10$$
$$= 3[(x + \text{.......})^2 - \text{.......}] - 10$$
$$= 3(x + \text{.......})^2 - \text{.......} - 10$$
$$= 3(x + \text{.......})^2 - \text{.......}$$

b) Use a similar method to write $2x^2 + 8x - 5$ in the form $k(x + m)^2 - n$, where k, m and n are positive integers.

Challenge 19.6

a) What is the smallest possible value of $(x - 3)^2$?
What would be the value of x in that case?

b) What is the smallest possible value of $(x - 3)^2 + 5$?

c) What is the smallest possible value of $(x - 3)^2 - 2$?

d) What is the smallest possible value of $(x + 2)^2 + 4$?
What would be the value of x in that case?

e) What is the smallest possible value of $(x - 1)^2 - 6$?
What would be the value of x in that case?

f) What is the smallest possible value of $(x + m)^2 + n$?
What would be the value of x in that case?

Solving quadratic equations by completing the square

Some quadratic equations do not factorise. However, if a quadratic equation is in the form of a completed square, it can easily be solved. Remember to do the same operation to the whole of both sides.

EXAMPLE 19.5

Solve each of these quadratic equations.

a) $(x - 3)^2 - 8 = 0$

b) $(x + 2)^2 - 11 = 0$

Solution

a) $(x - 3)^2 - 8 = 0$

$$(x - 3)^2 = 8 \qquad \text{Add 8 to each side.}$$
$$x - 3 = \pm\sqrt{8} \qquad \text{Take the square root of each side.}$$
$$x = \pm\sqrt{8} + 3 \qquad \text{Add 3 to each side.}$$
$$x = +5.83 \text{ or } x = 0.17 \text{ (to 2 d.p.)}$$

> **TIP**
>
> Since the number 8 is not a perfect square, the final answer to part **a)** is not exact and has to be rounded. The exact answer is seen in the previous line of the solution, where it is given in **surd form**: $x = \pm\sqrt{8} + 3$. Surds are covered in Chapter 9.

b) $(x + 2)^2 - 11 = 0$

$$(x + 2)^2 = 11 \qquad \text{Add 11 to each side.}$$
$$x + 2 = \pm\sqrt{11} \qquad \text{Take the square root of each side.}$$
$$x = \pm\sqrt{11} - 2 \qquad \text{Subtract 2 from each side.}$$
$$x = 1.32 \text{ or } x = -5.32 \text{ (to 2 d.p.)}$$

If the equation is not in the form of a completed square, you need to complete the square first.

EXAMPLE 19.6

Solve each of these quadratic equations.

a) $x^2 - 8x + 10 = 0$

b) $x^2 + 10x - 9 = 0$

Solution

a)
$$x^2 - 8x + 10 = 0$$
$$(x - 4)^2 - 16 + 10 = 0 \qquad \text{First complete the square.}$$
$$(x - 4)^2 - 6 = 0$$
$$(x - 4)^2 = 6 \qquad \text{Add 6 to each side.}$$
$$x - 4 = \pm\sqrt{6} \qquad \text{Take the square root of each side.}$$
$$x = \pm\sqrt{6} + 4 \qquad \text{Add 4 to each side.}$$
$$x = 6.45 \text{ or } x = 1.55 \text{ (to 2 d.p.)}$$

b)
$$x^2 + 10x - 9 = 0$$
$$(x + 5)^2 - 25 - 9 = 0 \qquad \text{First complete the square.}$$
$$(x + 5)^2 - 34 = 0$$
$$(x + 5)^2 = 34 \qquad \text{Add 34 to each side.}$$
$$x + 5 = \pm\sqrt{34} \qquad \text{Take the square root of each side.}$$
$$x = \pm\sqrt{34} - 5 \qquad \text{Subtract 5 from each side.}$$
$$x = 0.83 \text{ or } x = -10.83 \text{ (to 2 d.p.)}$$

> **TIP**
>
> If you are asked to solve a quadratic equation giving your solutions to a number of decimal places or significant figures, it almost certainly will not factorise.

EXERCISE 19.5

In this exercise, give all your answers correct to 2 decimal places.

1 Solve each of these quadratic equations.

a) $(x - 3)^2 - 10 = 0$ **b)** $(x + 1)^2 - 2 = 0$

c) $(x + 6)^2 - 28 = 0$ **d)** $(x + 4)^2 - 17 = 0$

2 For each of the quadratic equations, first complete the square and then solve the equation.

a) $x^2 - 4x - 10 = 0$ **b)** $x^2 + 8x + 11 = 0$ **c)** $x^2 - 2x - 7 = 0$

d) $x^2 - 10x + 19 = 0$ **e)** $x^2 - 12x + 21 = 0$ **f)** $x^2 + 4x - 6 = 0$

g) $x^2 + 8x + 13 = 0$ **h)** $x^2 + 20x + 50 = 0$ **i)** $x^2 - 3x - 11 = 0$

Solving quadratic equations by using the formula

The method of completing the square can be used to solve the general quadratic equation

$$ax^2 + bx + c = 0$$

This leads to the general solution of this quadratic equation, known as the **quadratic formula**.

$$x = \frac{-b \pm \sqrt{b^2 - 4ac}}{2a}$$

> **TIP**
>
> Remember that $\pm$ means 'plus or minus'. You obtain the two solutions of the equation by using the formula with $+$ and $-$ separately.

You do not need to know the proof of this formula, but it can be done by completing the square.

The formula can be used to solve quadratic equations which do not factorise.

EXAMPLE 19.7

Solve each of these quadratic equations. Give your answers correct to 2 decimal places.

a) $x^2 + 6x + 2 = 0$ **b)** $2x^2 - 9x + 5 = 0$ **c)** $3x^2 - 2x - 7 = 0$

Solution

a) $x^2 + 6x + 2 = 0$

$a = 1, b = 6$ and $c = 2$

> **TIP**
>
> If you write out the formula each time you do the first few questions, you will learn it much more easily.

$$x = \frac{-b \pm \sqrt{b^2 - 4ac}}{2a}$$

$$x = \frac{-6 \pm \sqrt{6^2 - 4 \times 1 \times 2}}{2 \times 1}$$

Substitute $a = 1, b = 6$ and $c = 2$ into the equation.

$$x = \frac{-6 \pm \sqrt{36 - 8}}{2} = \frac{-6 \pm \sqrt{28}}{2}$$

$$x = \frac{-6 \pm \sqrt{28}}{2} \quad \text{or} \quad x = \frac{-6 - \sqrt{28}}{2}$$

Find both solutions.

$$x = -0.35 \quad \text{or} \quad x = -5.65 \text{ (to 2 d.p.)}$$

b) $2x^2 - 9x + 5 = 0$

$a = 2, b = -9$ and $c = 5$

$$x = \frac{-b \pm \sqrt{b^2 - 4ac}}{2a}$$

$$x = 9 \pm \frac{\sqrt{(-9)^2 - 4 \times 2 \times 5}}{2 \times 2}$$ Substitute $a = 2, b = -9$ and $c = 5$ into the equation.

$$x = \frac{9 \pm \sqrt{81 - 40}}{4}$$

$$x = \frac{9 \pm \sqrt{41}}{4}$$

$$x = \frac{9 \pm \sqrt{41}}{4}$$ or $$x = \frac{9 - \sqrt{41}}{4}$$ Find both solutions.

$x = 3.85$ or $x = 0.65$ (to 2 d.p.)

> **TIP**
> When doing a calculation such as $\dfrac{9 - \sqrt{41}}{4}$ on your calculator, always press the $\boxed{=}$ button after working out the numerator and before dividing by 4, otherwise you will get the wrong answer.

c) $3x^2 - 2x - 7 = 0$

$a = 3, b = -2$ and $c = -7$

$$x = \frac{-b \pm \sqrt{b^2 - 4ac}}{2a}$$

$$x = 2 \pm \frac{\sqrt{(-2)^2 - 4 \times 3 \times (-7)}}{2 \times 3}$$ Substitute $a = 3, b = -2$ and $c = -7$ into the equation.

$$x = \frac{2 \pm \sqrt{4 + 84}}{6}$$

$$x = \frac{2 \pm \sqrt{88}}{6}$$

$$x = \frac{2 + \sqrt{88}}{6}$$ or $$x = \frac{2 - \sqrt{88}}{6}$$ Find both solutions.

$x = 1.90$ or $x = -1.23$ (to 2 d.p.)

> **TIP**
> If the square root works out exactly then the equation would have factorised.

Quadratic equations with no real solutions

If you square a number, the result will always be positive, because $+ \times + = +$ and $- \times - = +$. This means that the square root of a negative number is not a **real** number.

Therefore, if you are using the formula to solve a quadratic equation and the number inside the square root ($b^2 - 4ac$) works out to be negative, then there are no real solutions to the equation.

The graph of such a quadratic equation could look like this.

Notice that the curve does not cross the x-axis, so there is no point where $y = 0$.

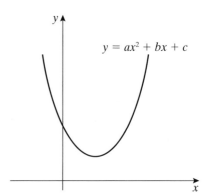

$y = ax^2 + bx + c$

> **TIP**
>
> In Exercise 19.6 there are a few equations with no real solutions. However, if you are asked to solve an equation it will usually have solutions. If the number inside the square root works out to be negative, you may have made a mistake: go back and check your working.

◎ EXERCISE 19.6

Solve each of these quadratic equations using the formula.
Give your answers correct to 2 decimal places.
If there are no real solutions, say so.

1 $x^2 + 8x + 5 = 0$ **2** $x^2 - 5x + 3 = 0$ **3** $x^2 + 7x - 5 = 0$

4 $x^2 - x - 3 = 0$ **5** $2x^2 + 10x + 5 = 0$ **6** $x^2 + 3x + 6 = 0$

7 $2x^2 + 4x - 7 = 0$ **8** $3x^2 + 8x + 1 = 0$ **9** $3x^2 - 6x - 4 = 0$

10 $2x^2 - 9x + 8 = 0$ **11** $6x^2 + 3x - 1 = 0$ **12** $5x^2 + 8x - 5 = 0$

13 $4x^2 - 12x + 7 = 0$ **14** $3x^2 - 5x + 4 = 0$ **15** $2x^2 + x - 20 = 0$

16 $7x^2 + 3x - 1 = 0$ **17** $x^2 - 2x - 100 = 0$ **18** $5x^2 + 11x + 3 = 0$

19 $4x^2 + 7x + 6 = 0$ **20** $3x^2 - 6x - 8 = 0$ **21** $x^2 + 10x + 11 = 0$

Challenge 19.7

Jane's Mum is twice as old as Jane.

Jane works out that in 5 years' time the product of their ages will be approximately 1000.

a) Write down an equation in x.

b) Solve your equation to find Jane's present age.

WHAT YOU HAVE LEARNED

- To solve a quadratic equation by factorising, collect all the terms on one side, factorise the quadratic expression and obtain the solutions by equating each factor to zero
- You can solve a quadratic equation which does not factorise by completing the square: write the equation in the form $(x \pm m)^2 = n$, where m is equal to half the coefficient of x and n is equal to m^2; then square root each side
- You can also solve a quadratic equation which does not factorise by using the formula $x = \dfrac{-b \pm \sqrt{b^2 - 4ac}}{2a}$, where a, b and c are the coefficients in the equation $ax^2 + bx + c = 0$
- If $b^2 - 4ac < 0$, then there are no real solutions

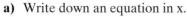

MIXED EXERCISE 19

1 Solve each of these quadratic equations.

    a) $x(x - 6) = 0$ **b)** $(x + 3)(x - 7) = 0$

    c) $(2x + 5)(x - 4) = 0$ **d)** $(3x + 4)(5x + 1) = 0$

2 Solve each of these quadratic equations by factorising.

    a) $x^2 + 9x = 0$ **b)** $x^2 - 49 = 0$ **c)** $x^2 + 5x + 4 = 0$

    d) $x^2 + 7x - 8 = 0$ **e)** $x^2 - 9x + 14 = 0$

3 Solve each of these quadratic equations by factorising.

    a) $3x^2 - 12x = 0$ **b)** $4x^2 - 25 = 0$ **c)** $2x^2 - 5x - 3 = 0$

    d) $3x^2 - 7x + 2 = 0$ **e)** $3x^2 - 5x - 12 = 0$

4 Solve each of these quadratic equations by rearranging and factorising.

    a) $x^2 = 7x$ **b)** $x^2 + 3x = 4$ **c)** $x^2 + 7 = 8x$

    d) $x^2 = 9x - 20$ **e)** $6x^2 = 1 - x$

5 Write each of these expressions in the form $(x + m)^2 + n$.

    a) $x^2 - 2x + 7$ **b)** $x^2 + 8x - 10$

    c) $x^2 - 4x - 9$ **d)** $x^2 - 5x + 1$

6 Solve each of these equations by completing the square.
Give your answers correct to 2 decimal places.

    a) $x^2 - 6x + 4 = 0$ **b)** $x^2 + 10x + 5 = 0$

    c) $x^2 - 4x - 3 = 0$ **d)** $x^2 + 3x - 1 = 0$

7 Solve each of these equations using the quadratic formula.
Give your answers correct to 2 decimal places.

    a) $x^2 - 4x + 2 = 0$ **b)** $x^2 + 9x + 5 = 0$

    c) $x^2 - 7x - 4 = 0$ **d)** $3x^2 + 6x - 5 = 0$

8 Which of these equations have no solutions? Show your calculations.

    a) $x^2 - 4x - 7 = 0$ **b)** $x^2 + 3x + 4 = 0$

    c) $x^2 - 5x + 8 = 0$ **d)** $4x^2 + 3x - 1 = 0$

9 The area of this rectangle is 33 cm².

Write down an equation in x and solve it to find
the dimensions of the rectangle.

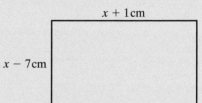

10 I think of a number. I double it and take away 3.
I then multiply by the number I first thought of.
The answer is 77.
Assume the number I started with was x.
Write down an equation in x and solve it to find the number I started with.

11 The surface area of a cuboid with height x cm, length $(x + 3)$ cm and width $(x + 2)$ cm
is 242 cm².

    a) Show that x satisfies the equation $5x^2 + 15x - 236 = 0$.

    b) Use the formula method to solve the equation $5x^2 + 15x - 236 = 0$, giving solutions
to 2 decimal places.

    c) Hence write down the dimensions of the cuboid.

12 **a)** Show that the hypotenuse of the right-angled triangle
shown is $\sqrt{2(x^2 + 4x + 8)}$.

    b) Write an equation in terms of x and solve it to
find the value of x correct to 1 decimal place,
given that the area of the triangle is 45.6 cm².

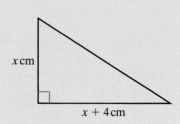

THIS CHAPTER IS ABOUT

- Solving simultaneous equations when the equations are both linear

YOU SHOULD ALREADY KNOW

- How to solve quadratic equations by factorisation

Solving simultaneous equations using the method of elimination

You learned how to solve **simultaneous equations** using the method of elimination in Chapter 15.

EXAMPLE 20.1

Solve each of these pairs of simultaneous equations.

a) $3x - 2y = 14$
$x + 2y = 10$

b) $2x + 3y = 21$
$2x + y = 11$

Solution

a)
$$3x - 2y = 14 \quad (1)$$
$$\underline{x + 2y = 10} \quad (2)$$
$$4x \quad\quad = 24 \quad (1) + (2)$$
$$x \quad\quad = 6$$

Remember to label the equations (1) and (2).
The coefficients of y in the two equations have the same magnitude, 2. Because the signs are different, the two y terms will be eliminated (cancel each other out) if the two equations are added.

Substitute $x = 6$ in (2).
$$6 + 2y = 10$$
$$2y = 4$$
$$y = 2$$

You could substitute $x = 6$ in equation (1) instead, but in this example it is easier to use equation (2).

Solution is $x = 6, y = 2$

Check in (1).
$$3x - 2y = 18 - 4$$
$$= 14 ✓$$

TIP

Always check your solution using the other equation.

b)

$$2x + 3y = 21 \quad (1)$$
$$\underline{2x + y = 11} \quad (2)$$
$$2y = 10 \quad (1) - (2)$$
$$y = 5$$

This time x has the same coefficient, 2, in both equations. Because the signs are the same, the two x terms will be eliminated if the two equations are subtracted.

Substitute $y = 5$ in (1).

$$2x + 15 = 21$$
$$2x = 6$$
$$x = 3$$

Solution is $x = 3, y = 5$

Check in (2).

$$2x + y = 6 + 5$$
$$= 11 \checkmark$$

The coefficients of either x or y must be the same magnitude, positive or negative. If the signs are different, add the equations. If the signs are the same, subtract the equations.

You also learned how to solve simultaneous equations where neither of the unknowns (x or y) has the same coefficient in both equations. In such cases you first need to multiply one of the equations.

EXAMPLE 20.2

Solve each of these pairs of simultaneous equations.

a) $5x + 3y = 4$
$2x - y = -5$

b) $4x + 5y = 30$
$2x - 3y = 4$

Solution

a)

$$5x + 3y = 4 \quad (1)$$
$$2x - y = -5 \quad (2)$$

$$6x - 3y = -15 \quad (2) \times 3 = (3)$$

In this example the coefficients of x and y are different in the two equations.

Multiply equation (2) by 3 to make the coefficient of y the same as in equation (1).
Remember to multiply *each* term in the equation by 3.
The signs are different, so add.

$$5x + 3y = 4 \quad (1)$$
$$\underline{6x - 3y = -15} \quad (3)$$
$$11x = -11 \quad (1) + (3)$$
$$x = -1$$

Substitute $x = -1$ in (1).

$$-5 + 3y = 4$$
$$3y = 9$$
$$y = 3$$

Solution is $x = -1, y = 3$

Check in (2).

$$2x - y = -2 - 3$$
$$= -5 \checkmark$$

b) $4x + 5y = 30$ (1)
     $2x - 3y = 4$ (2)
     $4x - 6y = 8$ (2) × 2 = (3)   Multiply equation (2) by 2 to make the coefficient of x the same as in equation (1).

     $4x + 5y = 30$ (1)   The signs are the same, so subtract.
     $\underline{4x - 6y = \ 8}$ (3)   Take care with the signs. $5y - (-6y) = 11y$.
           $11y = 22$ (1) − (3)
            $y = \ 2$

    Substitute $y = 2$ in (1).
    $4x + 10 = 30$
         $4x = 20$
          $x = 5$

    Solution is $x = 5, y = 2$

    Check in (2).
    $2x - 3y = 10 - 6$
           $= 4 ✓$

In both pairs of equations in Example 20.2, you can make the coefficient of one of the values (x or y) the same by multiplying one of the equations. Sometimes this is not possible, and you have to multiply each of the equations by a different number.

EXAMPLE 20.3

Solve the simultaneous equations $2x + 5y = -4$ and $3x + 4y = 1$.

Solution

$2x + 5y = -4$ (1)
$3x + 4y = 1$ (2)

In this example, you cannot make the coefficient of either the x terms or the y terms the same by multiplying just one equation. This is because 2 is not a factor of 3 (the coefficients of x) and 4 is not a factor of 5 (the coefficients of y).

Instead, look at the coefficients of the x terms. You know that $2 \times 3 = 6$. So if you multiply equation (1) by 3 and multiply equation (2) by 2, the x terms in both equations will be the same, $6x$.

Alternatively, you could make the y terms the same by multiplying equation (1) by 4 and equation (2) by 5, but these numbers are larger: for this pair of equations, the arithmetic will be easier if you make the x terms the same.

$$6x + 15y = -12 \quad (1) \times 3 = (3)$$
$$\underline{6x + 8y = 2} \quad (2) \times 2 = (4)$$
$$7y = -14 \quad (3) - (4)$$
$$y = -2$$

The signs are the same, so subtract.

Substitute $y = -2$ in (1).
$$2x - 10 = -4$$
$$2x = 6$$
$$x = 3$$

Check in (2).
$$3x + 4y = 9 - 8$$
$$ = 1 \checkmark$$

EXERCISE 20.1

Solve each of these pairs of simultaneous equations.

1 $\quad 5x + y = 7$
$ \quad 3x + y = 5$

2 $\quad 4x - 3y = 7$
$ \quad 2x + 3y = 17$

3 $\quad 4x + 2y = 16$
$ \quad x + 2y = 10$

4 $\quad 2x + 3y = 3$
$ \quad 2x - y = 7$

5 $\quad 3x + 2y = 7$
$ \quad 3x - y = -8$

6 $\quad 2x + 3y = 14$
$ \quad 4x - 3y = 1$

7 $\quad -3x + 2y = 0$
$ \quad 3x - 4y = 6$

8 $\quad x + 5y = 9$
$ \quad 2x + 3y = 11$

9 $\quad 5x - 2y = 19$
$ \quad 3x + y = 18$

10 $\quad 4x + y = 8$
$ \quad 7x + 3y = 9$

11 $\quad 2x - 3y = 8$
$ \quad x + 2y = -10$

12 $\quad 2x + 6y = 34$
$ \quad 4x - 2y = 5$

13 $\quad 2x + 3y = 10$
$ \quad 5x - 6y = 16$

14 $\quad 2x + 3y = 0$
$ \quad 8x + 9y = -1$

15 $\quad 7x + 8y = 19$
$ \quad 3x - 2y = -19$

16 $\quad 3x + 4y = 15$
$ \quad x - 6y = -6$

17 $\quad 3x - 4y = 14$
$ \quad 5x - 8y = 30$

18 $\quad 3x + 5y = 21$
$ \quad 4x + 3y = 17$

19 $\quad 3x - 2y = 17$
$ \quad 2x + 7y = 3$

20 $\quad 5x - 2y = 26$
$ \quad 3x - 5y = 27$

21 $\quad 2x + 4y = 5$
$ \quad 5x + 7y = 8$

Mr and Mrs Brown went to the cinema with their three children. Their total entrance fee was £20.

Mr and Mrs Khan also went to the cinema with their grown-up son and his wife and two grandchildren. Their total entrance fee was £28.

Let £x be the entrance fee for an adult and £y be the entrance fee for a child.

a) Write down two equations in x and y.

b) Solve your equations simultaneously to find the entrance fee for adults and for children.

Mrs Jones is buying cans of coke and lemonade for her daughter's birthday party.

She buys x cans of coke and y cans of lemonade. Altogether she buys 32 cans.

Coke costs 50p per can and lemonade 40p per can. Altogether she spends £14.80.

Write down a pair of simultaneous equations and solve them to find how many of each type of can Mrs Jones buys.

Hint: Take care with the units and simplify one of the equations before solving them.

Solving simultaneous equations by substitution

Simultaneous equations can also be solved algebraically by **substitution**.

This involves making x or y the subject of one equation, and substituting this into the other equation.

For some forms of equation this is the easier method, or the only method possible.

EXAMPLE 20.4

Solve the simultaneous equations $2x + 3y = 21$ and $2x + y = 11$.

Solution

$$2x + 3y = 21 \qquad (1)$$
$$2x + y = 11 \qquad (2)$$
$$y = 11 - 2x \qquad (2) \qquad \text{First make } y \text{ the subject of equation (2) by subtracting } 2x \text{ from each side.}$$

Substitute for y in equation (1).

$$2x + 3y = 21 \quad (1)$$
$$2x + 3(11 - 2x) = 21$$

$2x + 33 - 6x = 21$	Expand the brackets.
$33 - 4x = 21$	Collect the x terms together.
$33 = 21 + 4x$	Add $4x$ to each side.
$12 = 4x$	Subtract 21 from each side.
$3 = x$	Divide each side by 4.

Substitute $x = 3$ in the rearranged version of equation (2).
$$y = 11 - 2x$$
$$y = 11 - 6$$
$$y = 5$$

Solution is $x = 3, y = 5$

If you look back at Example 20.1 part **b)** you will see that this is the same solution as we got using the method of elimination.

Substitution is often a longer, more difficult method than elimination, especially as rearranging one of the equations often creates fractions.

If, however, x or y is the subject of one or both of the given equations, it is the easier method.

EXAMPLE 20.5

Solve each of these pairs of the simultaneous equations.

a) $2x + 5y = 4$
$x = 3y - 9$

b) $y = 3x - 17$
$y = 8 - 2x$

Solution

a) $2x + 5y = 4 \quad (1)$
$x = 3y - 9 \quad (2)$

Substitute for x from equation (2) in equation (1).
$$2(3y - 9) + 5y = 4$$

$6y - 18 + 5y = 4$	Expand the brackets.
$11y - 18 = 4$	Collect the y terms together.
$11y = 22$	Add 18 to each side.
$y = 2$	Divide each side by 11.

Substitute $y = 2$ in equation (2).
$$x = 6 - 9$$
$$x = -3$$

Solution is $x = -3, y = 2$

b) $y = 3x - 17$ (1)
    $y = 8 - 2x$ (2)

y is the subject of both equation (1) and equation (2), so put the two equations equal to each other.

$3x - 17 = 8 - 2x$

$5x - 17 = 8$ Add $2x$ to each side.

$\quad\quad 5x = 25$ Add 17 to each side.

$\quad\quad\quad x = 5$ Divide each side by 5.

Substitute $x = 5$ in equation (1).

$y = 15 - 17$

$y = -2$

Solution is $x = 5, y = -2$

EXERCISE 20.2

Solve each of these pairs of simultaneous equations by substitution.

1 $y = 2x - 3$
      $y = 3x - 5$

2 $y = 2x - 7$
      $y = 8 - 3x$

3 $y = x + 7$
      $2x + y = 1$

4 $y = 3x - 9$
      $5x + 2y = 4$

5 $7x - y = 10$
      $y = x + 2$

6 $y = 2x - 10$
      $7x - 2y = 29$

WHAT YOU HAVE LEARNED

- To solve simultaneous equations by elimination, make the coefficients of either x or y take the same value, positive or negative, in both equations by multiplying one equation (or occasionally both) by a number, then add or subtract the equations. If the signs are different, add the equations. If the signs are the same, subtract the equations. Then substitute the value of x or y you have found into either of the equations to find the value of the other unknown

- To solve simultaneous equations by substitution, first make x or y the subject of one of the equations, then substitute from this equation into the other equation. Solve the new equation to find y or x, then substitute this value into the transformed equation to find the value of the other unknown

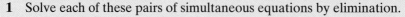

1 Solve each of these pairs of simultaneous equations by elimination.

a) $3x - y = 10$
$2x + y = 5$

b) $2x + 3y = -3$
$4x - y = 8$

c) $3x - 2y = -11$
$x + 5y = 2$

d) $4x - 3y = 7$
$3x + 2y = 18$

2 Solve each of these pairs of simultaneous equations by substitution.

a) $y = 3x + 5$
$x = y - 3$

b) $y = 4x + 1$
$y = x - 11$

3 Jamie and Sasha went to the stationery shop.

Jamie bought three pens and two exercise books. The total cost was £1.55.
Sasha bought four pens and one exercise book. The total cost was £1.40.

Let x pence be the cost of a pen and y pence be the cost of an exercise book.

a) Write down two equations in x and y to represent Jamie and Sasha's purchases.
Hint: Take care with the units.

b) Solve the equations to find the cost of a pen and the cost of an exercise book.

21 → SEQUENCES

THIS CHAPTER IS ABOUT

- **Using rules to find terms of sequences**
- **Seeing patterns in sequences**
- **Explaining how you have found another term in a sequence**
- **Recognising common integer sequences such as square or triangular numbers**
- **Finding the nth term of a linear sequence**

YOU SHOULD ALREADY KNOW

- **How to find terms of simple sequences using term-to-term rules and position-to-term rules**

Using rules to find terms of sequences

You already know how to find the next **term** in a **sequence** from Key Stage 3.

For example, for the sequence 3, 8, 13, 18, 23, 28, … , you find the next term by adding 5.

This is known as the **term-to-term** rule.

You also know how to find a term given its **position** in the sequence using a **position-to-term** rule.

For example, for the sequence 3, 8, 13, 18, 23, 28, … , taking the position number (n), multiplying it by 5 and then subtracting 2 gives the term.

The term-to-term rule and position-to-term rule for any sequence can be expressed as formulae using the following notation.

T_1 represents the first term of a sequence,
T_2 represents the second term of a sequence,
T_3 represents the third term of a sequence,
and so on.

T_n represents the nth term of a sequence.

EXAMPLE 21.1

Marie makes some matchstick patterns. Here are her first three patterns.

To get the next pattern from the previous one, Marie adds three more matchsticks to complete another square.

She makes this table.

Pattern	1	2	3
Number of matchsticks	4	7	10

a) Find the term-to-term rule.

b) The nth term of the sequence is $3n + 1$.
   Check that the first three terms are 4, 7 and 10.
   What is the 4th term?

Solution

a) First find the rule in words.

The first term is 4.

To find the next term, add 3.

You must always state the value of the first term when giving a term-to-term rule.

Then write the rule using the notation.

$T_1 = 4$

$T_{n+1} = T_n + 3$ Add 3 to each term to find the next one.

For example, $T_4 = T_3 + 3$

$$= 10 + 3 = 13$$

b) nth term $= 3n + 1$

First term $= 3(1) + 1 = 4$ ✓

Second term $= 3(2) + 1 = 7$ ✓

Third term $= 3(3) + 1 = 10$ ✓

Fourth term $= 3(4) + 1 = 13$

EXAMPLE 21.2

For a sequence, $T_1 = 10$ and $T_{n+1} = T_n - 4$.
Find the first four terms of the sequence.

Solution

$T_1 = 10$ $T_2 = T_1 - 4$ $T_3 = T_2 - 4$ $T_4 = T_3 - 4$

$= 10 - 4$ $= 6 - 4$ $= 2 - 4$

$= 6$ $= 2$ $= -2$

The first four terms are 10, 6, 2 and -2.

A position-to-term rule is very useful if you need to find a term a long way into the sequence, such as the 100th term. It means that you can find it straight away without having to find the previous 99 terms as you would using a term-to-term rule.

EXAMPLE 21.3

The nth term of a sequence is $5n + 1$.

a) Find the first four terms of the sequence.

b) Find the 100th term of the sequence.

Solution

a) $T_1 = 5 \times 1 + 1$ $T_2 = 5 \times 2 + 1$ $T_3 = 5 \times 3 + 1$ $T_4 = 5 \times 4 + 1$
    $= 6$ $= 11$ $= 16$ $= 21$

The first four terms are 6, 11, 16 and 21.

b) $T_{100} = 5 \times 100 + 1$
      $= 501$

EXERCISE 21.1

1 Look at this sequence of circles. The first four patterns in the sequence have been drawn.

a) Describe the position-to-term rule for this sequence.

b) How many circles are there in the 100th pattern?

2 Look at this sequence of matchstick patterns.

a) Copy and complete the table.

Pattern number	1	2	3	4	5
Number of matchsticks					

b) What patterns can you see in the numbers?

c) Find the number of matchsticks in the 50th pattern.

3 Here is a sequence of star patterns.

a) Draw the next pattern in the sequence.

b) Without drawing the pattern, find the number of stars in the 8th pattern. Explain how you found your answer.

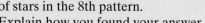

4 The numbers in a sequence are given by the rule 'Multiply the position number by 3, then subtract 5'.

   a) Show that the first term of the sequence is -2.

   b) Find the next four terms in the sequence.

5 Find the first four terms of the sequences with these nth terms.

   a) $6n - 2$ **b)** $4n + 1$ **c)** $6 - 2n$

6 Find the first five terms of the sequences with these nth terms.

   a) n^2 **b)** $n^2 + 2$ **c)** n^3

7 The first term of a sequence is 2.
The general rule for the sequence is 'Multiply a term by 2 to get to the next term'.
Write down the first five terms of the sequence.

8 For a sequence, $T_1 = 5$ and $T_{n+1} = T_n - 3$.
Write down the first four terms of the sequence.

9 Draw suitable patterns to represent this sequence.

         $1, 5, 9, 13, \dots$

10 Draw suitable patterns to represent this sequence.

         $1, 4, 9, 16, \dots$

Challenge 21.1

Work in pairs.

Find as many different sequences as you can where $T_1 = 1$ and $T_2 = 3$.

For each one, write down the first four terms on a separate piece of paper.
On the back of the paper, write the rule you have used.

Swap a sequence with your partner.
Try to find each other's rule.

Finding the nth term of a linear sequence

Look at this linear sequence.

To get from one term to the next, you add 5 each time.

Another way of saying this is that there is a **common difference** between the terms, 5.

A sequence like this, which has a common difference, is called a **linear sequence**.

If you plot the terms of a linear sequence on a graph, you get a straight line.

Term (n)	1	2	3	4	5
Value (y)	4	9	14	19	24

As 5 is added each time,

$$T_2 = T_1 + 5$$
$$= 4 + 5$$
$$T_3 = 4 + 5 \times 2$$
$$T_4 = 4 + 5 \times 3, \text{etc.}$$

So
$$T_n = 4 + 5(n - 1)$$
$$= 5n - 1$$

The nth term of this sequence is $5n - 1$.

Now look at some of the other linear sequences we have met so far in this chapter:

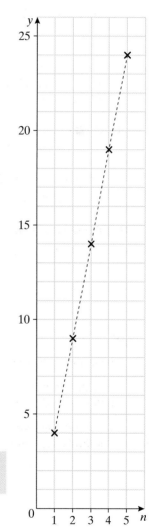

Sequence	Common difference	First term − common difference	nth term
$4, 7, 10, 13, \ldots$	3	$4 - 3 = 1$	$3n + 1$
$6, 11, 16, 21, \ldots$	5	$6 - 5 = 1$	$5n + 1$
$10, 6, 4, -2, \ldots$	-4	$10 - (-4) = 14$	$-4n + 14$
$2, 4, 6, 8, \ldots$	2	$2 - 2 = 0$	$2n$
$4, 10, 16, 22, \ldots$	6	$4 - 6 = -2$	$6n - 2$

Looking at the patterns in the table, you can see some evidence for the formula below.

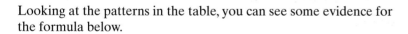

nth term of a linear sequence =
common difference $\times n +$ (first term − common difference)

This can be written as

$$n\text{th term} = An + b$$

where A represents the common difference and b is the first term minus A.

You can also find b by comparing An with any term in the sequence.

EXAMPLE 21.4

Find the nth term of this sequence: $4, 7, 10, 13, \ldots$.

Solution

4   7   10   13   ...
    +3   +3   +3

The common difference (A) is 3, so the formula contains $3n$.

When $n = 1, 3n = 3$. The first term is actually 4, which is 1 more.
So the nth term is $3n + 1$.

You can check your answer using a different term.

When $n = 2, 3n = 6$. The second term is actually 7, which is 1 more.

This confirms that the nth term is $3n + 1$.

You can use sequences and position-to-term rules to solve problems.

EXAMPLE 21.5

Lucy has ten CDs. She decides to buy three more CDs each month.

a) Copy and complete the table to show the number of CDs Lucy has after each of the first four months.

Number of months	1	2	3	4
Number of CDs				

b) Find the formula for the number of CDs she will have after n months.

c) After how many months will Lucy have 58 CDs?

Solution

a)

Number of months	1	2	3	4
Number of CDs	13	16	19	22

b) nth term $= An + b$
    $A = 3$ A is the common difference.
    $b = 10$ b is the first term minus the common difference.
    nth term $= 3n + 10$

c) $3n + 10 = 58$ Solve the equation to find n when the nth term is 58.
        $3n = 48$
         $n = 16$
Lucy will have 58 CDs after 16 months.

Some special sequences

You have already met some special sequences in the examples and in Exercise 21.1.

Discovery 21.1

Look at each of these sequences.

Even numbers $2, 4, 6, 8, \ldots$
Odd numbers $1, 3, 5, 7, \ldots$
Multiples of 4 $4, 8, 12, 16, \ldots$
Powers of 2 $2, 4, 8, 16, \ldots$
Square numbers $1, 4, 9, 16, \ldots$
Triangular numbers $1, 3, 6, 10, \ldots$

Look for different patterns in each of the sequences.

a) Describe the term-to-term rule.   **b)** Describe the position-to-term rule.

For triangular numbers, you may find it helpful to look at this diagram.

```
                                          *       *       *
                *       *                 *   *   *   *
*               *   *                     *   *   *   *
*               *   *                     *   *   *   *
```

EXERCISE 21.2

1 Find the nth term for each of these sequences.
  a) $5, 7, 9, 11, 13, \ldots$ **b)** $2, 5, 8, 11, 14, \ldots$ **c)** $7, 8, 9, 10, 11, \ldots$

2 Find the nth term for each of these sequences.
  a) $17, 14, 11, 8, 5, \ldots$ **b)** $5, 0, -5, -10, -15, \ldots$ **c)** $0, -1, -2, -3, -4, \ldots$

3 Which of these sequences are linear?
  Find the next two terms of each of the sequences that are linear.
  a) $5, 8, 11, 14, \ldots$ **b)** $2, 4, 7, 11, \ldots$ **c)** $6, 12, 18, 24, \ldots$ **d)** $2, 6, 18, 54, \ldots$

4 **a)** Write the first five terms of the sequence with nth term $12 - 6n$.
  b) Write the nth term of this sequence: $8, 2, -4, -10, -16, \ldots$

5 A theatre agency charges £15 per ticket, plus an overall booking charge of £2.
  a) Copy and complete the table.
  b) Write an expression for the cost, in pounds, of n tickets.
  c) Jenna pays £107 for her tickets. How many does she buy?

Number of tickets	1	2	3	4
Cost in £				

6 Write down the first ten triangular numbers.

7 The nth triangular number is $\dfrac{n(n+1)}{2}$. Find the 20th triangular number.

8 The nth term of a sequence is 10^n.
    a) Write down the first five terms of this sequence.
    b) Describe this sequence.

9 **a)** Write down the first five square numbers.
    b) **(i)** Compare the sequence below with the sequence of square numbers.
$$4, 7, 12, 19, 28, \ldots$$
      (ii) Write down the nth term of this sequence.
      (iii) Find the 100th term of this sequence.

10 **a)** Compare the sequence below with the sequence of square numbers.
$$3, 12, 27, 48, 75, \ldots$$
    b) Write down the nth term of this sequence.
    c) Find the 20th term of this sequence.

Challenge 21.2

Work in pairs.

You are going to use a spreadsheet to explore sequences. Don't let your partner see you input your formula. Make sure that they can't see the formula on the computer screen: click on View in the toolbar and make sure the Formula Bar is not checked.

1 Open a new spreadsheet.

2 Enter the number 1 in cell A1.
Select cell A1, and hold down the mouse button and drag down the column. Then click on Edit in the toolbar and select Fill, then Series to call up the Series dialogue box. Make sure that the Columns and Linear boxes are checked and that the Step value is 1. Click OK.

3 Enter a formula in cell B1. For example **=A1*3+5**. Press the enter key.
Click on cell B1, click on Edit in the toolbar and select Copy.
Click on cell B2, and hold down the mouse key and drag down the column. Then click on Edit in the toolbar and select Paste.

4 Ask your partner to try to work out the formula and generate the same sequence in column C.

If you have time, you could explore some non-linear sequences as well. For example, enter the formula **=A1^2+A1**.

- Sequences may be described by a list of numbers, diagrams in a pattern, a term-to-term rule (for example, $T_{n+1} = T_n + 3$ when $T_1 = 4$) or a position-to-term rule (for example, nth term $= 3n + 1$ or $T_n = 3n + 1$)
- The nth term of a linear sequence $= An + b$, where A is the common difference and b is the first term minus A
- Here are some important sequences:

Name	Sequence	nth term	Term-to-term rule
Even numbers	$2, 4, 6, 8, \ldots$	$2n$	Add 2
Odd numbers	$1, 3, 5, 7, \ldots$	$2n - 1$	Add 2
Multiples e.g. multiples of 6	$6, 12, 18, 24, \ldots$	$6n$	Add 6
Powers of 2	$2, 4, 8, 16, \ldots$	2^n	Multiply by 2
Square numbers	$1, 4, 9, 16, \ldots$	n^2	Add 3 then 5 then 7, etc. (the odd numbers)
Triangular numbers	$1, 3, 6, 10, \ldots$	$\dfrac{n(n + 1)}{2}$	Add 2 then 3 then 4, etc. (consecutive integers)

MIXED EXERCISE 21

1 Look at this sequence of circles.

The first four patterns in the sequence have been drawn.

a) How many circles are there in the 100th pattern?

b) Describe a rule for this sequence.

2 Here is a sequence of star patterns.

a) Draw the next pattern in the sequence.

b) Without drawing the pattern, find the number of stars in the 8th pattern. Explain how you found your answer.

3 The numbers in a sequence are given by the rule 'Multiply the position number by 6, then subtract 2'.

    a) Show that the first term of the sequence is 4.

    b) Find the next four terms in the sequence.

4 Find the first four terms of the sequences with these nth terms.

    a) $5n + 2$ **b)** $n^2 + 1$ **c)** $90 - 2n$

5 The first term of a sequence is 4.
The general rule for the sequence is 'Multiply a term by 2 to get to the next term'.
Write down the first five terms of the sequence.

6 Find the nth term for each of these sequences.

    a) $5, 8, 11, 14, 17, \ldots$ **b)** $1, 7, 13, 19, 25, \ldots$ **c)** $2, -3, -8, -13, -18, \ldots$

7 Which of these sequences are linear?
Find the next two terms of each of the sequences that are linear.

    a) $4, 9, 14, 19, \ldots$ **b)** $3, 6, 10, 15, \ldots$

    c) $5, 10, 20, 40, \ldots$ **d)** $12, 6, 0, -6, \ldots$

8 The nth term of a sequence is 3^n.

    a) Write down the first five terms of this sequence.

    b) Describe this sequence.

9 Draw suitable diagrams to show the first five triangular numbers.
Write the triangular numbers under your diagrams.

10 **a)** Write down the first five terms of the sequence with nth term n^2.

    b) Hence find the nth term of the sequence below.
           $5, 8, 13, 20, 29, \ldots$

22 → FORMULAE 1

THIS CHAPTER IS ABOUT

- **Using simple formulae**
- **Writing down and creating formulae**
- **Rearranging formulae**
- **Solving equations using trial and improvement**

YOU SHOULD ALREADY KNOW

- **How to substitute numbers into simple formulae**
- **How to simplify and solve linear equations**
- **How to simplify a formula by, for example, collecting together 'like' terms**

Using formulae

You should already know how to **substitute** numbers into formulae. Formulae are also used in other chapters: for example, in Chapter 31 the area of circles is found using the formula $A = \pi r^2$.

Check up 22.1

The area of a circle is given by the formula $A = \pi r^2$.
Find A when $r = 10$ cm. Use $\pi = 3.14$.

You also know how to write a formula in letters for a given situation.
You solve these formulae in the same way, by substitution.

EXAMPLE 22.1

To work out the cost of hiring a car for a certain number of days, multiply the number of days by the daily rate and add the fixed charge.

a) Write a formula using letters to work out the cost of hiring a car.

b) If the fixed charge is £20 and the daily rate is £55, find the cost of hiring a car for five days.

Solution

a) $c = nd + f$ This uses c to represent the cost, n to represent the number of days, d to represent the daily rate and f to represent the fixed rate.

b) $c = 5 \times 55 + 20$ Substitute the numbers into the formula.

$c = 275 + 20$

$c = 295$

Cost = £295

⊙ EXERCISE 22.1

1 To find the time needed, in minutes, to cook a piece of beef, multiply the weight of the beef in kilograms by 40 and add 10.
How many minutes are needed to cook a piece of beef weighing

    a) 2 kilograms?   **b)** 5 kilograms?

2 The further you go up a mountain, the colder it gets.
There is a simple formula which tells you roughly how much the temperature will drop.

      Temperature drop (°C) = height climbed in metres ÷ 200.

If you climb 800 m, by about how much will the temperature drop?

3 The total coach fare, £P, for a group going to the airport is given by the formula

$$P = 8A + 5C$$

where A is the number of adults and C is the number of children.
Calculate the cost for two adults and three children.

4 The average speed (s) of a journey is the distance (d) divided by the time (t).

    a) Write the formula for this.

    b) A car journey of 150 km took 2 hours 30 minutes.
        What was the average speed of the journey?

5 The distance, d, in metres, that a stone falls in t seconds when dropped is given by the formula

$$d = \frac{9.8t^2}{2}.$$

Find d when $t = 10$ seconds.

6 This formula tells you the number of heaters needed to heat an office.

$$\text{Number of heaters} = \frac{\text{length of office} \times \text{width of office}}{10}$$

An office measures 15 m by 12 m.
How many heaters are needed?

7 A bread shop calculates the number of sandwiches, S, needed for a party using the formula $S = 3P + 10$, where P is the number of people expected.

   a) How many sandwiches are needed when 15 people are expected?

   b) How many people are expected when 70 sandwiches are provided?

8 The diagram shows a rectangle.

   a) What is the perimeter of the rectangle in terms of x?

   b) What is the area of the rectangle in terms of x?

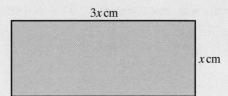

3x cm

x cm

9 The time, T minutes, needed to cook a leg of lamb is given by the formula

$$T = 50W + 30$$

where W is the weight of the leg in kilograms.

   a) How long, in hours and minutes, does it take to cook a leg of lamb weighing 2 kilograms?

   b) What is the weight of a leg of lamb that takes 105 minutes to cook?

10 The formula linking volume, area and length for a prism is $L = \dfrac{V}{A}$.

   If $V = 200$ and $A = 40$, find L.

11 Anne walks for 5 hours at an average speed of 3 mph.
   Use the formula $d = st$ to work out the distance she walked.
   d stands for distance in miles.
   s stands for average speed in mph.
   t stands for time in hours.

12 Connor is making an ornamental fence.
   He joins 3 posts with 6 chains as shown in the diagram.

   a) Connor fixes 5 posts into the ground. How many chains will he need?

   b) Copy and complete this table.

Number of posts, P	1	2	3	4	5	6
Number of chains, C	0	3	6			

   c) Write down the formula which gives the number of chains for any number of posts. Let C = total number of chains, and P = the number of posts.

   d) How many chains are needed for a fence with 30 posts?

13 The area of a parallelogram is equal to the base multiplied by the vertical height.
   What is the area of a parallelogram with these dimensions?

   a) Base = 6 cm and vertical height = 4 cm

   b) Base = 4.5 cm and vertical height = 5 cm

14 The volume of a cuboid is the length multiplied by the width multiplied by the height. What is the volume of a cuboid with these dimensions?

a) Length = 5 cm, width = 4 cm and height = 6 cm

b) Length = 4.5 cm, width = 8 cm and height = 6 cm

15 The cost of a long taxi journey is a fixed charge of £20 plus £1 per mile travelled.

a) What is the cost of a journey of 25 miles?

b) The cost of a journey was £63. How far was the journey?

Challenge 22.1

Write down the 'formula' you get by following each of these sets of instructions.

a) • Choose any number
   • Multiply it by two
   • Add five
   • Multiply by five
   • Subtract twenty-five

b) • Choose any number
   • Double it
   • Add nine
   • Add the original number
   • Divide by three
   • Subtract three

What answer do you get for each set if your starting number is ten?

Challenge 22.2

The Golden Ratio is achieved when a rectangle's length is (approximately) equal to 1.6 times its width.

A golden rectangle has an area of 230 cm².
Find its dimensions. Give your answers correct to the nearest millimetre.

Rearranging formulae

Sometimes you need to find the value of a letter which is not on the left-hand side of the formula. To find the value of the letter, you first need to **rearrange** the formula.

For example, the formula $d = st$ links distance (d), speed (s) and time (t). If you know the distance covered during a journey and the time it took, and you want to find the average speed, you need to get the s by itself.

The method used to rearrange a formula is similar to the method used to solve equations. This was covered in Chapter 12.

In this case, to get the s on its own, you need to divide by t. As with equations, you must do each operation to the whole of both sides of the formula.

$d = st$

$\dfrac{d}{t} = \dfrac{st}{t}$ Divide each side by t.

$\dfrac{d}{t} = s$

$s = \dfrac{d}{t}$ A formula is usually written with the single term (in this case, s) on the left-hand side.

s is now the **subject** of the formula.
The formula gives s in terms of d and t.

EXAMPLE 22.2

$y = mx + c$
Make x the subject.

Solution

$y = mx + c$

$y - c = mx + c - c$ Subtract c from each side.

$y - c = mx$

$\dfrac{y - c}{m} = \dfrac{mx}{m}$ Divide each side by m.

$\dfrac{y - c}{m} = x$

$x = \dfrac{y - c}{m}$ Turn the formula around so that x is on the left-hand side.

EXAMPLE 22.3

The formula for the volume, v, of a square-based pyramid of side a and vertical height h, is $v = \frac{1}{3}a^2h$.
Rearrange the formula to make h the subject.

Solution

$v = \frac{1}{3}a^2h$ Get rid of the fraction first.

$3v = a^2h$ Multiply each side by 3.

$\dfrac{3v}{a^2} = h$ Divide each side by a^2.

$h = \dfrac{3v}{a^2}$ Turn the formula around so that h is on the left-hand side.

EXAMPLE 22.4

Rearrange the formula $A = \pi r^2$ to make r the subject.

Solution

$A = \pi r^2$

$\dfrac{A}{\pi} = r^2$ Divide each side by π.

$r^2 = \dfrac{A}{\pi}$ Turn the formula around so that r^2 is on the left-hand side.

$r = \sqrt{\dfrac{A}{\pi}}$ Take the square root of each side.

EXERCISE 22.2

1 Rearrange each of these formulae to make the letter in brackets the subject.

     a) $a = b - c$ (b)   b) $4a = wx + y$ (x)   c) $v = u + at$ (t)

     d) $c = p - 3t$ (t)   e) $A = p(q + r)$ (q)   f) $p = 2g - 2f$ (g)

     g) $F = \dfrac{m + 4n}{t}$ (n)

2 Make u the subject of the formula $s = \dfrac{3uv}{bn}$.

3 Rearrange the formula $a = \dfrac{bh}{2}$ to give h in terms of a and b.

4 The formula for calculating simple interest is $I = \dfrac{PRT}{100}$.

    Make R the subject of this formula.

5 The volume of a cone is given by the formula $V = \dfrac{\pi r^2 h}{3}$, where V is the volume in cm^3,

    r is the radius of the base in cm and h is the height in cm.

     a) Rearrange the formula to make h the subject.

     b) Calculate the height of a cone with radius 5 cm and volume 435 cm^3.
        Use $\pi = 3.14$ and give your answer correct to 1 decimal place.

6 To change from degrees Celsius (°C) to degrees Fahrenheit (°F), you can use the formula

        $F = \frac{9}{5}(C + 40) - 40$.

     a) The temperature is 60°C. What is this in °F?

     b) Rearrange the formula to find C in terms of F.

7 Rearrange the formula $V = \dfrac{\pi r^2 h}{3}$ to make r the subject.

8 **a)** Make a the subject of the formula $v^2 = u^2 + 2as$.

　　b) Make u the subject of the formula $v^2 = u^2 + 2as$.

Solving equations by trial and improvement

Sometimes you will need to solve an equation by **trial and improvement**. This means that you substitute different values into the equation until you find a solution.

It is important that you work systematically and do not just choose the numbers you try at random.

First you need to find two numbers between which the solution lies. Next you try the number halfway between these two numbers. You continue this process until you find the answer to the required degree of accuracy.

EXAMPLE 22.5

Find a solution of the equation $x^3 - x = 40$.
Give your answer correct to 1 decimal place.

Solution

$x^3 - x = 40$

Try $x = 3$	$3^3 - 3 = 24$	Too small. Try a larger number.
Try $x = 4$	$4^3 - 4 = 60$	Too large. The solution must lie between 3 and 4.
Try $x = 3.5$	$3.5^3 - 3.5 = 39.375$	Too small. Try a larger number.
Try $x = 3.6$	$3.6^3 - 3.6 = 43.056$	Too large. The solution must lie between 3.5 and 3.6.
Try $x = 3.55$	$3.55^3 - 3.55 = 41.118\ldots$	Too large. The solution must lie between 3.5 and 3.55.

So the answer is $x = 3.5$, correct to 1 decimal place.

EXAMPLE 22.6

a) Show that $x^3 - 3x = 6$ has a solution between $x = 2$ and $x = 3$.

b) Find the solution correct to 1 decimal place.

Solution

a) $x^3 - 3x = 6$

    Try $x = 2$ $2^3 - 3 \times 2 = 2$ Too small.
    Try $x = 3$ $3^3 - 3 \times 3 = 18$ Too large.

    6 is between 2 and 18. Therefore there is a solution of $x^3 - 3x = 6$ between $x = 2$ and $x = 3$.

b) Try $x = 2.5$ $2.5^3 - 3 \times 2.5 = 8.125$ Too large. Try a smaller number.
    Try $x = 2.3$ $2.3^3 - 3 \times 2.3 = 5.267$ Too small. Try a larger number.
    Try $x = 2.4$ $2.4^3 - 3 \times 2.4 = 6.624$ Too large. Try a smaller number.
    Try $x = 2.35$ $2.35^3 - 3 \times 2.35 = 5.927 \ldots$ Very close, but still too small.

    x must be greater than 2.35. Therefore the solution, correct to 1 decimal place, is $x = 2.4$.

EXAMPLE 22.7

a) Show that the equation $x^3 - x = 18$ has a root between $x = 2.7$ and $x = 2.8$.

b) Find the solution correct to 2 decimal places.

Solution

a) $x^3 - x = 18$

    Try $x = 2.7$ $2.7^3 - 2.7 = 16.983$ Too small.
    Try $x = 2.8$ $2.8^3 - 2.8 = 19.152$ Too large.

    18 is between 16.983 and 19.152. Therefore there is a solution of $x^3 - x = 18$ between $x = 2.7$ and $x = 2.8$.

b) Try halfway between $x = 2.7$ and $x = 2.8$, that is, try $x = 2.75$.

    Try $x = 2.75$ $2.75^3 - 2.75 = 18.04688$ Too large. Try a smaller number.
    Try $x = 2.74$ $2.74^3 - 2.74 = 17.83082$ Too small. Try a larger number.

    18 is between 17.83082 and 18.04688. Therefore there is a solution of $x^3 - x = 18$ between $x = 2.74$ and $x = 2.75$.

    Try halfway between $x = 2.74$ and $x = 2.75$, that is, try $x = 2.745$.
    $2.745^3 - 2.745 = 17.93864$ Too small.

    x must be greater than 2.745. Therefore the solution, correct to 2 decimal places, is $x = 2.75$.

◉ EXERCISE 22.3

1 Find a solution, between $x = 1$ and $x = 2$, of the equation $x^3 = 5$.
    Give your answer correct to 1 decimal place.

2 **a)** Show that a solution of the equation $x^3 - 5x = 8$ lies between $x = 2$ and $x = 3$.
    b) Find the solution correct to 1 decimal place.

3 **a)** Show that a solution of the equation $x^3 - x = 90$ lies between $x = 4$ and $x = 5$.

b) Find the solution correct to 1 decimal place.

4 **a)** Show that the equation $x^3 - x = 50$ has a root between $x = 3.7$ and $x = 3.8$.

b) Find this root correct to 2 decimal places.

5 Find a solution of the equation $x^3 + x = 15$.
Give your answer correct to 1 decimal place.

6 Find a solution of the equation $x^3 + x^2 = 100$.
Give your answer correct to 2 decimal places.

7 Which whole number, when cubed, gives the value closest to 10 000?

8 Use trial and improvement to find which number, when squared, gives 1000.
Give your answer correct to 1 decimal place.

9 A number, added to the square of this number, gives 10.

a) Write this as a formula.

b) Find the number correct to 1 decimal place.

10 The product of two whole numbers, the difference of which is 4, is 621.

a) Write this as formula in terms of x.

b) Use trial and improvement to find the two numbers.

11 Use trial and improvement to find which number, when squared, gives 61.
Give your answer correct to 1 decimal place.

WHAT YOU HAVE LEARNED

- **How to substitute into formulae**
- **To rearrange a formula, do each operation to the whole of both sides of the formula until you get the required term on its own, on the left-hand side of the formula**
- **To find the solution of an equation by trial and improvement you first need to find two numbers between which the solution lies. You then try the number halfway between these two numbers and continue the process until you find the solution to the required degree of accuracy**

1 To convert temperatures on the Celsius (°C) scale to the Fahrenheit (°F) scale you can use the formula $F = 1.8C + 32$.

Calculate the Fahrenheit temperature when the temperature is

    a) 40°C.   **b)** 0°C.   **c)** −5°C.

2 The cost of a child's ticket on a bus is half that of an adult's ticket, plus 25p.
Find the cost of a child's ticket when the adult fare is £1.40.

3 The area of a rhombus is found by multiplying the lengths of the diagonals together and then dividing by 2.
Find the area of a rhombus with diagonals of length

    a) 4 cm and 6 cm.   **b)** 5.4 cm and 8 cm.

4 The Trenton bus company estimates the time, in minutes, for its local bus journeys by using the formula $T = 1.2m + 2s$, where m is the number of miles in a journey and s is the number of stops. Find T when

    a) $m = 5$ and $s = 14$.   **b)** $m = 6.5$ and $s = 20$.

5 Rearrange each of these formulae to make the letter in brackets the subject.

    a) $p = q + 2r$ (q) **b)** $x = s + 5r$ (r)

    c) $m = \dfrac{pqr}{s}$ (r) **d)** $A = t(x - 2y)$ (y)

6 The cooking time, T minutes, for w kg of meat is given by the formula
$$T = 45w + 40.$$

    a) Make w the subject of the formula.

    b) What is the value of w when the cooking time is 2 hours 28 minutes?

7 The area of a triangle is given by the formula $A = b \times h \div 2$, where b is the base and h is the height.

    a) Find the length of the base when $A = 12$ cm² and $h = 6$ cm.

    b) Find the height when $A = 22$ cm² and $b = 5.5$ cm.

8 The cost in £ of an advert in a local paper is given by the formula $C = 12 + \dfrac{w}{5}$ where w is the number of words in the advert.
How many words can you have if you are willing to pay

    a) £18?   **b)** £24?

9   **a)** Show that a solution of the equation $x^3 + 4x = 12$ lies between $x = 1$ and $x = 2$.

    b) Find the solution correct to 1 decimal place.

10   **a)** Show that a solution of the equation $x^3 - x^2 = 28$ lies between $x = 3$ and $x = 4$.

    b) Find the solution correct to 1 decimal place.

11 A number, added to the cube of this number, gives 100.

    a) Write this as a formula.   **b)** Find the number correct to 1 decimal place.

23 → FORMULAE 2

THIS CHAPTER IS ABOUT

- Rearranging formulae where the new subject occurs more than once
- Rearranging formulae where the new subject is raised to a power

YOU SHOULD ALREADY KNOW

- How to factorise simple expressions
- How to expand brackets and manipulate simple algebraic expressions
- How to rearrange simple formulae

Formulae where the new subject occurs more than once

In all the formulae you have rearranged so far, the new subject has occurred only once. For example, you already know how to make t the subject of $v = u + at$.

You also need to be able to rearrange formulae where the new subject occurs more than once. In such cases, all terms containing the new subject must be collected together.

EXAMPLE 23.1

Rearrange the formula $mxy = y + 4x$ to make x the subject.

Solution

$$mxy = y + 4x$$

$$mxy - 4x = y \qquad \text{Subtract } 4x \text{ from each side so that all the } x \text{ terms are together.}$$

$$x(my - 4) = y \qquad \text{Factorise the left-hand side, taking out the common factor, } x.$$

$$x = \frac{y}{my - 4} \qquad \text{Divide each side by } (my - 4).$$

Sometimes, it is not immediately obvious that the new subject occurs more than once.

EXAMPLE 23.2

Rearrange the formula $m = \dfrac{1}{x} + \dfrac{4}{y}$ to make x the subject.

Solution

$$m = \frac{1}{x} + \frac{4}{y}$$

$$mxy = y + 4x \qquad \text{Multiply each side by } xy \text{ to get rid of the fractions.}$$
$$mxy - 4x = y \qquad \text{The formula is now in the form it was in Example 23.1.}$$
$$x(my - 4) = y$$
$$x = \frac{y}{my - 4}$$

Formulae like the one in Example 23.2 can also be rearranged using a different approach.

EXAMPLE 23.3

Rearrange the formula $m = \dfrac{1}{x} + \dfrac{4}{y}$ to make x the subject.

Solution

$$m = \frac{1}{x} + \frac{4}{y}$$

$$m - \frac{4}{y} = \frac{1}{x} \qquad \text{Subtract } \frac{4}{y} \text{ from each side.}$$

$$\frac{1}{x} = m - \frac{4}{y} \qquad \text{Swap the sides so that the } x \text{ term is on the left.}$$

$$\frac{1}{x} = \frac{my - 4}{y} \qquad \text{Arrange the right-hand side over a common denominator.}$$

$$x = \frac{y}{my - 4} \qquad \text{Invert each side.}$$
$$\text{Remember that you can only do this if each side is a single fraction.}$$

Rearrange each of these formulae to make the letter in brackets the subject.

1 $pq - rs = rt$ $\qquad$ (r) $\qquad\qquad$ **2** $A = P + \dfrac{PRT}{100}$ $\qquad$ (P)

3 $3(x - 5) = y(4 - 3x)$ $\quad$ (x) $\qquad\qquad$ **4** $pq + r = rq - p$ $\qquad$ (p)

5 $pq + r = rq - p$ $\qquad$ (r) $\qquad\qquad$ **6** $y = x + \dfrac{px}{q}$ $\qquad$ (x)

7 $s = ut + \dfrac{at}{2}$ $\qquad$ (t) $\qquad\qquad$ **8** $\dfrac{a}{2x + 1} = \dfrac{b}{3x - 1}$ $\qquad$ (x)

9 $s = \dfrac{uv}{u + v}$ $\qquad$ (u) $\qquad\qquad$ **10** $\dfrac{1}{f} = \dfrac{1}{u} + \dfrac{1}{v}$ $\qquad$ (v)

11 $3 = \dfrac{4f + 5g}{2f + e}$ $\qquad$ (f) $\qquad\qquad$ **12** $\dfrac{2b + c}{3b - c} = 5a$ $\qquad$ (b)

Formulae where the new subject is raised to a power

You also need to be able to rearrange formulae where the new subject is raised to a power.

In these cases you will need, at the appropriate stage, to use the inverse operation. For example, for $\sqrt{x}$, you need to square; for x^2, you need to take the square root.

EXAMPLE 23.4

Rearrange the formula $v = y + \sqrt{\dfrac{p}{x}}$ to make x the subject.

Solution

$v = y + \sqrt{\dfrac{p}{x}}$ $\qquad$ When a formula involves a power or root, rearrange the formula to get that term by itself.

$v - y = \sqrt{\dfrac{p}{x}}$ $\qquad$ In this case, subtract y from each side so that $\sqrt{\dfrac{p}{x}}$ is by itself.

$(v - y)^2 = \dfrac{p}{x}$ $\qquad$ Square each side. In this case you do not need to expand $(v - y)^2$.

$x(v - y)^2 = p$ $\qquad$ Multiply each side by x.

$x = \dfrac{p}{(v - y)^2}$ $\qquad$ Divide each side by $(v - y)^2$.

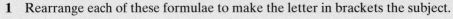

EXERCISE 23.2

1 Rearrange each of these formulae to make the letter in brackets the subject.

a) $y = 3x^2 - 4$ $\qquad$ (x) $\qquad$ **b)** $t = 2\pi \sqrt{\dfrac{l}{g}}$ $\qquad$ (l)

c) $A = \pi r \sqrt{h^2 + r^2}$ $\qquad$ (h) $\qquad$ **d)** $v^2 - u^2 = 2as$ $\qquad$ (u)

e) $V = \frac{1}{3}\pi r^2 h$ $\qquad$ (r) $\qquad$ **f)** $s = 15 - \frac{1}{2}at^2$ $\qquad$ (t)

2 The formula for the volume of a cylinder is $V = \pi r^2 h$, where r is the radius of the cylinder and h is its height.

a) Find the volume of a cylinder of radius 12 cm and height 20 cm.
Give your answer to 2 significant figures.

b) Rearrange the formula to make r the subject.

c) What is the radius of a cylinder of volume 500 cm³ and height 5 cm?
Give your answer to 3 significant figures.

3 The formula for finding the length, d, of the diagonal of a cuboid with dimensions x, y and z is $d = \sqrt{x^2 + y^2 + z^2}$.

a) Find d when $x = 2$, $y = 3$ and $z = 4$.

b) How long is the diagonal of a cuboid block of concrete with dimensions 2 m, 3 m and 75 cm?

c) Rearrange the formula to make x the subject.

d) Find x when $d = 0.86$ m, $y = 0.25$ m, and $z = 0.41$ m.

Challenge 23.1

a) The formula for the surface area of a closed cyliner is $A = 2\pi r(r + h)$, where r is the radius of the cylinder and h is its height.
What happens if you try to make r the subject?

b) The formula $S = \frac{1}{2}n(n + 1)$ gives the sum, S, of the first n positive integers.
Find n when $S = 325$.

- When rearranging a formula where the new subject occurs more than once, collect together all terms containing the new subject
- When a formula involves a power or root, rearrange the formula to get that term by itself
- When the new subject is raised to a power, use the inverse operation. For example, for $\sqrt{x}$, square; for x^2, take the square root

MIXED EXERCISE 23

1 Rearrange each of these formulae to make the letter in brackets the subject.

a) $r = \dfrac{a}{a + b}$ (a)

b) $t = 2\pi\sqrt{\dfrac{l}{g}}$ (g)

c) $\dfrac{1}{f} = \dfrac{1}{u} + \dfrac{1}{v}$ (u)

d) $x + a = \dfrac{x + b}{c}$ (x)

e) $a - b = \dfrac{a + 2}{b}$ (a)

f) $\dfrac{y + x}{y - x} = 3$ (y)

g) $ab + bc + a = 0$ (a)

h) $2x = \sqrt{x^2 + y}$ (x)

i) $\sqrt{\dfrac{x + 1}{x}} = y$ (x)

j) $p = 2\left[\dfrac{n - (r + 1)}{n - 1}\right]$ (n)

k) $y = \dfrac{x - np}{\sqrt{nqp}}$ (q)

l) $F = \dfrac{x^2}{1 - x^2}$ (x)

m) $s = \sqrt{\dfrac{x^2 + y^2}{n}}$ (y)

n) $m = \dfrac{ax + by}{a + b}$ (b)

2 The formula for the volume of a sphere is $V = \frac{4}{3}\pi r^3$.
   a) Find V if $r = 5$.
   b) Make r the subject of the formula.
   c) Find the radius of a sphere of volume 3500 m³.

24 → FUNCTIONS

Trigonometric graphs are covered in Chapter 33.

Function notation

$y = f(x)$ means that y is a **function** of x. You read $f(x)$ as 'f of x'. For any value of x, a function has only one value.

If $y = x^2 + 2$ then $f(x) = x^2 + 2$.
The value of the function when $x = 3$ is written $f(3)$.
So $f(3) = 3^2 + 2 = 9 + 2 = 11$.

EXAMPLE 24.1

$f(x) = 3x^2 + 2$. Find the value of each of these.

a) $f(4)$ **b)** $f(-1)$

Solution

a) $f(4) = 3 \times 4^2 + 2$
$= 3 \times 16 + 2$
$= 48 + 2$
$= 50$

b) $f(-1) = 3 \times (-1)^2 + 2$
$= 3 \times 1 + 2$
$= 3 + 2$
$= 5$

EXAMPLE 24.2

$g(x) = 5x + 6$

a) Solve $g(x) = 8$

b) Write an expression for each of these.

    (i) $g(3x)$ **(ii)** $3g(x)$

Solution

a) $5x + 6 = 8$

        $5x = 2$

        $x = \frac{2}{5}$

b) (i) To find $g(3x)$, you replace x with $3x$.

        $g(3x) = 5 \times (3x) + 6$

            $= 15x + 6$

   (ii) To find $3g(x)$, you multiply $g(x)$ by 3.

        $3g(x) = 3 \times (5x + 6)$

            $= 15x + 18$

EXAMPLE 24.3

$h(x) = x^2 - 6$. Write an expression for each of these.

a) $2h(x)$ **b)** $h(2x)$ **c)** $h(x) + 3$ **d)** $h(x + 3)$

Solution

a) $2h(x) = 2 \times (x^2 - 6)$ **b)** $h(2x) = (2x)^2 - 6$

         $= 2x^2 - 12$ $= 4x^2 - 6$

c) $h(x) + 3 = x^2 - 6 + 3$ **d)** $h(x + 3) = (x + 3)^2 - 6$

             $= x^2 - 3$ $= x^2 + 6x + 9 - 6$

                                          $= x^2 + 6x + 3$

EXERCISE 24.1

1 $f(x) = x^2 - 5$. Find the value of each of these.

    a) $f(3)$ **b)** $f(-2)$

2 $g(x) = 3x^2 - 2x + 1$. Find the value of each of these.

    a) $g(3)$ **b)** $g(-2)$ **c)** $g(0)$

3 $h(x) = 2x - 5$

   a) Solve $h(x) = 7$.

   b) Write an expression for each of these.

      (i) $h(x - 2)$ **(ii)** $h(2x)$

4 $f(x) = 7 - 3x$

   a) Solve $f(x) = 1$.

   b) Write an expression for each of these.

      (i) $3f(x)$ **(ii)** $f(x + 3)$

5 $g(x) = 4x - 3$

   a) Solve $g(x) = 0$.

   b) Write an expression for each of these.

      (i) $g(x + 5)$ **(ii)** $g(x) + 5$

6 $h(x) = x^2 - 2$

   a) Solve $h(x) = 7$.

   b) Write an expression for each of these.

      (i) $h(x + 1)$ **(ii)** $h(x) + 1$

7 $f(x) = 3x^2 - 2x$

   a) Find the value of $f(-4)$.

   b) Write an expression for each of these.

      (i) $f(x - 1)$ **(ii)** $f(2x)$

8 $g(x) = x^2 - 3x$

   a) Solve $g(x) = 4$.

   b) Write an expression for each of these.

      (i) $g(2x)$ **(ii)** $g(x + 1)$

Challenge 24.1

Function of a function

$fg(x)$ means first find $g(x) = y$ and then find $f(y)$.
You always start with the function nearest the brackets.
For example, $f(x) = 2x - 5$ and $g(x) = x^2 - 1$.
To find $fg(3)$ you find $g(3) = 8$, then $f(8) = 11$.
To find $fg(x)$ you find $f(x^2 - 1) = 2(x^2 - 1) - 5 = 2x^2 - 7$.
To find $gf(x)$ you find $g(2x - 5) = (2x - 5)^2 - 1 = 4x^2 - 20x + 24$.
Now try these.
Given $f(x) = 3x + 2$, $g(x) = x^2$ and $h(x) = x^2 - 2x$, work out these.

a) $fg(2)$ **b)** $gf(2)$ **c)** $gh(3)$ **d)** $gh(-2)$ **e)** $fgh(-1)$
f) $fg(x)$ **g)** $gf(x)$ **h)** $fh(x)$ **i)** $hg(x)$ **j)** $fgh(x)$

Transforming graphs

Translations parallel to the y-axis

The diagram shows the graphs of $y = x^2$, $y = x^2 + 4$ and $y = x^2 - 5$.

The graph of $y = x^2$ passes through the origin.

The graph of $y = x^2 + 4$ is the same shape and passes through the point $(0, 4)$.

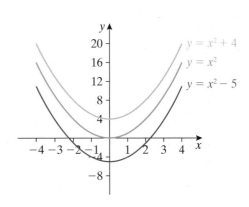

The graph of $y = x^2 - 5$ is also the same shape but passes through the point $(0, -5)$.

You can see that $y = x^2 + 4$ is the graph of $y = x^2$ translated by $\begin{pmatrix} 0 \\ 4 \end{pmatrix}$ and that $y = x^2 - 5$ is the graph of $y = x^2$ translated by $\begin{pmatrix} 0 \\ -5 \end{pmatrix}$.

This applies to all families of graphs. You can generalise the result.

> The graph of $y = f(x) + a$ is the graph of $y = f(x)$ translated by $\begin{pmatrix} 0 \\ a \end{pmatrix}$.

TIP

The graphs in this section are drawn accurately but you can sketch graphs unless you are told to do otherwise.

Translations parallel to the x-axis

The diagram shows the graphs of $y = x^2$, $y = (x + 3)^2$ and $y = (x - 2)^2$.

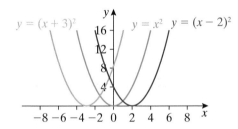

The graph of $y = x^2$ passes through the origin.

The graph of $y = (x + 3)^2$ is the same shape and passes through $(-3, 0)$.

The graph of $y = (x - 2)^2$ is also the same shape but passes through $(2, 0)$.

You can see that $y = (x + 3)^2$ is the graph of $y = x^2$ translated by $\begin{pmatrix} -3 \\ 0 \end{pmatrix}$ and that $y = (x - 2)^2$ is the graph of $y = x^2$ translated by $\begin{pmatrix} 2 \\ 0 \end{pmatrix}$.

This applies to all families of graphs. You can generalise the result.

> The graph of $y = f(x + a)$ is the graph of $y = f(x)$ translated by $\begin{pmatrix} -a \\ 0 \end{pmatrix}$.

TIP

It is useful to learn these results but you should also be able to work them out when needed.

EXAMPLE 24.4

a) Sketch these graphs on the same diagram.
   (i) $y = x^2$ **(ii)** $y = x^2 - 4$

b) State the transformation that maps $y = x^2$ on to $y = x^2 - 4$.

Solution

a)

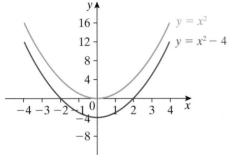

b) The transformation is a translation by $\begin{pmatrix} 0 \\ -4 \end{pmatrix}$.

EXAMPLE 24.5

This graph is a transformed sine curve. Find its equation.

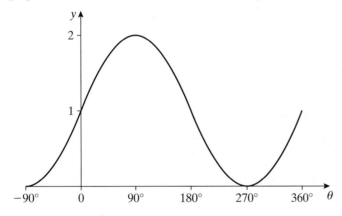

Solution

The graph of $y = \sin \theta$ passes through the origin.

The graph in the diagram has been translated by $\begin{pmatrix} 0 \\ 1 \end{pmatrix}$ so its equation is $y = \sin \theta + 1$.

Find the translation that maps $y = x^3$ on to $y = (x - 5)^3 + 6$.

Solution

The transformation that maps $y = x^3$ on to $y = (x - 5)^3$ is a translation by $\begin{pmatrix} 5 \\ 0 \end{pmatrix}$.

The transformation that maps $y = (x - 5)^3$ on to $y = (x - 5)^3 + 6$ is a translation by $\begin{pmatrix} 0 \\ 6 \end{pmatrix}$.

So the transformation that maps $y = x^3$ on to $y = (x - 5)^3 + 6$ is a translation by $\begin{pmatrix} 5 \\ 6 \end{pmatrix}$.

EXERCISE 24.2

1 **a)** Sketch these graphs on the same diagram.
   (i) $y = x^2$ **(ii)** $y = (x - 5)^2$
   b) State the transformation that maps $y = x^2$ on to $y = (x - 5)^2$.

2 **a)** Sketch these graphs on the same diagram.
   (i) $y = -x^2$ **(ii)** $y = -x^2 - 4$
   b) State the transformation that maps $y = -x^2$ on to $y = -x^2 - 4$.

3 **a)** Sketch these graphs on the same diagram.
   (i) $y = x^2$ **(ii)** $y = (x + 2)^2$ **(iii)** $y = (x + 2)^2 - 3$
   b) State the transformation that maps $y = x^2$ on to $y = (x + 2)^2 - 3$.

4 **a)** Sketch the result of translating the graph of $y = \sin \theta$ by $\begin{pmatrix} 0 \\ -1 \end{pmatrix}$ for $\theta = 0°$ to $\theta = 360°$.
   b) State the equation of the transformed graph.

5 State the equation of $y = x^2$ after it has been translated by
   a) $\begin{pmatrix} 0 \\ -5 \end{pmatrix}$.   **b)** $\begin{pmatrix} 2 \\ 0 \end{pmatrix}$.

6 The diagram shows the graph of $y = f(x)$.
   Copy the diagram and draw these graphs on
   the same axes.
   a) $y = f(x) - 2$
   b) $y = f(x - 2)$

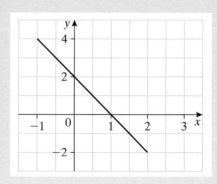

7 The diagram shows the graph of $y = g(x)$.

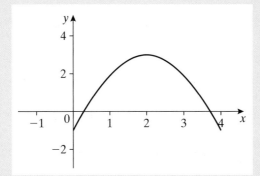

Copy the diagram and sketch these graphs on the same axes.

a) $y = g(x + 1)$ **b)** $y = g(x) + 1$

8 State the equation of the graph $y = x^2$ after it has been translated by $\begin{pmatrix} 1 \\ 2 \end{pmatrix}$.

9 The diagram shows the graph of a transformed cosine curve. State its equation.

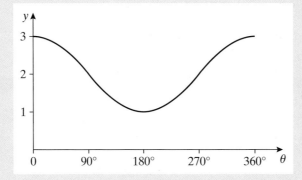

10 The graph of $y = x^2$ is translated by $\begin{pmatrix} -2 \\ 3 \end{pmatrix}$.

  a) State the equation of the transformed graph.

  b) Show that this equation may be written as $y = x^2 + 4x + 7$.

Challenge 24.2

a) Sketch the graphs of $y = \sin\theta$ and $y = \cos\theta$ on the same axes.

b) (i) By considering the transformation necessary, find a value for a so that
$\cos\theta = \sin(\theta + a)$.

   (ii) Investigate whether there are other possible values of a.

c) Investigate possible values of b so that $\sin\theta = \cos(\theta + b)$.

One-way stretches

Discovery 24.2

If possible, use graph-drawing software to draw the graphs in this task and print them out. Otherwise, draw the graphs on graph paper.

If your grid gets too crowded, start a new one.

Section 1

a) For the function $f(x) = x^2 - 2x$
- plot the graph of $y = f(x)$.
- on the same axes, plot the graphs of $y = 2f(x)$ and $y = 3f(x)$.

b) Experiment further until you can describe the transformation that maps the graph of $y = f(x)$ on to $y = kf(x)$, for any value of k.

Section 2

a) For the function $f(x) = x^2 - 2x$
- plot the graph of $y = f(x)$.
- on the same axes, plot the graph of $y = f(2x)$ (i.e. $y = 4x^2 - 4x$).
- also on the same axes, plot the graphs of $y = f\left(\frac{x}{2}\right)$ and $y = f(-3x)$.

b) Experiment further until you can describe the transformation that maps the graph of $y = f(x)$ on to $y = f(kx)$, for any value of k.

The diagram shows the graphs of $y = \sin\theta$ and $y = 3\sin\theta$.

To get from $y = \sin\theta$ to $y = 3\sin\theta$, the graph has been stretched parallel to the y-axis by a scale factor of 3.

This is an example of the following general principle.

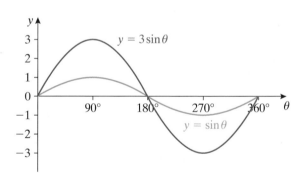

The graph of $y = kf(x)$ is a one-way stretch of the graph $y = f(x)$ parallel to the y-axis with a scale factor of k.

The diagram shows the graphs of $y = \cos\theta$ and $y = \cos 2\theta$.

The graph of $y = \cos\theta$ shows one period of the curve. By comparison, the graph of $y = \cos 2\theta$ shows two periods of the curve over the same range. This is a one-way stretch parallel to the x-axis with a scale factor of $\frac{1}{2}$.

This is an example of the following general principle.

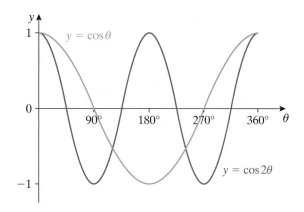

The graph of $y = f(kx)$ is a one-way stretch of the graph of $y = f(x)$ parallel to the x-axis with a scale factor of $\frac{1}{k}$.

A one-way stretch with a scale factor of $k = -1$ can be more simply described as a reflection.

If $y = f(x)$ is $y = x^2$ then $y = -f(x)$ is $y = -x^2$.

You can see that the graph of $y = -x^2$ is the reflection of $y = x^2$ in the x-axis.

This is an example of the following general principle.

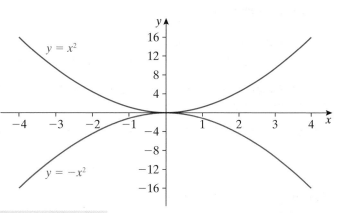

The graph of $y = -f(x)$ is a reflection of the graph of $y = f(x)$ in the x-axis.

If $y = f(x)$ is $y = 2x + 3$ then $y = f(-x)$ is $y = -2x + 3$.

You can see that $y = -2x + 3$ is a reflection of $y = 2x + 3$ in the y-axis.

This is an example of the following general principle.

The graph of $y = f(-x)$ is a reflection of the graph of $y = f(x)$ in the y-axis.

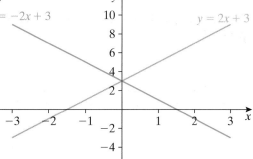

TIP

It is useful to learn these results but you should also be able to work them out when needed.

EXAMPLE 24.7

a) Sketch on the same diagram the graphs of $y = \cos\theta$ and $y = -\cos\theta$, for $0° \leqslant \theta \leqslant 360°$.

b) Describe the transformation that maps $y = \cos\theta$ on to $y = -\cos\theta$.

Solution

a)

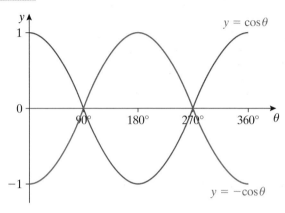

b) The transformation that maps $y = \cos\theta$ on to $y = -\cos\theta$ is a reflection in the x-axis.

EXAMPLE 24.8

Describe the transformation that maps the graph of $y = h(x)$ on to each of these graphs.

a) $y = h(x) - 2$ **b)** $y = 3h(x)$

c) $y = h(0.5x)$ **d)** $y = 4h(2x)$

Solution

a) A translation of $\begin{pmatrix} 0 \\ -2 \end{pmatrix}$.

b) A one-way stretch parallel to the y-axis with a scale factor of 3.

c) A one-way stretch parallel to the x-axis with a scale factor of 2.

d) A one-way stretch parallel to the x-axis of scale factor 0.5 and a one-way stretch parallel to the y-axis with a scale factor of 4.

EXAMPLE 24.9

The equation of this graph is $y = k \sin m\theta$. Find k and m.

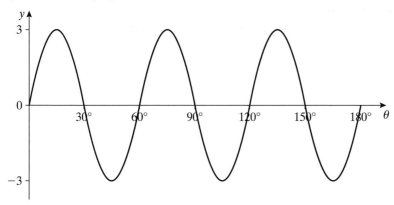

The graph is stretched parallel to the y-axis by a scale factor of 3, so $k = 3$.

The period of this sine curve is $60°$.

The graph of $y = \sin \theta$ has a period of $360°$. So the scale factor is $\frac{1}{6}$.

The graph is stretched parallel to the x-axis by a scale factor of $\frac{1}{6}$, so $m = 6$.

EXERCISE 24.3

1 a) Sketch on the same axes the graphs of $y = \sin \theta$ and $y = -\sin \theta$ for $0 \leqslant \theta \leqslant 360°$.

   b) Describe the transformation that maps $y = \sin \theta$ on to $y = -\sin \theta$.

2 a) Sketch on the same axes the graphs of $y = \sin \theta$ and $y = \sin \dfrac{\theta}{2}$ for $0 \leqslant \theta \leqslant 360°$.

   b) Describe the transformation that maps $y = \sin \theta$ on to $y = \sin \dfrac{\theta}{2}$.

3 Describe the transformation that maps

   a) $y = \cos \theta + 1$ on to $y = -\cos \theta - 1$.

   b) $y = x + 2$ on to $y = -x + 2$.

   c) $y = x^2$ on to $y = 5x^2$.

4 The graph of $y = \cos\theta$ is transformed by a one-way stretch parallel to the x-axis with a scale factor of $\frac{1}{3}$.

State the equation of the resulting graph.

5 State the equation of the graph $y = x^2 + 5$ after each of these transformations.
  a) A reflection in the y-axis
  b) A reflection in the x-axis

6 State the equation of the graph of $y = x + 2$ after each of these transformations.
  a) A one-way stretch parallel to the y-axis with a scale factor of 3
  b) A one-way stretch parallel to the x-axis with a scale factor of $\frac{1}{2}$

7 Describe the transformation that maps the graph of $y = f(x)$ on to each of these graphs.
  a) $y = f(x) + 1$ **b)** $y = 3f(x)$
  c) $y = f(2x)$ **d)** $y = 5f(3x)$

8 $y = -x^2 + 2x$. Find the equation of the graph after each of these transformations.
  a) A reflection in the x-axis
  b) A reflection in the y-axis
  c) A translation of 3 parallel to the x-axis

9 The graph of $y = x^2$ is stretched parallel to the x-axis by a scale factor of 2.
  a) State the equation of the resulting graph.
  b) What does the point $(1, 1)$ map on to under this transformation?
  c) What is the scale factor of a stretch parallel to the y-axis that maps $y = x^2$ on to the same graph?
  d) What does the point $(1, 1)$ map on to under this transformation?

10 The equation of this graph is $y = \sin k\theta$. Find k.

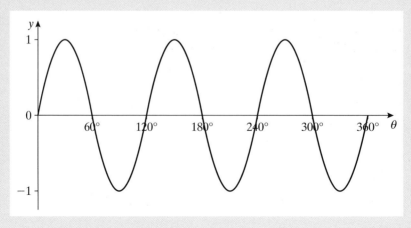

- To understand and use function notation
- The graph of f(x) + a is the graph of f(x) translated by $\begin{pmatrix} 0 \\ a \end{pmatrix}$
- The graph of y = f(x + a) is the graph of y = f(x) translated by $\begin{pmatrix} -a \\ 0 \end{pmatrix}$
- The graph of y = kf(x) is a one-way stretch of the graph y = f(x) parallel to the y-axis with a scale factor of k
- The graph of y = f(kx) is a one-way stretch of the graph of y = f(x) parallel to the x-axis with a scale factor of $\dfrac{1}{k}$
- The graph of y = −f(x) is a reflection of the graph of y = f(x) in the x-axis
- The graph of y = f(−x) is a reflection of the graph of y = f(x) in the y-axis

MIXED EXERCISE 24

1 $f(x) = 2x - 1$
   a) Solve $f(x) = 0$.
   b) Find an expression for each of these.
      (i) $f(x + 5)$ (ii) $f(x) + 5$

2 $g(x) = x^2 + 6$
   a) Solve $g(x) = 7$.
   b) Find an expression for each of these.
      (i) $g(x + 1)$ (ii) $g(2x) + 1$

3 $h(x) = x^2 - 2x$
   a) Find the value of $h(-4)$.
   b) Find an expression for each of these.
      (i) $h(x - 2)$ (ii) $h(2x) + 3$

4 a) On the same set of axes, sketch the graphs of $y = x^2$ and $y = x^2 - 4$.
   b) Describe the transformation that maps $y = x^2$ on to $y = x^2 - 4$.

5 The graph of $y = 2x - 3$ is translated by $\begin{pmatrix} 1 \\ 2 \end{pmatrix}$.

   Find the equation of the transformed graph.

6 Sketch the graph of $y = \sin \theta$ after it has been translated by $\begin{pmatrix} 0 \\ -1 \end{pmatrix}$.

7 Describe the transformation that maps the graph of $y = g(x)$ on to each of these graphs.

   a) $y = g(x + 1)$ **b)** $y = g(3x)$

   c) $y = 4g(x)$ **d)** $y = g(-x)$

8 State the equation of the graph of $y = \cos\theta$ after each of these transformations.

   a) A translation of $\begin{pmatrix} 0 \\ 3 \end{pmatrix}$

   b) A one-way stretch parallel to the x-axis with a scale factor of 0.25

9 State the equation of the graph of $y = x^3 + 5$ after it has been

   a) reflected in the x-axis.

   b) reflected in the y-axis.

10 The graph of $y = x^2$ is translated by $\begin{pmatrix} 3 \\ 1 \end{pmatrix}$ and then stretched with a scale factor of 2 parallel to the y-axis.

   a) Find the equation of the resulting curve and write it as simply as possible.

   b) Find the coordinates of the minimum point of this curve.

11 The diagram shows a sketch of $y = x^2$.
Copy the sketch and on the same diagram sketch each of these curves.

   a) $y = (x - 3)^2$

   b) $y = -(x - 3)^2$

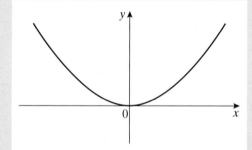

12 The diagram shows a sketch of $y = f(x)$.
Copy the sketch and on the same diagram sketch the curve $y = f(x) + 2$.
Mark clearly the coordinates of the point where the curve crosses the y-axis.

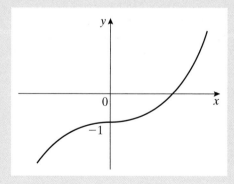

25 → ALGEBRAIC FRACTIONS

THIS CHAPTER IS ABOUT

- Adding and subtracting algebraic fractions
- Manipulating algebraic fractions
- Solving equations involving algebraic fractions

YOU SHOULD ALREADY KNOW

- How to carry out operations on numerical fractions
- How to factorise linear expressions
- How to factorise quadratic expressions
- How to solve equations

Adding and subtracting algebraic fractions

When adding numerical fractions, such as $\frac{3}{8} + \frac{7}{13}$, the first step is to write the fractions with a common denominator. In this example, the common denominator is 104 (8×13) and the numerator and denominator of each term need to be multiplied by the appropriate factor to give this denominator.

$$\frac{3}{8} + \frac{7}{13} = \frac{3 \times 13}{104} + \frac{7 \times 8}{104}$$
Write the fractions with a common denominator.

$$= \frac{39 + 56}{104}$$
Add the numerators.

$$= \frac{95}{104}$$

As you saw in Chapter 11, the rules for dealing with algebraic fractions are exactly the same as those for dealing with numerical fractions. In Example 25.1 the denominators are both numbers; in Example 25.2 one of the denominators is an algebraic expression but, as is shown, the technique is still the same.

EXAMPLE 25.1

Simplify $\dfrac{x+3}{2} + \dfrac{x+2}{3}$.

Solution

$\dfrac{x+3}{2} + \dfrac{x+2}{3}$　　　　Find the common denominator: 6.

$= \dfrac{3(x+3)}{6} + \dfrac{2(x+2)}{6}$　　　Multiply each numerator and denominator by the appropriate factor.

$= \dfrac{3x+9+2x+4}{6}$　　　Expand the brackets and combine the fractions (since the denominators are now the same).

$= \dfrac{5x+13}{6}$　　　Simplify by collecting like terms.

EXAMPLE 25.2

Simplify $\dfrac{2}{x-3} + \dfrac{x+2}{5}$.　　Sometimes, instead of the word 'simplify', you may be asked to 'write as a single fraction'.

Solution

$\dfrac{2}{x-3} + \dfrac{x+2}{5}$　　　Find the common denominator by multiplying the two denominators together: $5(x-3)$.

$= \dfrac{2 \times 5}{5(x-3)} + \dfrac{(x+2)(x-3)}{5(x-3)}$　　Multiply each numerator and denominator by the appropriate factor.

$= \dfrac{10 + x^2 - 3x + 2x - 6}{5(x-3)}$　　Expand the brackets in the numerator and combine the fractions.

$= \dfrac{x^2 - x + 4}{5(x-3)}$　　　Simplify by collecting like terms.

The numerator will not factorise and therefore the expression cannot be simplified further.

Simplify these.

1 $\dfrac{x+1}{2} + \dfrac{x-2}{3}$

2 $\dfrac{x+3}{5} - \dfrac{x+1}{2}$

3 $\dfrac{x-2}{4} + \dfrac{x-5}{3}$

4 $\dfrac{3x+2}{4} - \dfrac{2x-5}{2}$

5 $\dfrac{x+2}{3} + \dfrac{2}{x-1}$

6 $\dfrac{1}{x-2} - \dfrac{x+3}{4}$

7 $\dfrac{2x+3}{3} - \dfrac{2}{x-5}$

8 $\dfrac{3x+2}{4x-1} + \dfrac{2}{7}$

9 $\dfrac{5+3x}{2x-3} + \dfrac{1}{4}$

Adding and subtracting more complex algebraic fractions

In the previous section you learned how to add and subtract algebraic fractions when both denominators are numbers and when just one of the denominators is an algebraic expression.

In this section, both of the denominators are algebraic expressions. The key to dealing with these fractions is still to find the common denominator and, as before, the way to do this is to multiply the denominators.

EXAMPLE 25.3

Simplify $\dfrac{2}{x} + \dfrac{x+2}{3}$.

Solution

$\dfrac{2}{x} + \dfrac{x+2}{3}$

Find the common denominator: $x(x+2)$.

$= \dfrac{2(x+2)}{x(x+2)} + \dfrac{3(x)}{x(x+2)}$

Multiply each numerator and denominator by the appropriate factor.

$= \dfrac{2x+4+3x}{x(x+2)}$

Expand the brackets and combine the fractions.

$= \dfrac{5x+4}{x(x+2)}$

Simplify by collecting like terms.

EXAMPLE 25.4

Simplify $\dfrac{x-3}{x+2} + \dfrac{x-4}{x+1}$.

Solution

$$\dfrac{x-3}{x+2} + \dfrac{x-4}{x+1}$$

Find the common denominator: $(x+2)(x+1)$.

$$= \dfrac{(x-3)(x+1)}{(x+2)(x+1)} + \dfrac{(x-4)(x+2)}{(x+2)(x+1)}$$

Multiply each numerator and denominator by the appropriate factor.

$$= \dfrac{x^2 + x - 3x - 3 + x^2 + 2x - 4x - 8}{(x+2)(x+1)}$$

Expand the brackets in the numerators and combine the fractions.

$$= \dfrac{2x^2 - 4x - 11}{(x+2)(x+1)}$$

Simplify by collecting like terms.

In the next example an extra step is required because the numerator will factorise.

EXAMPLE 25.5

Simplify $\dfrac{8-9x}{2x-1} + \dfrac{6x+23}{x+4}$.

Solution

$$\dfrac{8-9x}{2x-1} + \dfrac{6x+23}{x+4}$$

Find the common denominator: $(2x-1)(x+4)$.

$$= \dfrac{(8-9x)(x+4)}{(2x-1)(x+4)} + \dfrac{(6x+23)(2x-1)}{(2x-1)(x+4)}$$

Multiply each numerator and denominator by the appropriate factor.

$$= \dfrac{-9x^2 - 28x + 32 + 12x^2 + 40x - 23}{(2x-1)(x+4)}$$

Expand the brackets in the numerators and combine the fractions.

$$= \dfrac{3x^2 + 12x + 9}{(2x-1)(x+4)}$$

Simplify by collecting like terms.

$$= \dfrac{3(x+1)(x+3)}{(2x-1)(x+4)}$$

Factorise the numerator.

Simplify these.

1 $\dfrac{2}{x+1} + \dfrac{1}{x+2}$ 　　　**2** $\dfrac{2}{x+3} + \dfrac{x+1}{x}$ 　　　**3** $\dfrac{x+2}{3x} + \dfrac{x}{x+3}$

4 $\dfrac{x+3}{x-4} + \dfrac{x-3}{x+4}$ 　　　**5** $\dfrac{3}{x+2} - \dfrac{2}{x-1}$ 　　　**6** $\dfrac{1}{x-2} - \dfrac{x+3}{x-4}$

7 $\dfrac{2x+3}{x-3} - \dfrac{x-2}{x-5}$ 　　　**8** $\dfrac{x+2}{3x-4} + \dfrac{x-3}{x+2}$ 　　　**9** $\dfrac{2}{2x+1} + \dfrac{3x+5}{x+2}$

10 $\dfrac{4x+17}{x+3} - \dfrac{2x-15}{x-3}$ 　　　**11** $\dfrac{2x-5}{3x-2} - \dfrac{3x+2}{5x-4}$ 　　　**12** $\dfrac{3x-4}{x+1} - \dfrac{x+2}{5x+3}$

Challenge 25.1

Simplify these.

a) $\dfrac{3}{x+2} + \dfrac{x}{x+3} + \dfrac{3}{4}$ 　　**b)** $\dfrac{x+2}{3x} + \dfrac{x-3}{x+2} - \dfrac{4}{x-1}$ 　　**c)** $\dfrac{x+1}{x-6} - \dfrac{x-3}{3x+4} - \dfrac{x+5}{5x-2}$

Solving equations involving algebraic fractions

The rules for dealing with equations involving algebraic fractions are exactly the same as those for dealing with equations involving numerical fractions, as is shown in the next examples.

EXAMPLE 25.6

Solve $\dfrac{2x-3}{4} - \dfrac{x+1}{3} = 1$.

Solution

$\dfrac{2x-3}{4} - \dfrac{x+1}{3} = 1$ 　　　　Find a common denominator: 12.

$3(2x-3) - 4(x+1) = 12 \times 1$ 　　　Multiply every term of the equation by the common denominator.

$6x - 9 - 4x - 4 = 12$ 　　　　Expand the brackets.

$2x = 25$ 　　　　Simplify by collecting like terms.

$x = 12.5$ 　　　　Solve the equation.

EXAMPLE 25.7

Solve $\dfrac{x + 4}{x - 1} = \dfrac{x}{x - 3}$.

Solution

$$\dfrac{x + 4}{x - 1} = \dfrac{x}{x - 3}$$ Find the common denominator: $(x - 1)(x - 3)$.

$$(x + 4)(x - 3) = x(x - 1)$$ Multiply each side of the equation by the common denominator.

$$x^2 + 4x - 3x - 12 = x^2 - x$$ Expand the brackets.

$$2x = 12$$ Simplify by collecting like terms.

$$x = 6$$ Solve the equation.

In the next example, the equation that results after you have eliminated the fractions is a quadratic equation (meaning that it has a term in x^2). You learned how to solve quadratic equations in Chapter 19.

EXAMPLE 25.8

Solve $\dfrac{2}{x} - \dfrac{3}{2x - 1} = \dfrac{1}{3x + 2}$.

Solution

$$\dfrac{2}{x} - \dfrac{3}{2x - 1} = \dfrac{1}{3x + 2}$$ Find the common denominator: $x(2x - 1)(3x + 2)$.

$$2(2x - 1)(3x + 2) - 3(x)(3x + 2) = x(2x - 1)$$ Multiply each term in the equation by the common denominator.

$$12x^2 + 2x - 4 - 9x^2 - 6x = 2x^2 - x$$ Expand the brackets.

$$x^2 - 3x - 4 = 0$$ Simplify by collecting like terms.

$$(x - 4)(x + 1) = 0$$ Factorise the resulting expression.

$$x = 4 \text{ or } x = -1$$ Solve by finding *both* solutions.

EXAMPLE 25.9

Solve $\dfrac{2x + 3}{x - 1} = \dfrac{x + 1}{2x + 3}$.

Solution

$$\dfrac{2x + 3}{x - 1} = \dfrac{x + 1}{2x + 3}.$$

Find the common denominator:
$(x - 1)(2x + 3)$.

$$(2x + 3)^2 = (x - 1)(x + 1)$$

Multiply each term in the equation by the common denominator.

$$4x^2 + 12x + 9 = x^2 - 1$$

Expand the brackets.

$$3x^2 + 12x + 10 = 0$$

Simplify by collecting like terms.

$$x = \dfrac{-12 \pm \sqrt{12^2 - 4 \times 3 \times 10}}{2 \times 3}$$

Solve the equation using the formula, since the expression will not factorise.

$x = -1.18$ or $x = -2.82$, correct to 2 decimal places.

EXERCISE 25.3

Solve these.

1 $\dfrac{x + 3}{2} - \dfrac{x + 4}{3} = 1$

2 $\dfrac{6}{x - 4} = \dfrac{5}{x - 3}$

3 $\dfrac{1}{2x + 3} = \dfrac{1}{3x - 2}$

4 $\dfrac{4}{x} + \dfrac{1}{x - 3} = 1$

5 $\dfrac{x}{x + 3} - \dfrac{x - 2}{8} = \dfrac{1}{4}$

6 $\dfrac{2x}{x - 3} - \dfrac{x}{x - 2} = 3$

7 $\dfrac{2}{x} + \dfrac{1}{x + 1} = 5$

8 $\dfrac{x}{x - 2} - 2x = 3$

9 $\dfrac{2x}{x - 5} + \dfrac{x - 1}{3x} = 2$

Challenge 25.2

The cost of hiring a minibus for a trip one Wednesday was £120.

Each of the people travelling paid the same amount.

On Saturday, the cost of hiring the minibus remained the same but two fewer people travelled and each person had to pay £2 more than they had paid on Wednesday.

How many people made the journey on Wednesday?

MIXED EXERCISE 25

1 Simplify these.

  a) $\dfrac{x-3}{3} + \dfrac{x+2}{2}$

  b) $\dfrac{x-3}{2} - \dfrac{x+5}{3}$

  c) $\dfrac{x+4}{5} - \dfrac{3}{2x}$

  d) $\dfrac{3}{4x-1} + \dfrac{1}{3x+2}$

2 Simplify these.

  a) $\dfrac{5}{x+4} - \dfrac{3}{x+2}$

  b) $\dfrac{x}{x+2} + \dfrac{x-1}{3}$

  c) $\dfrac{x-2}{x} + \dfrac{3x}{x+2}$

  d) $\dfrac{2x+3}{x-1} + \dfrac{4-x}{3x-5}$

3 Solve these.

  a) $\dfrac{x+2}{3} - \dfrac{x-3}{2} = 2$

  b) $\dfrac{3}{2(2x-1)} = \dfrac{4}{3x+2}$

  c) $\dfrac{2x-1}{3} + \dfrac{x-2}{6} = \dfrac{3x}{4}$

  d) $2 - \dfrac{2}{x-3} = \dfrac{8}{x}$

4 Solve these. Give your answers correct to 3 decimal places.

  a) $\dfrac{1}{3x-2} = \dfrac{2x+3}{x-1}$

  b) $\dfrac{3x+1}{x+2} = \dfrac{1-2x}{x+1}$

26 → PROPERTIES OF SHAPES

The area of a triangle

You already know that

Area of a rectangle = length × width

or

Area of a rectangle = $l \times w$.

Look at this diagram.

Area of rectangle ABCD = $l \times w$

You can see that

Area of triangle ABC = $\frac{1}{2} \times$ area of ABCD

$= \frac{1}{2} \times l \times w$.

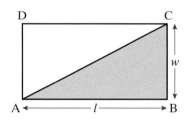

Now look at a different triangle.
From the diagram you can see that

Area of triangle ABC = $\frac{1}{2}$ area of BEAF + $\frac{1}{2}$ area of FADC

$= \frac{1}{2}$ area of BEDC

$= \frac{1}{2} \times l \times w$.

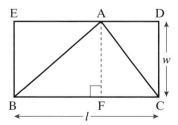

This shows that the area of any triangle can be found using the formula below.

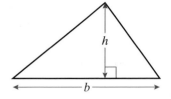

Area of a triangle $= \frac{1}{2} \times$ base $\times$ height

or

$$A = \frac{1}{2} \times b \times h$$

Note that the height of a triangle, h, is measured at right angles to the base. It is the **perpendicular height** or **altitude** of the triangle.

TIP You can use any of the sides as the base provided that you use the perpendicular height that goes with it.

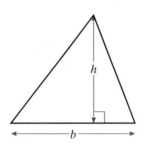 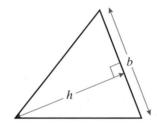 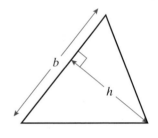

EXAMPLE 26.1

Find the area of this triangle.

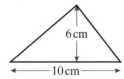

Solution

Use the formula.
$$A = \frac{1}{2} \times b \times h$$
$$= \frac{1}{2} \times 10 \times 6$$
$$= 30 \text{ cm}^2$$

 Do not forget the units, but notice how you only need to put the units in the answer. Remember that the two measurements used must have the same units.

Find the area of each of these triangles.

1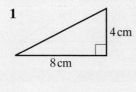

2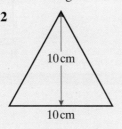

3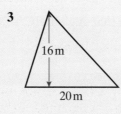

4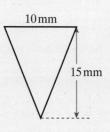

5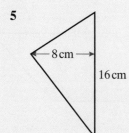

6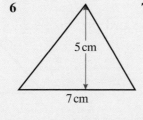

7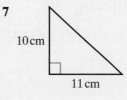

8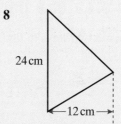

9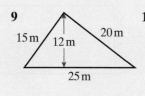

10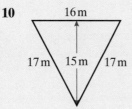

Challenge 26.1

The vertices of a triangle are at $(2, 2)$, $(7, 2)$ and $(4, 6)$.

Draw the triangle on squared paper and calculate its area.

Challenge 26.2

A triangle has area 12 cm². Its base is 4 cm long.

Work out the perpendicular height of this triangle.

In triangle ABC, AB = 6 cm, BC = 8 cm and AC = 10 cm.
Angle ABC = 90°.

a) Find the area of triangle ABC.

H is the point on AC such that angle BHC = 90°.

b) Find the length of BH.

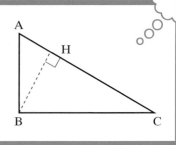

The area of a parallelogram

There are two ways to find the area of a parallelogram.

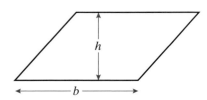

It can be cut and rearranged to form a rectangle. So,

$A = b \times h.$

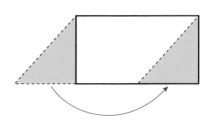

Or it can be split into two congruent triangles along
a diagonal.
(Remember that congruent means identical or exactly the same.)

The area of each triangle is $A = \frac{1}{2} \times b \times h$, so the total
area of the parallelogram is

$A = 2 \times \frac{1}{2} \times b \times h$
$\quad = b \times h.$

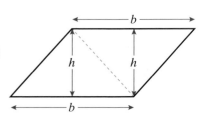

Note that the height of the parallelogram is the *perpendicular*
height, just as in the formula for the area of a triangle.

> Area of a parallelogram = $b \times h$
>
> or
>
> $A = b \times h$

Find the area of this parallelogram.

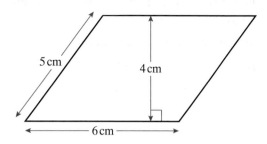

Use the formula. Make sure you choose the correct measurement for the height.

$A = b \times h$
$ = 6 \times 4$
$ = 24 \text{ cm}^2$

EXERCISE 26.2

Find the area of each of these parallelograms.

1
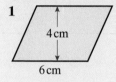
4 cm
6 cm

2
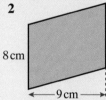
8 cm
9 cm

3
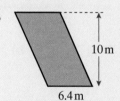
10 m
6.4 m

4
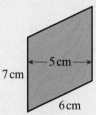
7 cm
5 cm
6 cm

5
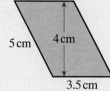
5 cm
4 cm
3.5 cm

6
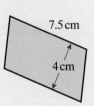
7.5 cm
4 cm

7
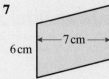
7 cm
6 cm

8
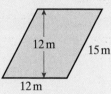
12 m
15 m
12 m

9
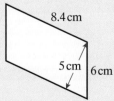
8.4 cm
5 cm
6 cm

10
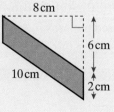
8 cm
6 cm
10 cm
2 cm

Use appropriate measurements to find the area of each of these parallelograms.

a)

b)

Find the lengths *x* and *y*.

a)

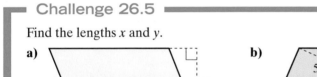

Area = 36 cm²

x

9 cm

b)

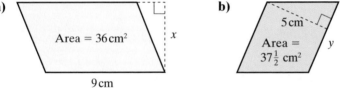

5 cm

Area = $37\frac{1}{2}$ cm²

y

OABC is a parallelogram with vertices at O(0, 0), A(4, 2), B(6, 0) and C(2, −2).

a) Draw the parallelogram OABC on squared paper.

b) Work out the area of OABC.

Angles made with parallel lines

Discovery 26.1

This is a map of part of New York.

a) Find Broadway and W 32nd Street
   on the map.
   Find some more angles equal
   to the angle between Broadway
   and W 32nd Street.

Two angles that add up to 180° are
called **supplementary** angles.

b) Find an angle that is supplementary
   to the angle between Broadway and
   W 32nd Street.

c) Explain your results.

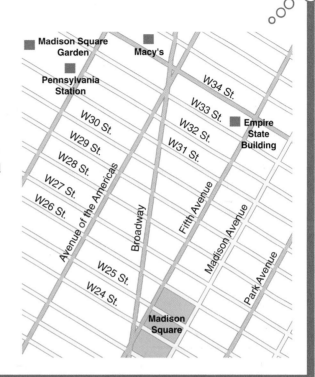

In Discovery 26.1 you should have found three sorts of angles made with
parallel lines.

Corresponding angles

The diagrams show equal angles made by a line cutting across a pair
of parallel lines. These equal angles are called **corresponding** angles.
Corresponding angles occur in an F-shape.

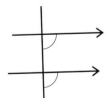 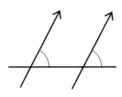 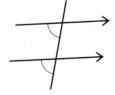 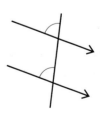

Alternate angles

These diagrams also show equal angles made by a line cutting across a pair of parallel lines. These equal angles are called **alternate** angles. Alternate angles occur in a Z-shape.

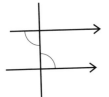 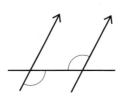 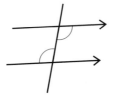 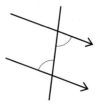

Allied angles

You can see that the two angles marked in these diagrams are not equal. Instead, they are supplementary. (Remember that supplementary angles add up to 180°.) These angles are called **allied** angles or **co-interior** angles and occur in a C-shape.

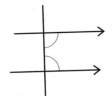 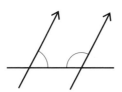 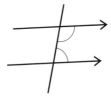 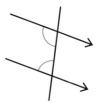

> **TIP**
>
> Questions about finding the size of angles often ask you to give reasons for your answers. This means that you must say why, for example, angles are equal. Stating that the angles are alternate angles or corresponding angles would be possible reasons. This is shown in the next example.

EXAMPLE 26.3

Work out the size of the lettered angles.
Give a reason for each answer.

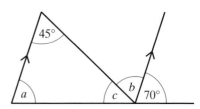

Solution

$a = 70°$ Corresponding angles
$b = 45°$ Alternate angles
$c = 65°$ Angles on a straight line add up to 180°
                         or Allied angles add up to 180°
                         or Angles in a triangle add up to 180°

Find the size of the lettered angles. Give a reason for each answer.

1

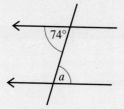

2

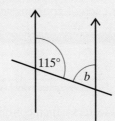

3

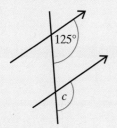

4

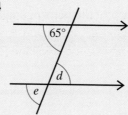

5

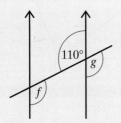

6

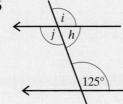

7

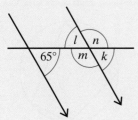

8

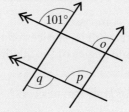

9

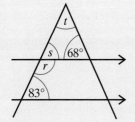

10

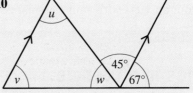

Challenge 26.7

One of the angles in a parallelogram is 125°.

a) Make a sketch of the parallelogram and mark this angle.

b) Calculate the other three angles and mark them on your diagram.
Give a reason for each of your answers.

Challenge 26.8

Make a list of capital letters which contain parallel lines.

Mark pairs of equal angles on each sketch.

Give a reason for each of the pairs of angles.

Challenge 26.9

ABCD is a parallelogram.

Show that triangles ABD and CDB contain the
same angles.

Give a reason for each step of your work.

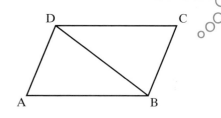

Challenge 26.10

Draw a circle using compasses.
Draw a quadrilateral with its vertices on the edge of the circle.
Measure the opposite angles of the quadrilateral and add each
pair together.

What did you get for your two answers?

Try this again with a different sized circle and a different
shaped quadrilateral.

Check your answers with your classmates. Did they get the same?

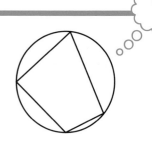

The angles in a triangle

You already know that the angles in a triangle add up to 180°. You can use
the properties of angles associated with parallel lines to prove this fact.

Proof 26.1

You draw a line parallel to the base of a triangle.

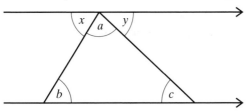

$$x + a + y = 180°$$ Angles on a straight line add up to 180°
$$b = x$$ Alternate angles
$$c = y$$ Alternate angles

So $$b + a + c = 180°$$ Since $b = x$ and $c = y$

This proves that the three angles in any triangle add up to 180°.

The angles inside a triangle (or any polygon) are called **interior** angles. If you extend a side of the triangle, there is an angle between the extended side and the next side. This angle is called an **exterior** angle.

The exterior angle of a triangle is equal to the sum of the opposite, interior angles.

Here is a proof of this fact.

Proof 26.2

One side is extended, as shown in the diagram. The exterior angle is marked x.

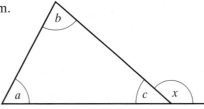

$$x + c = 180°$$ Angles on a straight line add up to 180°
$$a + b + c = 180°$$ Angles in a triangle add up to 180°

So $$x = a + b$$ Since $x = 180° - c$ and $a + b = 180° - c$

This proves that the exterior angle of a triangle is equal to the sum of the opposite, interior angles.

There is another way to prove that the exterior angle of a triangle is equal to the sum of the opposite, interior angles. It uses angle facts associated with parallel lines.

Complete a proof for this diagram. Remember to give a reason for each step.

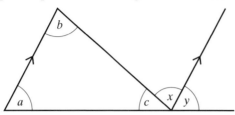

You can use the angle facts associated with triangles to work out missing angles. This is shown in the next example.

EXAMPLE 26.4

Work out the size of the lettered angles.
Give a reason for each answer.

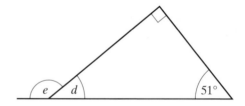

Solution

$d = 180° - (51° + 90°)$ Angles in a triangle add up to $180°$
$d = 39°$
$e = 51° + 90°$ Exterior angle of a triangle equals the sum
$e = 141°$ of the opposite interior angles

The angles in a quadrilateral

A quadrilateral is a four-sided shape.

> The angles in a quadrilateral add up to $360°$.

You can divide a quadrilateral into two triangles. You can then use the fact that angles in a triangle add up to $180°$ to prove this fact.

Proof 26.3

A quadrilateral is divided into two triangles, as shown in the diagram.

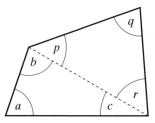

$a + b + c = 180°$ Angles in a triangle add up to 180°
$p + q + r = 180°$ Angles in a triangle add up to 180°
$a + b + c + p + q + r = 360°$, so

the interior angles of a quadrilateral add up to 360°.

You can use this fact to work out missing angles in quadrilaterals.

EXAMPLE 26.5

Work out the size of angle x.
Give a reason for your answer.

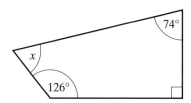

Solution

$x = 360° - (126° + 90° + 74°)$ Angles in a quadrilateral add up to 360°
$x = 70°$

EXERCISE 26.4

Find the size of each lettered angle. Give a reason for each answer.

1

2

3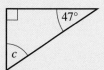

4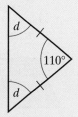

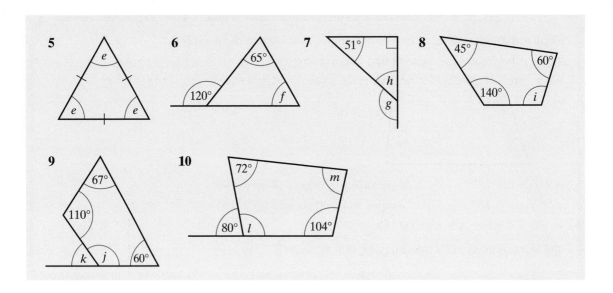

Special quadrilaterals

You already know about the special quadrilaterals: square, rectangle, parallelogram, rhombus, kite, arrowhead and trapezium. There is also a special type of trapezium called an isosceles trapezium.

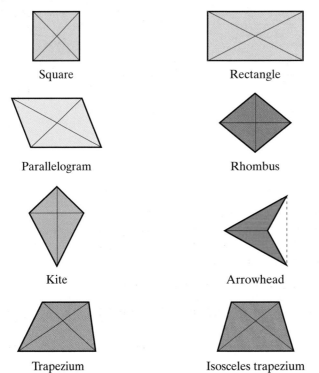

Square	Rectangle
Parallelogram	Rhombus
Kite	Arrowhead
Trapezium	Isosceles trapezium

Copy this decision tree.

Take each of the eight special quadrilaterals in turn.

Work through the decision tree with each shape and fill in the boxes at the bottom.

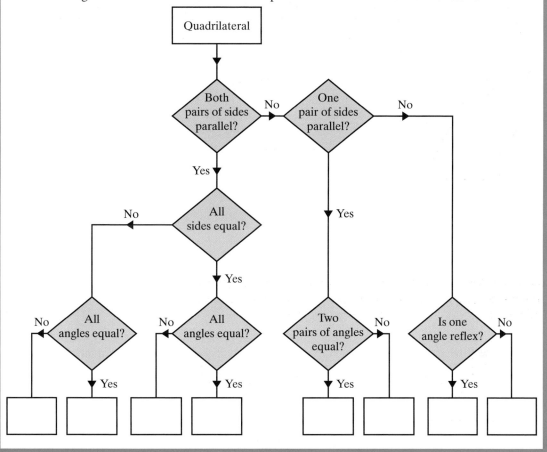

Look again at the diagram of the kite. It has one pair of opposite angles equal. Sides next to each other are called **adjacent** sides. A kite has two pairs of adjacent sides equal. Look at the diagonals. They cross at right angles and one of the diagonals is cut into two equal parts, or **bisected**, by the other.

Copy and complete this table for each of the special quadrilaterals.

Name	Diagram	Angles	Length of sides	Parallel sides	Diagonals

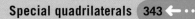

1 Name each of these quadrilaterals.

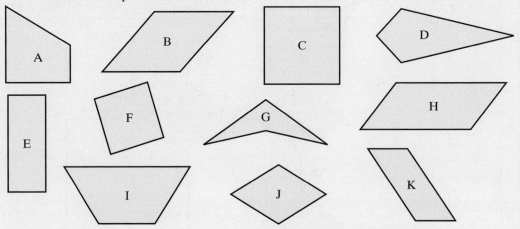

2 Name the quadrilateral or quadrilaterals which have the following properties.

a) All sides the same length

b) Two pairs of sides equal in length

c) Opposite sides the same length but not all four sides the same length

d) Just two sides parallel

e) Diagonals that cross at 90°

3 Plot each set of points on squared paper and join them in order to make a quadrilateral. Use a different grid for each part. Write down the special name of each quadrilateral.

a) $(3, 0), (5, 4), (3, 8), (1, 4)$

b) $(8, 1), (6, 3), (2, 3), (1, 1)$

c) $(1, 2), (3, 1), (7, 2), (5, 3)$

d) $(6, 2), (2, 3), (1, 2), (2, 1)$

4 A rectangle is a special type of parallelogram.
What extra properties does a rectangle have?

5 A quadrilateral has angles of 70°, 70°, 110° and 110°.
Which special quadrilaterals could have these as their angles?
Draw each of these quadrilaterals and mark on the angles.

The angles in a polygon

A polygon is a closed shape made with straight sides.

Earlier in this chapter you saw how, by dividing a quadrilateral into two triangles, you could see that the interior angles of a quadrilateral add up to 360°.

In the same way you can divide a polygon into triangles to find the sum of the interior angles of any polygon.

Copy and complete this table which will help you find the formula for the sum of the interior angles of a polygon with *n sides*.

Number of sides	Diagram	Name	Sum of interior angles
3		Triangle	$1 \times 180° = 180°$
4		Quadrilateral	$2 \times 180° = 360°$
5			$3 \times 180° =$
6			
7			
8			
9			
10			
n			

At each **vertex**, or corner, of a polygon there is an interior angle and an exterior angle.

Since these form a straight line you know the sum of the angles.

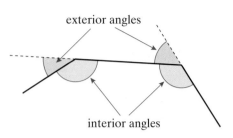
exterior angles

interior angles

> Interior angle + exterior angle = 180°

Discovery 26.3

Here is a pentagon showing all its exterior angles.

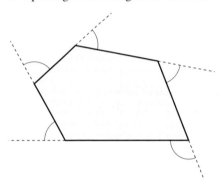

a) Measure each of the exterior angles and find the total.
Check with your neighbour. Do you both have the same total?

b) Draw another polygon.
Extend its sides and measure each of the exterior angles.
Is the total of these angles the same as for the pentagon?

You may have noticed that the five exterior angles of the pentagon go round in a full circle. This gives you another fact about the angles of a polygon.

> The sum of the exterior angles of a polygon is 360°.

EXAMPLE 26.6

Two of the exterior angles of this pentagon are equal.
Find their size.

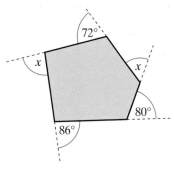

Solution

$x + x + 72° + 80° + 86° = 360°$ Exterior angles of a polygon add up to 360°

$$2x = 360° - 238°$$
$$2x = 122°$$
$$x = 61°$$

Regular polygons

You already know that, for a regular polygon, the sides are all the same
length and the interior angles are all the same size. You can see now that
the exterior angles are also all the same size.

EXAMPLE 26.7

Find the size of the exterior and interior angles of a regular octagon.

Solution

An octagon has eight sides.

Exterior angle $= \dfrac{360°}{8}$ Since the sum of the exterior angles of any
polygon is 360°

Exterior angle $= 45°$

Interior angle $= 180° - 45°$ Since the interior angle and the exterior
angle add up to 180°

Interior angle $= 135°$

Check up 26.2

Calculate the exterior and interior angles of each of these regular polygons:
triangle (equilateral triangle), quadrilateral (square), pentagon, hexagon, heptagon,
octagon, nonagon (nine sides) and decagon.

Remember that, for any regular polygon,

$$\text{The angle at the centre} = \frac{360°}{\text{number of sides}}$$

Note that the size of the angle at the centre of a polygon is the same as the exterior angle of the polygon. Can you see why?

Check up 26.3

Lines are drawn from the centre of a regular pentagon to each of its vertices.

a) What can you say about the triangles formed?

b) What sort of triangles are they?

c) Calculate the size of the lettered angles. Give a reason for each of your answers.

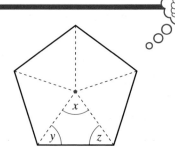

EXERCISE 26.6

1 A polygon has 15 sides.
Work out the sum of the interior angles of this polygon.

2 A polygon has 20 sides.
Work out the sum of the interior angles of this polygon.

3 Three of the exterior angles of a quadrilateral are 94°, 50° and 85°.
a) Work out the size of the fourth exterior angle.
b) Work out the sizes of the interior angles of the quadrilateral.

4 Four of the exterior angles of a pentagon are 90°, 80°, 57° and 75°.
a) Work out the size of the other exterior angle.
b) Work out the sizes of the interior angles of the pentagon.

5 A regular polygon has 12 sides.
Find the size of the exterior and interior angles of this polygon.

6 A regular polygon has 100 sides.
Find the size of the exterior and interior angles of this polygon.

7 A regular polygon has an exterior angle of 24°.
Work out the number of sides that the polygon has.

8 A regular polygon has an interior angle of 162°.
Work out the number of sides that the polygon has.

WHAT YOU HAVE LEARNED

- The area of a triangle $= \frac{1}{2} \times$ base $\times$ perpendicular height or $A = \frac{1}{2} \times b \times h$
- The area of a parallelogram $=$ base $\times$ perpendicular height or $A = b \times h$
- When a line crosses a pair of parallel lines, corresponding angles are equal
- When a line crosses a pair of parallel lines, alternate angles are equal
- When a line crosses a pair of parallel lines, allied angles add up to $180°$
- The interior angles of a triangle add up to $180°$
- The exterior angle of a triangle is equal to the sum of the opposite, interior angles
- The interior angles of a quadrilateral add up to $360°$
- The properties of special quadrilaterals
- The sum of the interior angles of a polygon with n sides is $180° \times (n - 2)$
- The interior angle and the exterior angle of a polygon add up to $180°$
- The sum of the exterior angles of a polygon is $360°$
- The angle at the centre of a regular polygon with n sides is $\dfrac{360°}{n}$

MIXED EXERCISE 26

1 Find the area of each of these triangles.

a)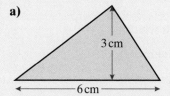

b)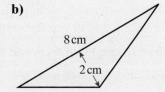

c)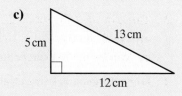

2 Find the area of each of these parallelograms.

a)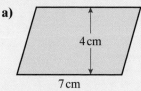

b)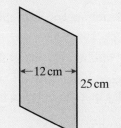

c)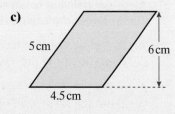

3 Find the size of the lettered angles. Give a reason for each answer.

a)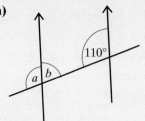

b)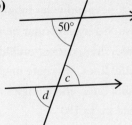

c)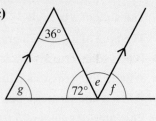

4 Find the size of the lettered angles. Give a reason for each answer.

a)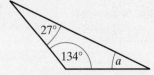

b)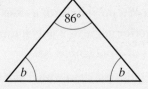

c)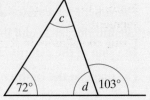

5 Find the size of the lettered angles. Give a reason for each answer.

a)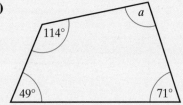

b)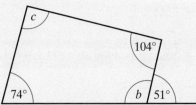

6 Write down the name of the quadrilateral or quadrilaterals which

   a) can be made using lengths of 10 cm, 10 cm, 5 cm and 5 cm.

   b) can be made using angles of 110°, 110°, 70° and 70°.

   c) have both diagonals equal in length.

7 **a)** A polygon has seven sides.
   Work out the sum of the interior angles of this polygon.

   b) A regular polygon has 15 sides.
   Find the size of the exterior and interior angles of this polygon.

   c) A regular polygon has an exterior angle of 10°.
   Work out the number of sides that the polygon has.

27 → PYTHAGORAS' THEOREM

THIS CHAPTER IS ABOUT

- **Calculating the length of a side of a right-angled triangle when you know the other two**
- **Deciding whether or not a triangle is right-angled**
- **Finding the coordinates of the midpoint of a line segment**
- **Using coordinates in three dimensions (3-D)**

YOU SHOULD ALREADY KNOW

- **How to find squares and square roots on your calculator**
- **The formula for the area of a triangle**
- **How to use coordinates in two dimensions**

Pythagoras' theorem

Discovery 27.1

Measure all three sides of the right-angled triangle in the diagram.

Use the lengths to work out the area of each of the three coloured squares.

What do you notice?

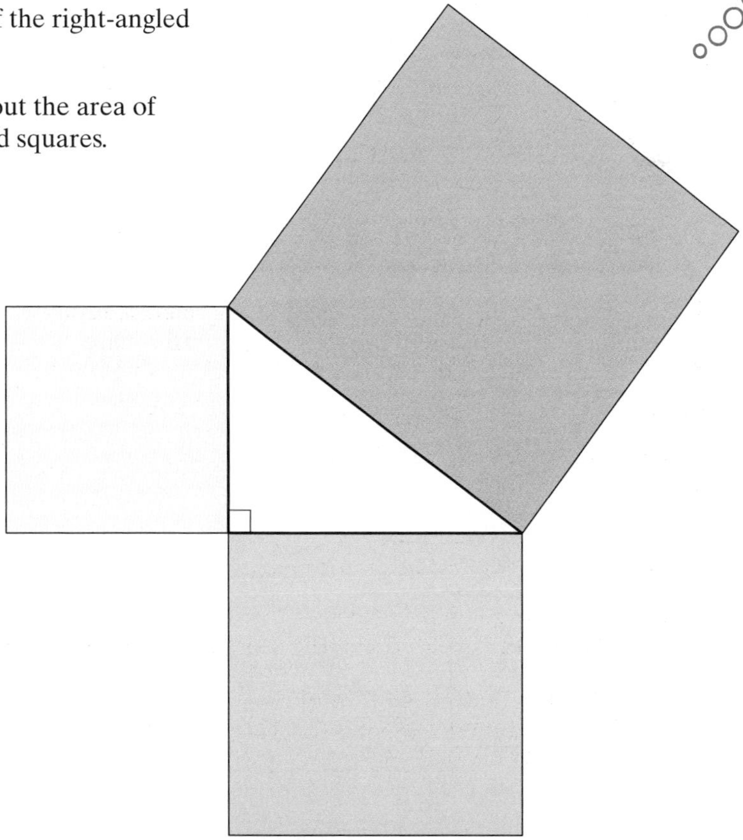

The area of the yellow square added to the area of the blue square is equal to the area of the red square.

The longest side of a right-angled triangle is called the **hypotenuse**. This is the side opposite the right angle.

What you discovered in Discovery 27.1 is true for all right-angled triangles. It was first 'discovered' by Pythagoras, a Greek mathematician, who lived around 500 BC.

Pythagoras' theorem states that:

The area of the square on the hypotenuse of a right-angled triangle is equal to the sum of the areas of the squares on the other two sides.

That is:

P + Q = R

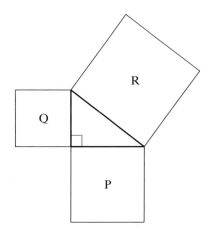

For each of these diagrams, find the area of the third square.

1

?

4 cm²

8 cm²

2

?

15 cm²

70 cm²

3

60 cm²

10 cm²

?

4

?

22 cm²

50 cm²

Using Pythagoras' theorem

Although the theorem is based on area it is usually used to find the length of a side.

If you drew squares on the three sides of this triangle their areas would be a^2, b^2 and c^2.

So Pythagoras' theorem can also be written as

$$a^2 + b^2 = c^2$$

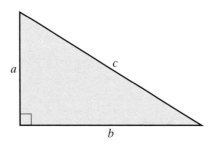

EXAMPLE 27.1

For each of these triangles, find the length marked x.

a)

5 cm

x

8 cm

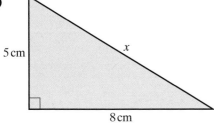

b)

x

8.7 cm

6.9 cm

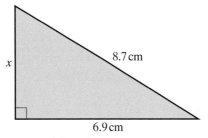

Solution

a) $c^2 = a^2 + b^2$ The length marked x is the hypotenuse, or c.
$x^2 = 8^2 + 5^2$ Substitute the numbers into the formula.
$x^2 = 64 + 25$
$x^2 = 89$
$x = \sqrt{89}$ Take the square root of both sides.
$x = 9.43$ cm (to 2 d.p.)

b) $a^2 + b^2 = c^2$ This time the length marked x is the shortest side, or a.

$x^2 + 6.9^2 = 8.7^2$

$\quad\quad x^2 = 8.7^2 - 6.9^2$ Subtract 6.9^2 from each side.

$\quad\quad x^2 = 75.69 - 47.61$

$\quad\quad x^2 = 28.08$

$\quad\quad x = \sqrt{28.08}$ Take the square root of both sides.

$\quad\quad x = 5.30$ (to 2 d.p.)

TIP

Always check whether you are finding the longest side (the hypotenuse) or one of the shorter sides.

If you are finding the longest side, add the squares.

If you are finding a shorter side, subtract the squares.

EXERCISE 27.2

1 For each of these triangles, find the length marked x.
Where the answer is not exact, give your answer correct to 2 decimal places.

a)

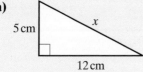

5 cm
12 cm
x

b)

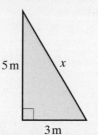

5 m
x
3 m

c)

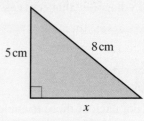

5 cm
8 cm
x

d)

25 cm
7 cm
x

e)

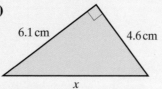

6.1 cm
4.6 cm
x

f)

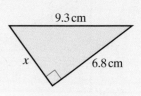

9.3 cm
x
6.8 cm

g)

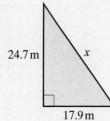

24.7 m
x
17.9 m

h)

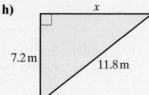

x
7.2 m
11.8 m

i)

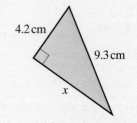

4.2 cm
9.3 cm
x

2 The diagram shows a ladder standing on horizontal ground and leaning against a vertical wall.

The ladder is 4.8 m long and the foot of the ladder is 1.6 m away from the wall.

How far up the wall does the ladder reach?

Give your answer correct to 2 decimal places.

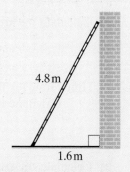

3 The size of a television screen is the length of the diagonal.

The screen size of this television is 27 inches.

If the height of the screen is 13 inches, what is the width? Give your answer correct to 2 decimal places.

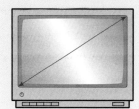

We can also write Pythagoras' theorem in terms of the letters used to name a triangle.

ABC is a triangle right-angled at B.

Pythagoras' theorem is written as

$$AC^2 = AB^2 + BC^2$$

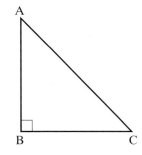

EXAMPLE 27.2

ABCD is a rectangle with length 8 cm and width 6 cm.
Find the length of its diagonals.

Solution

Using Pythagoras' theorem

$$DB^2 = DA^2 + AB^2$$
$$= 36 + 64$$
$$= 100$$
$$DB = \sqrt{100} = 10 \text{ cm}$$

The diagonals of a rectangle are equal, therefore AC = DB = 10 cm.

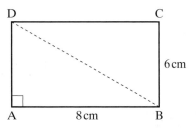

Challenge 27.1

a) Calculate the area of the isosceles triangle ABC.

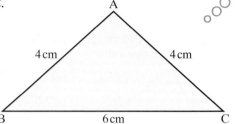

Hint: Draw the height AD of the triangle.
Calculate the length of AD.

b) Calculate the area of each of these isosceles triangles.
Give your answers correct to 1 decimal place.

(i)

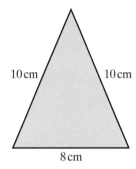

(ii)

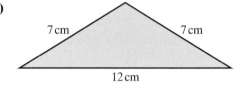

Pythagorean triples

Look again at the answers to question **1** parts **a)** and **d)** in Exercise 27.2.

The answers were exact.

In part **a)** $5^2 + 12^2 = 13^2$

In part **d)** $7^2 + 24^2 = 25^2$

These are examples of **Pythagorean triples**, or three numbers that exactly fit the Pythagoras relationship.

Another Pythagorean triple is $3, 4, 5$.

You saw this in the diagram at the start of the chapter.

$3, 4, 5$ $5, 12, 13$ and $7, 24, 25$ are the most well-known Pythagorean triples.

You can also use Pythagoras' theorem in reverse.

If the lengths of the three sides of a triangle form a Pythagorean triple then the triangle is right-angled.

Work out whether or not each of these triangles is right-angled.
Show your working.

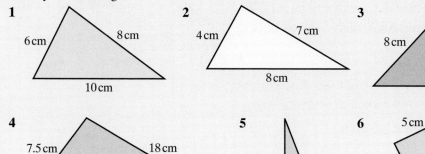

1 6 cm 8 cm 10 cm

2 4 cm 7 cm 8 cm

3 8 cm 8 cm 11 cm

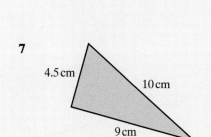

4 7.5 cm 18 cm 19.5 cm

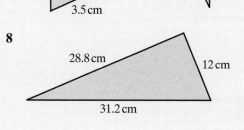

5 12.5 cm 12 cm 3.5 cm

6 5 cm 10 cm 11 cm

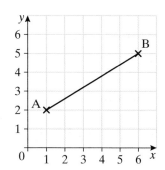

7 4.5 cm 10 cm 9 cm

8 28.8 cm 12 cm 31.2 cm

Coordinates and midpoints

Coordinates

You can use Pythagoras' theorem to find the distance between two points on a grid.

EXAMPLE 27.3

Find the length AB.

Solution

First make a right-angled triangle by drawing across from A and down from B.

By counting squares, you can see that the lengths of the short sides are 5 and 3.

You can then use Pythagoras' theorem to work out the length of AB.

$AB^2 = 5^2 + 3^2$
$AB^2 = 25 + 9$
$AB^2 = 34$
$AB = \sqrt{34}$
$AB = 5.83$ units (correct to 2 d.p.)

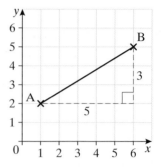

EXAMPLE 27.4

A is the point $(-5, 4)$ and B is the point $(3, 2)$.
Find the length AB.

Solution

Plot the points and complete the right-angled triangle.

Then use Pythagoras' theorem to work out the length of AB.

$AB^2 = 8^2 + 2^2$
$AB^2 = 64 + 4$
$AB^2 = 68$
$AB = \sqrt{68}$
$AB = 8.25$ units (correct to 2 d.p.)

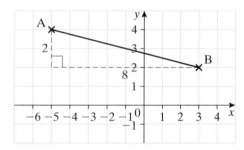

It is also possible to answer Example 27.4 without drawing a diagram.

From $A(-5, 4)$ to $B(3, 2)$ the x value has been increased from -5 to 3. That is, it has been increased by 8.

From $A(-5, 4)$ to $B(3, 2)$ the y value has been decreased from 4 to 2. That is, it has been decreased by 2.

So the short sides of the triangles are 8 units and 2 units.

As before, you can now use Pythagoras' theorem to work out the length of AB.

You may prefer, however, to draw the diagram first.

Midpoints

Discovery 27.2

For each of these pairs of points:

- Draw a diagram on squared paper.
  The first one is done for you.
- Find the middle point of the line joining the
  two points and label it M.
- Write down the coordinates of M.

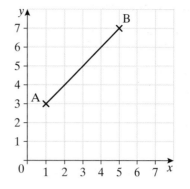

a) A(1, 3) and B(5, 7)

b) C(1, 5) and D(7, 1)

c) E(2, 5) and F(6, 6)

d) G(3, 7) and H(6, 0)

What do you notice?

TIP

'Middle point' is often shortened to **midpoint**.

The coordinates of the midpoint of a line are the means of the coordinates
of the two endpoints.

$$\text{Midpoint of line with coordinates } (a, b), (c, d) = \left(\frac{a + c}{2}, \frac{b + d}{2}\right)$$

EXAMPLE 27.5

Find the coordinates of the midpoints of these pairs of points without drawing the graph.

a) A(2, 1) and B(6, 7) **b)** C(−2, 1) and D(2, 5)

Solution

a) A(2, 1) and B(6, 7)
$a = 2, b = 1, c = 6, d = 7$

$\text{Midpoint} = \left(\frac{a + c}{2}, \frac{b + d}{2}\right)$

$\text{Midpoint} = \left(\frac{2 + 6}{2}, \frac{1 + 7}{2}\right)$

$= (4, 4)$

b) C(−2, 1) and D(2, 5)
$a = -2, b = 1, c = 2, d = 5$

$\text{Midpoint} = \left(\frac{a + c}{2}, \frac{b + d}{2}\right)$

$\text{Midpoint} = \left(\frac{-2 + 2}{2}, \frac{1 + 5}{2}\right)$

$= (0, 3)$

You can check your answers by drawing the graph of the line.

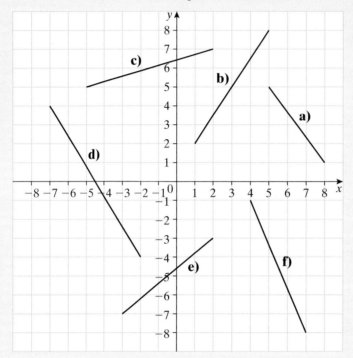

1 Find the coordinates of the midpoint of each of the lines in the diagram.

2 Find the coordinates of the midpoint of the line joining each of these pairs of points.
Try to do them without plotting the points.

   a) A(1, 4) and B(1, 8)   **b)** C(1, 5) and D(7, 3)
   c) E(2, 3) and F(8, 6)   **d)** G(3, 7) and H(8, 2)
   e) I(−2, 3) and J(4, 1)   **f)** K(−4, −3) and L(−6, −11)

Challenge 27.2

a) The midpoint of AB is (5, 3).
A is the point (2, 1).
What are the coordinates of B?

b) The midpoint of CD is (−1, 2).
C is the point (3, 6).
What are the coordinates of D?

Three-dimensional (3-D) coordinates

You already know how to describe a point using coordinates in two dimensions.
You use x- and y-coordinates.

If you are working in three dimensions, you need a third coordinate. This is known as the z-coordinate.

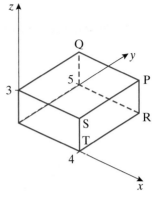

> **TIP**
> Notice that the x- and y-axes lie flat with the z-axis vertical.

The coordinates of point P are $(4, 5, 3)$.
That is, 4 in the x direction, 5 in the y direction and 3 in the z direction.

> **TIP**
> As with 2-D coordinates, 3-D coordinates are written in alphabetical order: x then y then z.

Check up 27.1

What are the coordinates of the points Q, R, S and T in the diagram above?

EXAMPLE 27.6

A is the point $(4, 2, 3)$ and B is the point $(2, 6, 9)$.
What are the coordinates of the midpoint of AB?

Solution

The coordinates of the midpoint of a line are the means of the coordinates of the two endpoints.

In three dimensions:

$$\text{Midpoint of line with coordinates } (a, b, c) \text{ and } (d, e, f) = \left(\frac{a + d}{2}, \frac{b + e}{2}, \frac{c + f}{2}\right)$$

For A$(4, 2, 3)$ and B$(2, 6, 9)$ the coordinates of the midpoint of the line AB are $\left(\frac{4 + 2}{2}, \frac{2 + 6}{2}, \frac{3 + 9}{2}\right) = (3, 4, 6)$.

1 The diagram shows the outline of a cuboid.
   The coordinates of point A are $(5, 0, 0)$.
   The coordinates of point B are $(0, 3, 0)$.
   The coordinates of point C are $(0, 0, 2)$.
   Write down the coordinates of

   a) point D.

   b) point E.

   c) point F.

   d) point G.

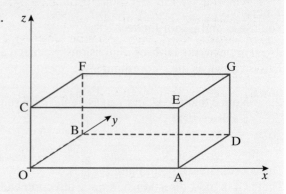

2 VOABC is a square-based pyramid.
   A is the point $(6, 0, 0)$.
   N is the centre of the base.
   The perpendicular height VN of the pyramid is 5 units.
   Write down the coordinates of

   a) point C.

   b) point B.

   c) point N.

   d) point V.

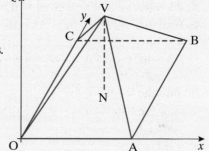

3 The diagram shows the outline of a cuboid.
   The coordinates of point A are $(8, 0, 0)$.
   The coordinates of point B are $(0, 6, 0)$.
   The coordinates of point C are $(0, 0, 4)$.
   L is the midpoint of AD.
   M is the midpoint of EG.
   N is the midpoint of FG.
   Write down the coordinates of

   a) point D.   **b)** point L.

   c) point M.   **d)** point N.

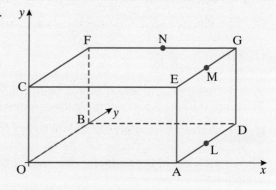

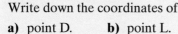

Challenge 27.3

a) For the pyramid in Exercise 27.5, question **2**, calculate these lengths.
   (i) OB   **(ii)** VB

b) For the cuboid in Exercise 27.5, question **1**, calculate the length of OG.

The diagram shows a mast supported by three wires.

The mast is vertical and standing on flat horizontal ground.

The mast is 48 metres tall.
Each wire is attached to a point on the ground 17 metres from the foot of the mast.

Find the total length of the three wires.

Give your answer correct to the nearest metre.

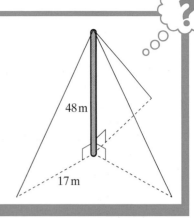

48 m

17 m

WHAT YOU HAVE LEARNED

- The longest side of a right-angled triangle is called the hypotenuse
- Pythagoras' theorem states that the area of the square on the hypotenuse of a right-angled triangle is equal to the sum of the areas of the squares on the other two sides, that is, if the hypotenuse of a right-angled triangle is c and the other sides are a and b, $a^2 + b^2 = c^2$

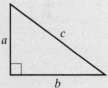

- To find the length of the longest side using Pythagoras' theorem, add the squares
- To find the length of one of the shorter sides using Pythagoras' theorem, subtract the squares
- If the lengths of the three sides of a triangle satisfy $a^2 + b^2 = c^2$, then the triangle is right-angled
- The three most well-known Pythagorean triples are 3, 4, 5; 5, 12, 13 and 7, 24, 25
- The coordinates of the midpoint of the line joining (a, b) to (c, d) are $\left(\dfrac{a + c}{2}, \dfrac{b + d}{2}\right)$
- In three dimensions a point has three coordinates, the third being the z-coordinate

MIXED EXERCISE 27

1 For each of these diagrams, find the area of the third square.

a)

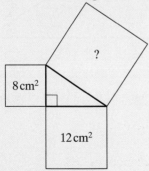

8 cm²

?

12 cm²

b)

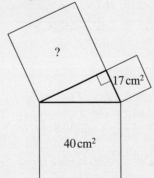

?

17 cm²

40 cm²

2 For each of these triangles, find the length marked x.
Give your answers correct to 2 decimal places.

a)

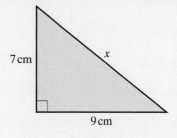

7 cm

x

9 cm

b)

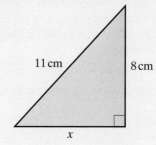

11 cm

8 cm

x

c)

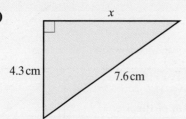

x

4.3 cm

7.6 cm

d)

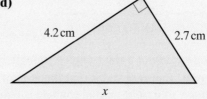

4.2 cm

2.7 cm

x

3 Work out whether or not each of these triangles is right-angled.
Show your working.

a)

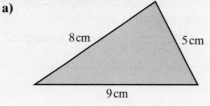

8 cm

5 cm

9 cm

b)

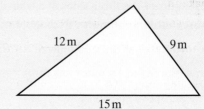

12 m

9 m

15 m

c)

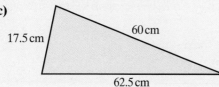

17.5 cm

60 cm

62.5 cm

d)

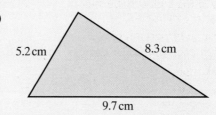

5.2 cm

8.3 cm

9.7 cm

4 Find the coordinates of the midpoint of the line joining each of these pairs of points.
Try to do them without plotting the points.

a) A(2, 1) and B(4, 7)

b) C(2, 3) and D(6, 8)

c) E(2, 0) and F(7, 9)

5 The diagram shows the end of a shed.
The vertical sides are of height 2.8 m and 2.1 m.
The width of the shed is 1.8 m.
Calculate the sloping length of the roof.
Give your answer correct to 2 decimal places.

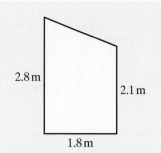

2.8 m

2.1 m

1.8 m

6 Find the area of this isosceles triangle.
Give your answer correct to 1 decimal place.

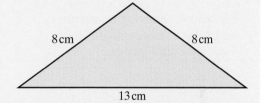

8 cm   8 cm

13 cm

7 The diagram shows a farm gate made from
seven pieces of metal.

The gate is 2.6 m wide and 1.2 m high.

Calculate the total length of metal used to
make the gate.

Give your answer correct to 2 decimal places.

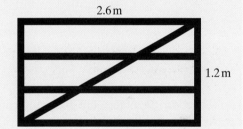

2.6 m

1.2 m

8 The diagram shows the outline of a
house. All the measurements
are in metres.
All the walls are vertical.
E, F, G, H and N are in the
same horizontal plane.

a) Using the axes on the
diagram, write down
the coordinates of these
points.
  (i) B
  (ii) H
  (iii) G

b) The coordinates of Q are
(11, 5, 9).
N is vertically below Q.
Write down the coordinates of N.

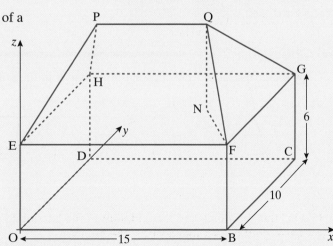

28 → TRANSFORMATIONS

Reflections

In a reflection, the object and image are **congruent**. Congruent means exactly the same shape and size.

Drawing reflections

Check up 28.1

Copy these diagrams. Reflect each of the shapes in the mirror line.

a)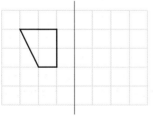

b)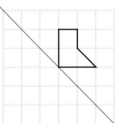

> **TIP**
> When you have drawn a reflection in a sloping line, check it by turning the page so that the line is vertical. You can also use a mirror or tracing paper.

Recognising and describing reflections

You also need to be able to recognise and describe reflections.

You can tell whether a shape has been reflected using tracing paper: if you trace the **object**, you will have to turn the tracing paper over to fit the tracing on to the **image**.

You also need to find the **mirror line**. You do this by measuring the distance between points on the object and image.

EXAMPLE 28.1

Describe the single transformation that maps shape ABC on to shape A′B′C′.

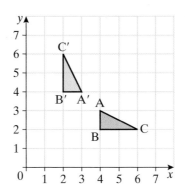

Solution

You can probably tell just by looking that the transformation is a reflection, but you could check using tracing paper.

To find the mirror line, put a ruler between two corresponding points (B and B′) and mark the midpoint of the line between them. The midpoint is $(3, 3)$.

Do the same for two other corresponding points (C and C′).

The midpoint is $(4, 4)$.

Join the points to find the mirror line. The mirror line passes through $(1, 1), (2, 2), (3, 3), (4, 4) \dots$.
It is the line $y = x$.

The transformation is a reflection in the line $y = x$.

> TIP
>
> You must both state that the transformation is a reflection and give the mirror line.

> TIP
>
> Check the line is correct by turning the page so that the mirror line is vertical.

The mirror line can be any straight line.

Draw a pair of axes and label them −4 to 4 for x and y.
Draw these lines on the graph and label them.

a) $x = 2$ **b)** $y = -3$ **c)** $y = x$ **d)** $y = -x$

EXERCISE 28.1

1 Draw a pair of axes and label them −4 to 4 for x and y.
  a) Draw a triangle with vertices at $(1, 0), (1, -2)$ and $(2, -2)$. Label it A.
  b) Reflect triangle A in the line $y = 1$. Label the image B.
  c) Reflect triangle B in the line $y = x$. Label the image C.

2 Draw a pair of axes and label them −4 to 4 for x and y.
  a) Draw a triangle with vertices at $(1, 1), (2, 3)$ and $(3, 3)$. Label it A.
  b) Reflect triangle A in the line $y = 2$. Label the image B.
  c) Reflect triangle A in the line $y = -x$. Label the image C.

3 For each part
  • copy the diagram carefully, making it larger if you wish.
  • reflect the shape in the mirror line.

  a) 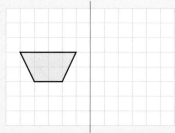   **b)** 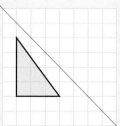   **c)**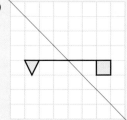

4 Describe fully the single transformation
  that maps
  a) flag A on to flag B.
  b) flag A on to flag C.
  c) flag B on to flag D.

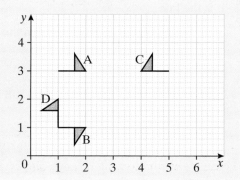

5 Describe fully the single
transformation that maps

    a) triangle A on to triangle B.

    b) triangle A on to triangle C.

    c) triangle C on to triangle D.

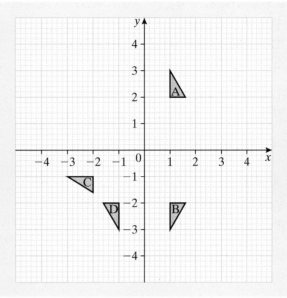

Rotations

In a rotation, the object and image are congruent.

Drawing rotations

Check up 28.3

Rotate each of these shapes as described.

a) Rotation of 90° anticlockwise
about the origin.

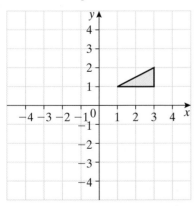

b) Rotation of 270° anticlockwise
about its centre, A.

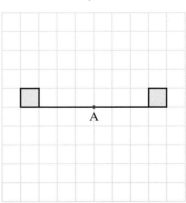

The centre of rotation can be any point. It does not have to be either the
origin or the centre of the shape.

EXAMPLE 28.2

Rotate the shape through 90° anticlockwise about the point C(1, 2).

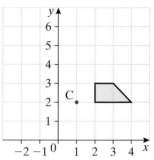

Solution

You can rotate the shape using tracing paper or you can count squares.

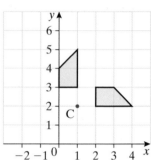

EXAMPLE 28.3

Rotate triangle ABC through 90° clockwise about C.

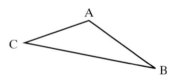

Solution

Because the triangle is not drawn on squared paper, you have to use a different method.

Measure an angle of 90° clockwise at C from the line AC, and draw a line.

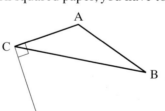

Trace the shape ABC.
Put a pencil or pin at C to hold the tracing to the diagram at that point.
Rotate the tracing paper until AC coincides with the line drawn.
Use a pin or the point of your compasses to prick through the other corners, A and B.
Join up the new points to make the image.

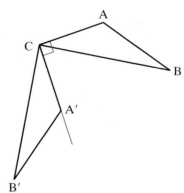

When the centre of rotation is not on the shape the method is slightly different.

EXAMPLE 28.4

Rotate triangle ABC through 90° clockwise about point P.

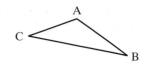

Solution

Join P to a point C on the object. Measure an angle of 90° clockwise from PC and draw a line.

Trace the triangle ABC and the line PC.

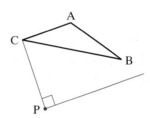

Put a pencil or pin at P to hold the tracing to the diagram at that point.

Rotate the tracing paper until PC coincides with the line drawn.

Use a pin or the point of your compasses to prick through the corners A, B and C.

Join up the new points to make the image.

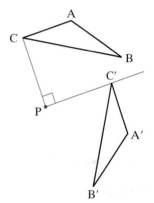

> **TIP**
>
> Corresponding points will be the same distance from the centre of rotation.

Recognising and describing rotations

Challenge 28.1

Which of the triangles B, C, D, E, F and G are reflections of triangle A and which are rotations of triangle A?

Hint: For reflections the tracing paper needs to be turned over, for rotations it does not.

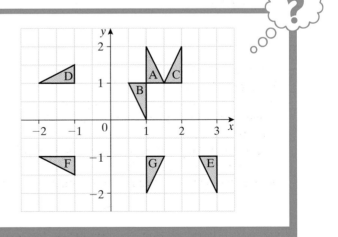

To describe a rotation, you need to know the **angle of rotation** and the **centre of rotation**.

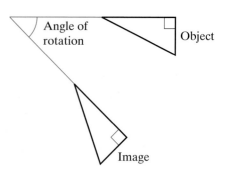

Sometimes you can tell the angle of rotation just by looking at the diagram.

If you can't, you need to identify a pair of sides that correspond in the object and image and measure the angle between them. You may need to extend the lines.

You can usually find the centre of rotation by counting squares or using tracing paper.

EXAMPLE 28.5

Describe fully the single transformation that maps flag A on to flag B.

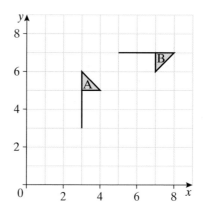

Solution

It is clear that the transformation is a rotation and that the angle is 90° clockwise. This is a rotation of 90° clockwise.

Use tracing paper and a pencil or compass point to find the centre of rotation.

Trace flag A and use the pencil or compass point to hold the tracing to the diagram at a point.
Rotate the tracing paper and see if the tracing fits over flag B.
Keep trying different points until you find the centre of rotation.
Here, the centre of rotation is (6, 4).

The transformation is a rotation of 90° clockwise about the point (6, 4).

> **TIP**
> You must state that the transformation is a rotation and give the angle of rotation, the direction of rotation and the centre of rotation.

1 Copy the diagram.

a) Rotate shape A through 90° clockwise about the origin. Label the image B.

b) Rotate shape A through 180° about the point (1, 2). Label the image C.

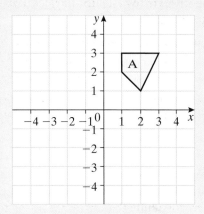

2 Copy the diagram.

a) Rotate flag A through 90° anticlockwise about the origin. Label the image B.

b) Rotate flag A through 90° clockwise about the point (1, 2). Label the image C.

c) Rotate flag A through 180° about the point (2, 0). Label the image D.

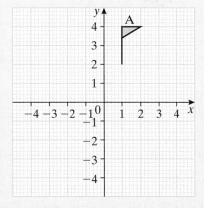

3 Draw a pair of axes and label them −4 to 8 for x and y.

a) Draw a triangle with vertices (0, 1), (0, 4) and (2, 3). Label it A.

b) Rotate triangle A through 180° about the origin. Label the image B.

c) Rotate triangle A through 90° anticlockwise about the point (0, 1). Label the image C.

d) Rotate triangle A through 90° clockwise about the point (2, −1). Label the image D.

4 Copy the diagram.

Rotate the triangle through 90° clockwise about point C.

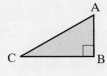

5 Copy the diagram.

Rotate the triangle through 90° clockwise about the point O.

×O

6 Copy the diagram.

Rotate the triangle through 120° clockwise about the point C.

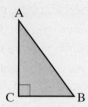

7 Describe fully the single transformation that maps

   a) triangle A on to triangle B.

   b) triangle A on to triangle C.

   c) triangle A on to triangle D.

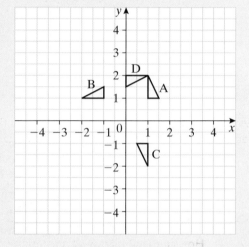

8 Describe fully the single transformation that maps

   a) flag A on to flag B.

   b) flag A on to flag C.

   c) flag A on to flag D.

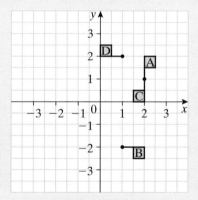

9 Describe fully the single transformation that maps triangle A on to triangle B.

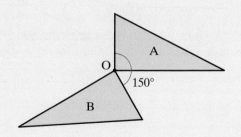

10 Describe fully the single transformation that maps

    a) triangle A on to triangle B.

    b) triangle A on to triangle C.

    c) triangle A on to triangle D.

    d) triangle A on to triangle E.

    e) triangle B on to triangle E.

    Hint: Some of these transformations are reflections.

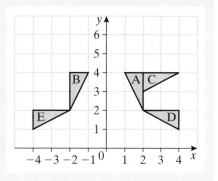

Translations

A **translation** moves all points of an object the same distance in the same direction. The object and image are congruent.

Discovery 28.1

Triangle B is a translation of triangle A.

a) How can you tell it is a translation?

b) How far across has it moved?

c) How far down has it moved?

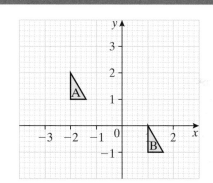

> **TIP**
>
> Take care with the counting. Choose a point on both the object and image and count the squares from one to the other.

How far a shape moves in a translation is written as a **column vector**.

The *top* number tells you how far the shape moves *across*, or in the *x* direction.
The *bottom* number tells you how far the shape moves *up or down*, or in the *y* direction.

A *positive* top number is a move to the *right*. A *negative* top number is a move to the *left*.
A *positive* bottom number is a move *up*. A *negative* bottom number is a move *down*.

A translation of 3 to the right and 2 down is written as $\begin{pmatrix} 3 \\ -2 \end{pmatrix}$.

EXAMPLE 28.6

Translate the triangle by $\begin{pmatrix} -3 \\ 4 \end{pmatrix}$.

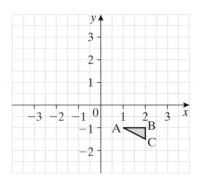

Solution

$\begin{pmatrix} -3 \\ 4 \end{pmatrix}$ means move 3 units left, and 4 units up.

Point A moves from $(1, -1)$ to $(-2, 3)$.
Point B moves from $(2, -1)$ to $(-1, 3)$.
Point C moves from $(2, -1.5)$ to $(-1, 2.5)$.

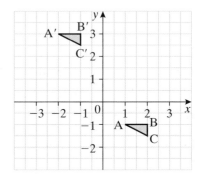

EXAMPLE 28.7

Describe fully the single transformation that maps shape A on to shape B.

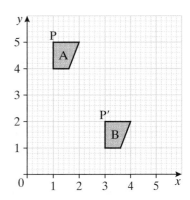

Solution

It is clearly a translation as the shape stays the same way up.

To find the movement choose one point on the object and image and count the squares moved.
For example, P moves from $(1, 5)$ to $(3, 2)$. This is a movement of 2 to the right and 3 down.

The transformation is a translation of $\begin{pmatrix} 2 \\ -3 \end{pmatrix}$.

TIP

You must both state that the transformation is a translation and give the column vector.

TIP

Try not to confuse the words *transformation* and *translation*.

Transformation is the general name for all changes made to shapes.

Translation is the particular transformation where all points of an object move the same distance in the same direction.

EXERCISE 28.3

1 Draw a pair of axes and label them −2 to 6 for x and y.

   a) Draw a triangle with vertices at $(1, 2), (1, 4)$, and $(2, 4)$. Label it A.

   b) Translate triangle A by vector $\begin{pmatrix} 2 \\ 1 \end{pmatrix}$. Label the image B.

   c) Translate triangle A by vector $\begin{pmatrix} 4 \\ -2 \end{pmatrix}$. Label the image C.

   d) Translate triangle A by vector $\begin{pmatrix} -2 \\ -3 \end{pmatrix}$. Label the image D.

2 Draw a pair of axes and label them −2 to 6 for x and y.

   a) Draw the trapezium with vertices at $(2, 1), (4, 1), (3, 2)$ and $(2, 2)$. Label it A.

   b) Translate trapezium A by vector $\begin{pmatrix} 2 \\ 3 \end{pmatrix}$. Label the image B.

   c) Translate trapezium A by vector $\begin{pmatrix} -4 \\ 0 \end{pmatrix}$. Label the image C.

   d) Translate trapezium A by vector $\begin{pmatrix} -3 \\ 2 \end{pmatrix}$. Label the image D.

3 Describe the single transformation that maps

   a) triangle A on to triangle B.

   b) triangle A on to triangle C.

   c) triangle A on to triangle D.

   d) triangle B on to triangle D.

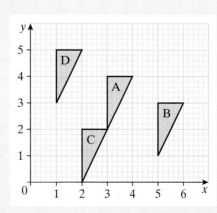

4 Describe the single transformation that maps

  a) flag A on to flag B.

  b) flag A on to flag C.

  c) flag A on to flag D.

  d) flag A on to flag E.

  e) flag A on to flag F.

  f) flag E on to flag G.

  g) flag B on to flag E.

  h) flag C on to flag D.

  Hint: Not all the transformations are translations.

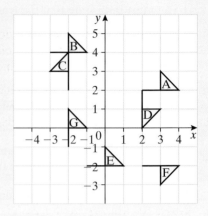

Challenge 28.2

Draw a pair of axes and label them −6 to 6 for x and y.

a) Draw a shape in the positive region near the origin. Label it A.

b) Translate shape A by vector $\begin{pmatrix} 2 \\ 1 \end{pmatrix}$. Label the image B.

c) Translate shape B by vector $\begin{pmatrix} 3 \\ -2 \end{pmatrix}$. Label the image C.

d) Translate shape C by vector $\begin{pmatrix} -6 \\ -1 \end{pmatrix}$. Label the image D.

e) Translate shape D by vector $\begin{pmatrix} 1 \\ 2 \end{pmatrix}$. Label the image E.

f) What do you notice about shapes A and E? Can you suggest why this happens? Try to find other combinations of translations for which this happens.

Challenge 28.3

Sometimes you can apply different transformations to an object and get the same image.

a) Describe all the different single transformations that will map shape A on to shape B.

b) Try to find other pairs of shapes where the object, A, can be mapped on to the image, B, by more than one single transformation.

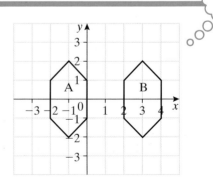

Enlargements

There is another kind of transformation: **enlargement**. In an enlargement, the object and image are not congruent, but they are **similar**. The lengths change but the angles in the object and the image are the same.

You will have done some work on enlargements with positive integer scale factors at Key Stage 3. In this section you will learn about fractional scale factors and in Chapter 35 you will learn about negative scale factors.

Check up 28.4

Draw a pair of axes and label them 0 to 6 for x and y.

a) Draw a triangle with vertices at $(1, 2)$, $(3, 2)$ and $(3, 3)$. Label it A.

b) Enlarge the triangle by scale factor 2, with the origin as the centre of enlargement. Label the image B.

Discovery 28.2

a) (i) Think about what happens to the lengths of the sides of an object when it is enlarged by scale factor 2.
    What do you think will happen to the lengths of the sides of an object if it is enlarged by scale factor $\frac{1}{2}$?

   (ii) Think about the position of the image when an object is enlarged by scale factor 2. What happens to the distance between the centre of enlargement and the object? What do you think will be the position of the image if an object is enlarged by scale factor $\frac{1}{2}$?

b) Draw a pair of axes and label them 0 to 6 for x and y.
   (i) Draw a triangle with vertices at $(2, 4)$, $(6, 4)$ and $(6, 6)$. Label it A.
   (ii) Enlarge the triangle by scale factor $\frac{1}{2}$, with the origin as the centre of enlargement. Label the image B.

c) Compare your diagram with the diagram you drew in Check up 28.4. What do you notice?

An enlargement with scale factor $\frac{1}{2}$ is the **inverse** of an enlargement with scale factor 2.

> **TIP**
> Although the image is smaller than the object, an enlargement with scale factor $\frac{1}{2}$ is still called an enlargement.

You can also draw enlargements with other fractional scale factors.

EXAMPLE 28.8

Draw a pair of axes and label them 0 to 8 for both x and y.
a) Draw a triangle with vertices at P(5, 1), Q(5, 7) and R(8, 7).
b) Enlarge the triangle PQR by scale factor $\frac{1}{3}$, centre C(2, 1).

Solution

The sides of the enlargement are $\frac{1}{3}$ the lengths of the original.

The distance from the centre of enlargement, C, to P is 3 across.
So the distance from C to P' is $3 \times \frac{1}{3} = 1$ across.

The distance from C to Q is 3 across and 6 up.
So the distance from C to Q' is $3 \times \frac{1}{3} = 1$ across and $6 \times \frac{1}{3} = 2$ up.

The distance from C to R is 6 across and 6 up.
So the distance from C to R' is $6 \times \frac{1}{3} = 2$ across and $6 \times \frac{1}{3} = 2$ up.

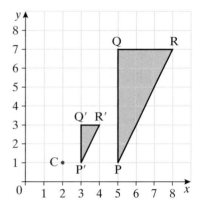

EXAMPLE 28.9

Describe fully the single transformation that maps triangle PQR on to triangle P'Q'R'.

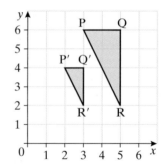

Solution

It is obvious that the shape has been enlarged.
The length of each side of triangle P'Q'R' is half the length of the corresponding side of triangle PQR, so the scale factor is $\frac{1}{2}$.

To find the centre of enlargement, join the corresponding corners of the two triangles and extend the lines until they cross.
The point where they cross is the centre of enlargement, C. Here, C is at (1, 2).

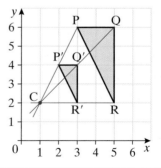

> **TIP**
>
> To check the centre of enlargement, find the distances from the centre of enlargement to a corresponding pair of points. For example, the distance from C to P is 2 across and 4 up, so the distance from C to P' should be 1 across and 2 up.

The transformation is an enlargement with scale factor $\frac{1}{2}$, centre $(1, 2)$.

TIP
You must state that the transformation is an enlargement and give the scale factor and centre of enlargement.

1 Draw a pair of axes and label them 0 to 6 for both x and y.

    a) Draw a triangle with vertices at $(4, 2), (6, 2)$ and $(6, 6)$. Label it A.

    b) Enlarge triangle A by scale factor $\frac{1}{2}$, with the origin as the centre of enlargement. Label the image B.

    c) Describe fully the single transformation that maps triangle B on to triangle A.

2 Draw a pair of axes and label them 0 to 8 for both x and y.

    a) Draw a triangle with vertices at $(4, 5), (4, 8)$ and $(7, 8)$. Label it A.

    b) Enlarge triangle A by scale factor $\frac{1}{3}$, with centre of enlargement $(1, 2)$. Label the image B.

    c) Describe fully the single transformation that maps triangle B on to triangle A.

3 Draw a pair of axes and label them 0 to 8 for both x and y.

    a) Draw a triangle with vertices at $(0, 2), (1, 2)$ and $(2, 1)$. Label it A.

    b) Enlarge triangle A by scale factor 4, with the origin as the centre of enlargement. Label the image B.

    c) Describe fully the single transformation that maps triangle B on to triangle A.

4 Draw a pair of axes and label them 0 to 8 for both x and y.

    a) Draw a triangle with vertices at $(4, 3), (4, 5)$ and $(6, 2)$. Label it A.

    b) Enlarge triangle A by scale factor $1\frac{1}{2}$, with centre of enlargement $(2, 1)$. Label the image B.

    c) Describe fully the single transformation that maps triangle B on to triangle A.

5 Describe fully the single transformation that maps

    a) triangle A on to triangle B.

    b) triangle B on to triangle A.

    c) triangle A on to triangle C.

    d) triangle C on to triangle A.

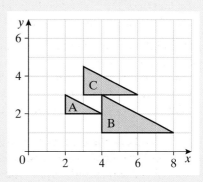

Hint: In questions **6** to **8**, not all the transformations are enlargements.

6 Describe fully the single transformation that maps

a) triangle A on to triangle B.

b) triangle A on to triangle C.

c) triangle C on to triangle D.

d) triangle A on to triangle E.

e) triangle A on to triangle F.

f) triangle G on to triangle A.

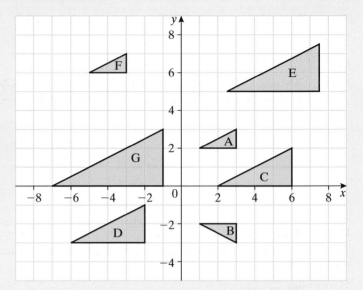

7 Describe fully the single transformation that maps

a) flag A on to flag B.

b) flag A on to flag C.

c) flag A on to flag D.

d) flag A on to flag E.

e) flag F on to flag E.

f) flag E on to flag G.

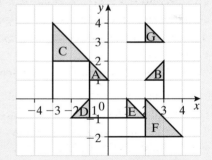

8 Describe fully the single transformation that maps

a) triangle A on to triangle B.

b) triangle A on to triangle C.

c) triangle B on to triangle D.

d) triangle C on to triangle E.

e) triangle F on to triangle G.

f) triangle H on to triangle G.

g) triangle G on to triangle H.

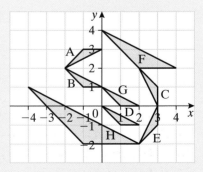

Challenge 28.4

A triangle ABC has sides AB = 9 cm, AC = 7 cm and BC = 6 cm.

A line XY is drawn parallel to BC through a point X on AB and a point Y on AC.

AX = 5 cm.

a) Draw a sketch of the triangle.

b) **(i)** Describe fully the transformation that maps triangle ABC on to triangle AXY.
  (ii) Work out the length of XY, correct to 2 decimal places.

WHAT YOU HAVE LEARNED

- In reflections, rotations and translations, the object and image are congruent
- To describe a reflection, you must state that the transformation is a reflection and give the mirror line
- How to find the mirror line
- To describe a rotation, you must state that the transformation is a rotation and give the centre of rotation, the angle of rotation and the direction of rotation
- How to find the centre of rotation and the angle of rotation
- In a translation, the object and image are the same way around
- To describe a translation, you must state that the transformation is a translation and give the column vector
- What the column vector $\begin{pmatrix} a \\ b \end{pmatrix}$ represents
- In an enlargement, the object and the image are similar. If the scale factor is fractional, the image will be smaller than the object
- To describe an enlargement, you must state that the transformation is an enlargement and give the scale factor and the centre of enlargement
- How to find the scale factor and the centre of enlargement

MIXED EXERCISE 28

1 Draw a pair of axes and label them −4 to 4 for x and y.

  a) Draw a triangle with vertices at $(2, -1), (4, -1)$ and $(4, -2)$. Label it A.

  b) Reflect triangle A in the line $y = 0$. Label the image B.

  c) Reflect triangle A in the line $y = -x$. Label the image C.

 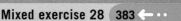

2 Copy these diagrams, making them larger if you wish.
Reflect each shape in the mirror line.

a)

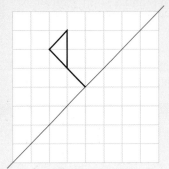

b)

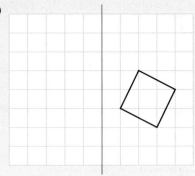

3 Copy the diagram.

  a) Rotate shape A through 90° anticlockwise about the origin. Label the image B.

  b) Rotate shape A through 180° about the point $(2, -1)$. Label the image C.

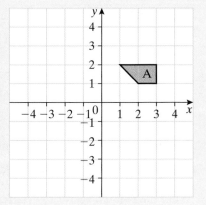

4 Copy the diagram.

  a) Translate shape A by vector $\begin{pmatrix} 1 \\ -6 \end{pmatrix}$. Label the image B.

  b) Translate shape A by vector $\begin{pmatrix} -3 \\ 0 \end{pmatrix}$. Label the image C.

  c) Translate shape A by vector $\begin{pmatrix} -5 \\ -4 \end{pmatrix}$. Label the image D.

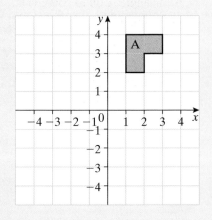

5 Draw a pair of axes and label them 0 to 9 for both x and y.

  a) Draw a triangle with vertices at $(6, 3)$, $(6, 6)$ and $(9, 3)$. Label it A.

  b) Enlarge the triangle by scale factor 13, with centre of enlargement $(0, 0)$. Label the image B.

6 Draw a pair of axes and label them 0 to 8 for both x and y.

   a) Draw a triangle with vertices at $(6, 4)$, $(6, 6)$ and $(8, 6)$. Label it A.

   b) Enlarge the triangle by scale factor $\frac{1}{2}$, with centre of enlargement $(2, 0)$. Label the image B.

7 Describe fully the single transformation that maps

   a) flag A on to flag B.

   b) flag A on to flag C.

   c) flag A on to flag D.

   d) flag B on to flag E.

   e) flag F on to flag C.

   f) flag C on to flag G.

   g) flag B on to flag C.

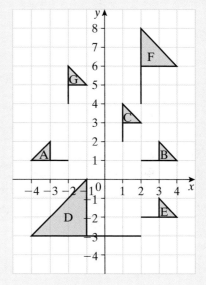

8 Describe fully the single transformation that maps

   a) triangle A on to triangle B.

   b) triangle A on to triangle C.

   c) triangle B on to triangle D.

   d) triangle C on to triangle E.

   e) triangle F on to triangle C.

   f) triangle A on to triangle G.

   g) triangle H on to triangle G.

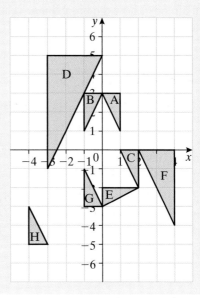

29 → MEASURES

THIS CHAPTER IS ABOUT

- Converting between measures, especially measures of area and volume
- Accuracy of measurement
- Giving answers to a sensible degree of accuracy
- Using compound measures such as speed and density

YOU SHOULD ALREADY KNOW

- The common metric units for length, area, volume and capacity

Converting between measures

You already know the basic **linear** relationships between metric measures. Linear means 'of length'.

You can use these relationships to work out the relationships between metric units of area and volume.

For example:

$1 \text{ cm} = 10 \text{ mm}$ $1 \text{ m} = 100 \text{ cm}$

$1 \text{ cm}^2 = 1 \text{ cm} \times 1 \text{ cm}$ $1 \text{ m}^2 = 1 \text{ m} \times 1 \text{ m}$
$1 \text{ cm}^2 = 10 \text{ mm} \times 10 \text{ mm}$ $1 \text{ m}^2 = 100 \text{ cm} \times 100 \text{ cm}$
$1 \text{ cm}^2 = 100 \text{ mm}^2$ $1 \text{ m}^2 = 10\,000 \text{ cm}^2$

$1 \text{ cm}^3 = 1 \text{ cm} \times 1 \text{ cm} \times 1 \text{ cm}$ $1 \text{ m}^3 = 1 \text{ m} \times 1 \text{ m} \times 1 \text{ m}$
$1 \text{ cm}^3 = 10 \text{ mm} \times 10 \text{ mm} \times 10 \text{ mm}$ $1 \text{ m}^3 = 100 \text{ cm} \times 100 \text{ cm} \times 100 \text{ cm}$
$1 \text{ cm}^3 = 1000 \text{ mm}^3$ $1 \text{ m}^3 = 1\,000\,000 \text{ cm}^3$

EXAMPLE 29.1

Change these measures to the units shown.

a) 5 m^3 to cm^3 **b)** 5600 cm^2 to m^2

Solution

a) $5 \text{ m}^3 = 5 \times 1\,000\,000 \text{ cm}^3$ Convert 1 m^3 to cm^3 and multiply by 5.
   $= 5\,000\,000 \text{ cm}^3$

b) $5600 \text{ cm}^2 = 5600 \div 10\,000 \text{ m}^2$ To convert m^2 to cm^2 you multiply, so to convert cm^2 to m^2
   $= 0.56 \text{ m}^2$ you divide. Make sure you have done the right thing by
          checking your answer makes sense. If you had multiplied
          by 10 000 you would have got $56\,000\,000 \text{ m}^2$, which is
          obviously a much larger area than 5600 cm^2.

EXERCISE 29.1

1 Change these measures to the units shown.

   a) 25 m to cm   **b)** 42 cm to mm   **c)** 2.36 m to cm   **d)** 5.1 m to mm

2 Change these measures to the units shown.

   a) 3 m^2 to cm^2 **b)** 2.3 cm^2 to mm^2 **c)** 9.52 m^2 to cm^2 **d)** 0.014 cm^2 to mm^2

3 Change these measures to the units shown.

   a) $90\,000 \text{ mm}^2$ to cm^2 **b)** 8140 mm^2 to cm^2 **c)** $7\,200\,000 \text{ cm}^2$ to m^2 **d)** $94\,000 \text{ cm}^2$ to m^2

4 Change these measures to the units shown

   a) 3.2 m^3 to cm^3 **b)** 42 cm^3 to m^3 **c)** 5000 cm^3 to m^3 **d)** 6.42 m^3 to cm^3

5 Change these measures to the units shown.

   a) 2.61 litres to cm^3 **b)** 9500 ml to litres   **c)** 2.4 litres to ml   **d)** 910 ml to litres

6 What is wrong with this statement?

   The trench I have just dug is 5 m long, 2 m wide and 50 cm deep.
   To fill it in, I would need 500 m^3 of concrete.

7 A pane of glass measures 32 cm by 65 cm and is 0.3 cm thick.
   What is the volume of the glass?

Challenge 29.1

Cleopatra is reputed to have had a bath filled with asses' milk.
Today her bath might be filled with cola!

Assuming a can of drink holds 33 centilitres, approximately how many cans would she
need to have a bath in cola?

Accuracy in measurement

As you saw in Chapter 8, all measurements are **approximations**.
Measurements are given to the nearest practical unit.

Measuring a value to the nearest unit means deciding that it is nearer to
one mark on a scale than another; in other words, that the value is within
half a unit of that mark.

Look at this diagram.

Any value within the shaded area is 5 to the nearest unit.

The boundaries for this interval are 4.5 and 5.5. This would be written as
$4.5 \leqslant x < 5.5$.

> **TIP**
>
> Remember that $4.5 \leqslant x < 5.5$ means all values, x, which are
> greater than or equal to 4.5 but less than 5.5.
>
> $x < 5.5$ because if $x = 5.5$ it would round up to 6.
> Even though x cannot equal 5.5, it can be as close to it as you like.
> So 5.5 is the upper bound. Do not use 5.499 or 5.4999, etc.

4.5 is the lower bound and 5.5 the upper bound.

Any value less than 4.5 is closer to 4 (4 to the nearest unit).
Any value greater than or equal to 5.5 is closer to 6 (6 to the nearest unit).

EXAMPLE 29.2

a) Tom won the 100 m race with a time of 12.2 seconds, to the nearest
tenth of a second.
What are the upper and lower bounds for this time?

b) Copy and complete this statement.

> A mass given as 46 kg, to the nearest kilogram, lies between
> .... kg and .... kg.

Solution

a) Lower bound = 12.15 seconds, upper bound = 12.25 seconds

b) A mass given as 46 kg, to the nearest kilogram, lies between 45.5 kg
and 46.5 kg.

1 Copy and complete each of these statements.

a) A height given as 57 m, to the nearest metre, is between .... m and .... m.

b) A volume given as 568 ml, to the nearest millilitre, is between .... ml and .... ml.

c) A winning time given as 23.93 seconds, to the nearest hundredth of a second, is between .... seconds and .... seconds.

2 Copy and complete each of these statements.

a) A mass given as 634 g, to the nearest gram, is between .... g and .... g.

b) A volume given as 234 ml, to the nearest millilitre, is between .... ml and .... ml.

c) A height given as 8.3 m, to 1 decimal place, is between .... m and .... m.

3

Gloss Paint

Coverage from 7 to 8 square metres
depending on surface

750 ml

a) What are the least and greatest surface areas that 3 litres of the paint will cover?

b) How many cans of paint are needed to ensure coverage of an area of 100 m²?

4 Jessica measures the thickness of a metal sheet with a gauge.
The reading is 4.97 mm, accurate to the nearest hundredth of a millimetre.

a) What is the minimum thickness the sheet could be?

b) What is the maximum thickness the sheet could be?

5 Gina is fitting a new kitchen.
She has an oven which is 595 mm wide, to the nearest millimetre.
Will it definitely fit in a gap which is 60 cm wide, to the nearest centimetre?

6 Two metal blocks are placed together as shown.
The left-hand block is 6.3 cm long and the right-hand block is 8.7 cm.
Both blocks are 2 cm wide and 2 cm deep.
All measurements are correct to the nearest millimetre.

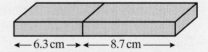

←—6.3 cm —→ ←—— 8.7 cm ——→

a) What are the least and greatest combined lengths of the two blocks?

b) What are the least and greatest depths of the blocks?

c) What are the least and greatest widths of the blocks?

7 A company manufactures components for the car industry. One component consists of a metal block with a hole drilled into it. A plastic rod is fixed into the hole.

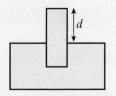

The hole is drilled to a depth of 20 mm, to the nearest millimetre.
The length of the rod is 35 mm, to the nearest millimetre.
What are the maximum and minimum values of d (the height of the rod above the block)?

8 The diagram shows a rectangle, ABCD.
AB = 15 cm and BC = 9 cm.
All measurements are correct to the nearest centimetre.
Work out the least and greatest values for the perimeter of the rectangle.

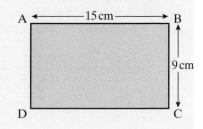

Working to a sensible degree of accuracy

Measurements and calculations should not be too accurate for their purpose.

It is obviously silly to claim that:

	a car journey took 4 hours, 56 minutes and 13 seconds
or	the distance between two houses is 93 kilometres, 484 metres and 78 centimetres.

Answers such as these would be more sensible rounded to 5 hours and 93 km respectively.

When calculating a measurement, you need to give your answer to a **sensible degree of accuracy**.
As a general rule your final answer should not be given to a greater degree of accuracy than any of the values used in the calculation.

EXAMPLE 29.3

A table is 1.8 m long and 1.3 m wide. Both measurements are correct to 1 decimal place.

Work out the area of the table.
Give your answer to a sensible degree of accuracy.

Solution

Area = length × width
     = 1.8 × 1.3
     = 2.34   The answer has 2 decimal places.
     = 2.3 m² (to 1 d.p.) However, the answer should not be more accurate than the original measurements. So you need to round the answer to 1 decimal place.

1 Rewrite each of these statements using sensible values for the measurements.

   a) It takes 3 minutes and 24.8 seconds to boil an egg.

   b) It will take me 2 weeks, 5 days, 3 hours and 13 minutes to paint your house.

   c) Helen's favourite book weighs 2.853 kg.

   d) The height of the classroom door is 2 metres, 12 centimetres and 54 millimetres.

2 Give your answer to each of these questions to a sensible degree of accuracy.

   a) Find the length of the side of a square field whose area is 33 m².

   b) Three friends share £48.32 equally. How much will each receive?

   c) It takes 1.2 hours to fly between two cities at 554 km/h.
      How far apart are they?

   d) A strip of card is 2.36 cm long and 0.041 cm wide.
      Calculate the area of the card.

Challenge 29.2

The health provision in different countries can be compared using various measures.

One measure is the number of under-one-year-olds who die per 1000 live births. Another measure could be the number of people surviving beyond the age of 65 years expressed as a percentage of the total population.

Use this table, showing figures for 2010, to compare the health provision in these countries.

Country	Population in millions	Live births in a year	Infant deaths in a year	Number of people aged 65 years and over
Australia	21.5	266 400	1 250	2 950 000
China	1 330	16 180 000	2 671 300	114 000 000
Burma	53.4	1 041 000	52 840	2 670 000
UK	61.2	653 000	3 120	10 040 000

Compound measures

Some measures are calculated using the same type of measurements. Area, for example, is calculated using length and width, which are both measures of length.

Compound measures are calculated using two different types of measure.
Speed is calculated using distance and time.

$$\text{Speed} = \frac{\text{distance}}{\text{time}}$$

Compound measures are written with **compound units**.
The units for speed are written in the form distance per unit of time.
For example, if the distance is in kilometres and the time is in hours, speed
is written as kilometres per hour, or km/h.

Another compound measure is **density**. Density is linked to **mass** and
volume.

$$\text{Density} = \frac{\text{mass}}{\text{volume}}$$

EXAMPLE 29.4

a) Calculate the average speed of a car travelling 80 km in 2 hours.
b) Gold has a density of 19.3 g/cm^3.
 Calculate the mass of a gold bar with a volume of 30 cm^3.

Solution

a) Speed $= \dfrac{\text{distance}}{\text{time}}$

$= \dfrac{80}{2} = 40$ km/h

b) Density $= \dfrac{\text{mass}}{\text{volume}}$ First rearrange the formula to make mass the subject.

Mass $=$ density $\times$ volume
$= 19.3 \times 30 = 579$ g

EXERCISE 29.4

1 A train covers a distance of 1250 metres in a time of 20 seconds.
 Calculate its average speed.

2 Caroline's car travels 129 miles in 3 hours. Calculate her average speed.

3 Paula jogs at a steady 6 miles per hour.
 How far does she run in one and a quarter hours?

4 How long will it take a boat sailing at 12 km/h to travel 60 km?

5 The density of aluminium is 2.7 g/cm³. What is the volume of a block of aluminium with a mass of 750 g? Give your answer to the nearest whole number.

6 Find the average speed of a car which travelled 150 miles in two and a half hours.

7 Calculate the density of a rock of mass 780 g and volume 84 cm³.
Give your answer to a suitable degree of accuracy.

8 A car travels 20 km in 12 minutes. What is the average speed in km/h?

9 Calculate the density of a stone of mass 350 g and volume 45 cm³.

10 **a)** Calculate the density of a 3 cm³ block of copper of mass 26.7 g.

   b) What would be the mass of a 17 cm³ block of copper?

11 Gold has a density of 19.3 g/cm³. Calculate the mass of a gold bar of volume 1000 cm³.
Give your answer in kilograms.

12 Air at normal room temperature and pressure has a density of 1.3 kg/m³.

   a) What mass of air is there in a room which is a cuboid measuring 3 m by 5 m by 3 m?

   b) What volume of air would have a mass of
      (i) 1 kg?   **(ii)** 1 tonne?

13 **a)** Find the speed of a car which travels 75 km in 1 hour 15 minutes.

   b) A car travels 15 km in 14 minutes. Find its speed in km/h.
      Give your answer correct to 1 decimal place.

14 Calculate the density of a stone of mass 730 g and volume 69 cm³.
Give your answer correct to 1 decimal place.

15 A town has a population of 74 000 and covers an area of 64 square kilometres.
Calculate the population density (number of people per square kilometre) of the town.
Give your answer correct to 1 decimal place.

WHAT YOU HAVE LEARNED

- **How to convert between metric measures of length, area and volume**
- **That all measurements are approximations**
- **When calculating a measurement, you need to give your answer to a sensible degree of accuracy. As a general rule your final answer should not be given to a greater degree of accuracy than any of the values used in the calculation**
- **Compound measures are calculated from two other measurements. Examples include speed, which is calculated using distance and time and expressed in units such as m/s, and density, which is calculated using mass and volume and expressed in units such as g/cm³**

1 Change these measures to the units shown.

    a) 12 m^2 to cm^2 **b)** 3.71 cm^2 to mm^2

    c) 0.42 m^2 to cm^2 **d)** 0.05 cm^2 to mm^2

2 Change these measures to the units shown.

    a) 3 m^2 to mm^2 **b)** $412\,500 \text{ cm}^2$ to m^2

    c) 9400 mm^2 to cm^2 **d)** 0.06 m^2 to cm^2

3 Change these measures to the units shown.

    a) 2.13 litres to cm^3 **b)** 5100 ml to litres

    c) 421 litres to ml   **d)** 91.7 ml to litres

4 Give the lower and upper bounds of each of these measurements.

    a) 27 cm to the nearest centimetre

    b) 5.6 cm to the nearest millimetre

    c) 1.23 m to the nearest centimetre

5 A policeman timed a car travelling along a 100 m section of road.
The time taken was 6 seconds.
The length of the road was measured accurate to the nearest 10 cm, and the time was measured accurate to the nearest second.
What is the greatest speed the car could have been travelling at?

6 a) A machine produces pieces of wood.
      The length of each piece is 34 mm, measured correct to the nearest millimetre.
      Between what limits does the actual length lie?

    b) Three of the pieces of wood are put together to make a triangle.
      What is the greatest possible perimeter of the triangle?

7 In a 10 km road race, one runner started at 11:48 and finished at 13:03.

    a) How long did it take this runner to complete the race?

    b) What was his average speed?

8 a) Eleanor drives to Birmingham on a motorway.
      She travels 150 miles in 2 hours 30 minutes. What is her average speed?

    b) Eleanor drives to Cambridge at an average speed of 57 mph.
      The journey takes 3 hours 20 minutes. How many miles is the journey?

9 The length of a field is 92.43 m and the width is 58.36 m.
Calculate the area of the field.
Give your answer to a sensible degree of accuracy.

30 → CONSTRUCTIONS AND LOCI

THIS CHAPTER IS ABOUT

- **Constructing the perpendicular bisector of a line**
- **Constructing the perpendicular from a point on a line**
- **Constructing the perpendicular from a point to a line**
- **Constructing the bisector of an angle**
- **Knowing that a locus is a line, a curve or a region of points**
- **Knowing the four basic loci results**
- **Locating a locus of points that follow a given rule or rules**

YOU SHOULD ALREADY KNOW

- **How to use a protractor and compasses**
- **How to make scale drawings**
- **How to construct a triangle given three facts about its sides and angles**

Constructions

You already know how to **construct** angles and triangles from Key Stage 3. You can use these skills in other constructions.

Four important constructions

You need to know four important constructions.

Construction 1: The perpendicular bisector of a line

Use the following method to construct the **perpendicular bisector** of line AB.

Perpendicular means 'at right angles to'.

A *bisector* is something that divides into 'two equal parts'.

1 Open your compasses to a radius more than half the length of the line AB. Put the point of your compasses on A. Draw one arc above the line, and one arc below.

2 Keep your compasses open to the same radius. Put the compass point on B. Draw two more arcs, cutting the first arcs at P and Q.

3 Join the points P and Q. This line divides AB into two equal parts and is at right angles to AB.

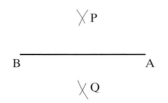

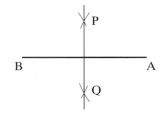

Discovery 30.1

a) (i) Draw a triangle. Make it big enough to fill about half of your page.
   (ii) Construct the perpendicular bisector of each of the three sides.
   (iii) If you have drawn them accurately enough, the bisectors should meet at one point. Put your compass point on this point, and the pencil on one of the corners of the triangle. Draw a circle.

b) You have drawn the **circumcircle** of the triangle. What do you notice about this circle?

Construction 2: The perpendicular from a point on a line

Use the following method to construct the perpendicular from point P on the given line.

1 Open your compasses to any radius. Put the compass point on P. Draw an arc on each side of P, cutting the line at Q and R.

2 Open your compasses to a larger radius. Put the compass point on Q. Draw an arc above the line. Now put the compass point on R and draw another arc, with the same radius, cutting the first arc at X.

3 Join the points P and X. This line is at right angles to the original line.

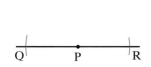

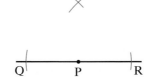

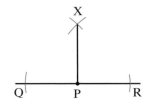

Discovery 30.2

a) Draw a line 10 cm long. Label it AB.

b) At A, draw a circle of radius 5 cm.
Label the point where the circle crosses the line P.

c) Construct the perpendicular from P.

This perpendicular is the **tangent** to the circle at P.

Construction 3: The perpendicular from a point to a line

Use the following method to construct the perpendicular from point P to the given line.

1 Open your compasses to any radius.
Put the compass point on P.
Draw two arcs, cutting the line at Q and R.

2 Keep your compasses open to the same radius.
Put the compass point on Q.
Draw an arc below the line.
Now put the compass point on R and draw another arc, cutting the first arc at X.

3 Line up your ruler with points P and X.
Draw the line PM.
This line is at right angles to the original line.

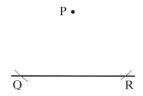

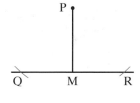

Discovery 30.3

a) (i) Draw a line across your page. Put a cross on one side of the line and label it P.

(ii) Construct the perpendicular from P to the line. Make sure you keep your compasses open to the same radius all the time.
This time, join P to X, don't stop at the original line.

b) (i) Measure PM and XM.

(ii) What do you notice?
What can you say about P and X?

Construction 4: The bisector of an angle

Use the following method to construct the bisector of the given angle.

1 Open your compasses to any radius.
Put the compass point on A. Draw two arcs, cutting the 'arms' of the angle at P and Q.

2 Open your compasses to any radius.
Put the compass point on P. Draw an arc inside the angle. Now put the compass point on Q and draw another arc, with the same radius, cutting the first arc at X.

3 Join the points A and X. This line divides the given angle into two equal parts.

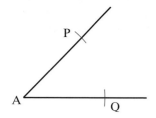 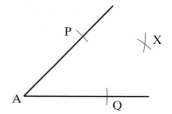 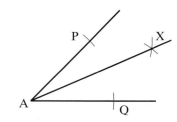

Discovery 30.4

a) (i) Draw a triangle. Make it big enough to fill about half of your page.
   (ii) Construct the bisector of each of the three angles.
   (iii) If you have drawn them accurately enough, the bisectors should meet at one point. Label this point A.
   Construct the perpendicular from A to one of the sides of the triangle. Label the point where the perpendicular meets the side of the triangle B.
   (iv) Put your compass point on A, and the pencil on point B. Draw a circle.
b) You have drawn the **incircle** of the triangle.
   (i) What do you notice about this circle?
   (ii) What can you say about the side of the triangle to which you have drawn the perpendicular from A?
   Hint: Look at your diagram from Discovery 30.2.
   (iii) What can you say about each of the other two sides of the triangle?

Loci

Constructing a locus

A **locus** is a line, curve or region of points that satisfy a certain rule.
The plural of locus is **loci**.

Four important loci

You need to know four important loci.

Locus 1: The locus of points that are the same distance from a given point

The locus of points that are 2 cm from point A is a circle, with centre A and radius 2 cm.

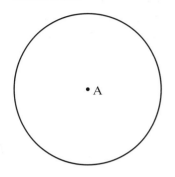

TIP

The locus of points that are less than 2 cm from A is the region inside the circle.

The locus of points that are more than 2 cm from A is the region outside the circle.

TIP

When trying to identify a particular locus, find several points that satisfy the required rule and see what sort of line, curve or region they form.

Locus 2: The locus of points that are the same distance from two given points

The locus of points that are the same distance from the points A and B is the perpendicular bisector of the line joining A and B.

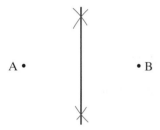

TIP

You learned how to construct a perpendicular bisector earlier in this chapter.

Locus 3: The locus of points that are the same distance from two given intersecting lines

The locus of points that are the same distance from AB and AC is the bisector of the angle BAC.

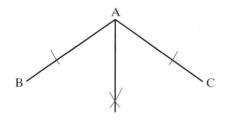

TIP

You learned how to construct the bisector of an angle earlier in this chapter.

Proof 30.1

a) Prove that locus 2 works.

b) Prove that locus 3 works.

Locus 4: The locus of points that are the same distance from a given line

The locus of points that are 3 cm from the line AB is a pair of lines, parallel to AB and 3 cm away from it on either side; at each end of the line there is a semicircle with centre A or B and radius 3 cm.

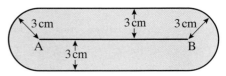

You can use constructions and loci to solve problems.

EXAMPLE 30.1

Two towns, P and Q, are 5 km apart.
Toby lives exactly the same distance from P as from Q.
a) Construct the locus of where Toby could live. Use a scale of 1 cm to 1 km.
Toby's school is closer to P than to Q.
b) Shade the region where Toby's school could be.

Solution

a) The locus of where Toby lives is the perpendicular bisector of the line joining P and Q.

b) Any point to the left of the line drawn in part **a)** is closer to point P than to point Q. Any point to the right of the line is closer to point Q than to point P.

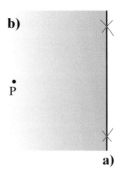

EXAMPLE 30.2

A security light is attached to a wall.
The light illuminates an area up to 20 m.
Construct the region illuminated by the light. Use a scale of 1 cm to 5 m.

Solution

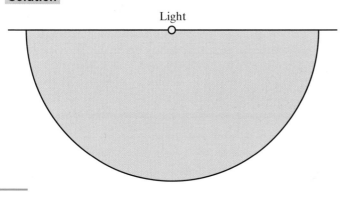

Remember that the light cannot illuminate the area behind the wall.

EXAMPLE 30.3

The diagram shows a port, P, and some rocks.

To leave the port safely, a boat must keep the same distance from each set of rocks.

Copy the diagram and construct the path of a boat from the port.

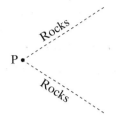

Solution

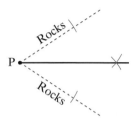

EXERCISE 30.1

1 Draw the locus of points that are less than 5 cm from a fixed point A.

2 Two rocks are 100 m apart.
A boat passes between the rocks so that it is always the same distance from each of them.
Construct the locus of the path of the boat.
Use a scale of 1 cm to 20 m.

3 In a farmer's field, a tree is 60 m from a long hedge.
The farmer decides to build a fence between the tree and the hedge.
The fence must be as short as possible.
Make a scale drawing of the tree and the hedge.
Construct the locus of where the fence must be built.
Use a scale of 1 cm to 10 m.

Hedge

● Tree

4 Draw an angle of 60°.
Construct the bisector of the angle.

5 Draw a square, ABCD, with side 5 cm.
Draw the locus of points inside the square that are less than 3 cm from corner C.

6 A rectangular shed measures 4 m by 2 m.
A path, 1 m wide and perpendicular to the shed, is to
be built from the door of the shed.
Construct the locus showing the edges of the path.
Use a scale of 1 cm to 1 m.

Door
◄1 m►

7 Construct a compass like the one on the right.
  • Draw a circle with radius 4 cm.
  • Draw a horizontal diameter of the circle.
  • Construct the perpendicular bisector of the diameter.
  • Bisect each of the four angles.

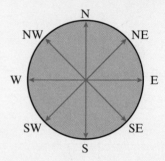

8 Draw a line 7 cm long.
Construct the region of points that are less than 3 cm from the line.

9 Construct a triangle ABC with AB = 8 cm, AC = 7 cm and BC = 6 cm.
Shade the locus of points inside the triangle that are closer to AB than to AC.

10 The owner of a theme park decides to build a moat 20 m wide around a castle.
The castle is a rectangle measuring 80 m long by 60 m wide.
Draw accurately an outline of the castle and the moat.

Challenge 30.1

Sketch the locus for each of the following.
a) The tip of the hour hand of a clock.
b) A ball thrown vertically up.
c) A ball thrown at an angle.
d) The middle of a bicycle wheel as it is ridden along a level road.
e) A nail stuck in the tyre of a bicycle as it is ridden along a level road.
f) A door handle as the door opens.
g) The lower edge of an 'up-and-over' garage door as it is opened.

Intersecting loci

Often a locus is defined by more than one rule.

EXAMPLE 30.4

Two points, P and Q, are 5 cm apart.
Find the locus of points that are less than
3 cm from P and equidistant from P and Q.

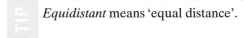

Equidistant means 'equal distance'.

Solution

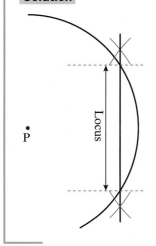

The locus of points that are less than 3 cm
from P is within a circle with centre P and
radius 3 cm.

The locus of points that are equidistant from
P and Q is the perpendicular bisector of the
line joining P and Q.

The points that satisfy both rules lie within
the circle *and* on the line.

EXAMPLE 30.5

A rectangular garden is 25 m long and 15 m wide.
A tree is to be planted in the garden so that it is more than 2.5 m
from the boundary and less than 10 m from the south-west corner.
Using a scale of 1 cm to 5 m, make a scale drawing of the garden
and find the region where the tree can be planted.

Solution

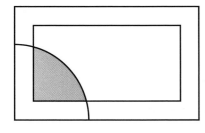

Draw a rectangle measuring 5 cm by 3 cm to represent the
garden.

The locus of points more than 2.5 m from the boundary is
a smaller rectangle inside the first. Each side of the smaller
rectangle is 0.5 cm inside the sides of the larger rectangle.

The locus of points less than 10 m from the south-west
corner is an arc with the corner as its centre and radius 2 cm.

The points that satisfy both rules lie within the smaller
rectangle *and* the arc.

1 Draw a point and label it A.
Draw the region of points that are more than 3 cm from A but less than 6 cm from A.

2 Two towns, P and Q, are 7 km apart.
Amina wants to buy a house that is within 5 km of P and also within 4 km of Q.
Make a scale drawing to indicate the region where Amina could buy a house.
Use a scale of 1 cm to 1 km.

3 Draw a square ABCD with side 5 cm.
Find the region of points that are more than 3 cm from AB and AD.

4 Steve is using an old map to locate treasure.
He is searching a rectangular plot of land, EFGH, which
measures 8 m by 5 m.
The map says that the treasure is hidden 6 m from E on
a line equidistant from F and H.
Using a scale of 1 cm to 1 m, make a scale drawing to
locate the treasure.
Mark the position where the treasure is hidden with
the letter T.

5 Construct triangle ABC with AB = 11 cm, AC = 7 cm and BC = 9 cm.
Construct the locus of points inside the triangle that are closer to AB than to AC
and equidistant from A and B.

6 The diagram shows the corner of a farm building, which stands
in a field.
A donkey is tethered at a point D with a rope 5 m long.
Shade the region of the field that the donkey can graze.

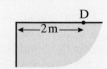

7 A sailor is shipwrecked at night. She is 140 m from the straight coastline.
She swims straight for the shore.

a) Make a scale drawing of the path she swims.

The coastguard is standing on the beach exactly where the sailor will come ashore.
He has a searchlight that can illuminate up to a distance of 50 m.

b) Mark on your diagram the part of the sailor's swim that will be lit up.

8 The positions of three radio stations, A, B and C, form a triangle such that AB = 7 km,
BC = 9.5 km and angle ABC = 90°.
The signal from each radio station can be received up to 5 km away.
Make a scale drawing to locate the region where none of the three radio stations can be
received.

9 A rectangular garden measures 20 m by 14 m.
The house wall is along one of the shorter sides of the garden.
Pete is going to plant a tree. It must be more than 10 m from the house and more than 8 m from any corner of the garden.
Find the region of the garden where the tree can be planted.

10 Two towns, H and K, are 20 miles apart.
A new leisure centre is to be built within 15 miles of H, but nearer to K than to H.
Using a scale of 1 cm to 5 miles, draw a diagram to show where the leisure centre could be built.

Challenge 30.2

a) Describe the 3-D locus produced when each of the following 2-D shapes is rotated 360° about the edge XY.

(i) X 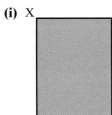 Y

(ii) X 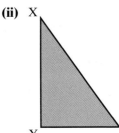 Y

(iii) X 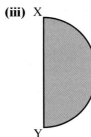 Y

b) In 3-D, P and Q are points 16 cm apart.
  (i) Describe the 3-D locus of a point that is always 10 cm from P.
  (ii) Describe the locus of a point that is always 10 cm from P and 10 cm from Q.

WHAT YOU HAVE LEARNED

- **How to construct the perpendicular bisector of a line**
- **How to construct the perpendicular from a point on a line**
- **How to construct the perpendicular from a point to a line**
- **How to construct the bisector of an angle**
- **That the locus of points that are the same distance from a given point is a circle**
- **That the locus of points that are the same distance from two points is the perpendicular bisector of the line joining the two points**
- **That the locus of points which are the same distance from two intersecting lines is the bisector of the angle formed by the two lines**
- **That the locus of points that are the same distance from a given line is a pair of parallel lines with a semicircle at each end**
- **That some loci must satisfy more than one rule**

1 Draw an angle of 100°.
Construct the bisector of the angle.

2 Draw a line 8 cm long.
Construct the locus of the points that are the same distance from each end of the line.

3 Draw a line 7 cm long.
Draw the locus of points which are 3 cm away from this line.

4 Draw the triangle ABC with AB = 9 cm, BC = 8 cm and CA = 6 cm.
Construct the perpendicular line from C to AB.
Measure the length of this line and hence work out the area of the triangle.

5 Draw the rectangle PQRS with PQ = 8 cm and QR = 5 cm.
Shade the region of points which are nearer to QP than to QR.

6 Two radio stations are 40 km apart. Each station can transmit signals up to 30 km.
Construct a scale drawing to show the region which can receive signals from both radio stations.

7 A garden is a rectangle ABCD with AB = 5 m and BC = 3 m.
A tree is planted so that it is within 5 m of A and within 3 m of C.
Indicate the region where the tree could be planted.

8 Draw triangle EFG with EF = 8 cm, EG = 6 cm and angle E = 70°.
Construct the point which is equidistant from F and G and is also 5 cm from G.

9 A lawn is a square with side 5 m.
A water sprinkler covers a circle of radius 3 m.
If the gardener puts the sprinkler at each corner, will the whole lawn get watered?

10 The gardener in question **9** lends his water sprinkler to a neighbour.
The neighbour has a large garden with a rectangular lawn measuring 10 m by 8 m.
She moves the sprinkler slowly around the edge of the lawn. Draw a scale diagram to show the region of the lawn which will be watered.

31 → AREAS, VOLUMES AND 2-D REPRESENTATION

THIS CHAPTER IS ABOUT

- The meaning of terms related to circles
- Circumference and area of a circle
- Area and volume of complex shapes
- Volume of prisms
- Volume and surface area of a cylinder
- Plans and elevations

YOU SHOULD ALREADY KNOW

- The meaning of *circumference, diameter* and *radius*
- The meaning of *area* and *volume*
- How to find the area of rectangles and triangles
- How to find the volume of cuboids

Circles

The distance all the way round a circle is its **circumference**.

A **diameter** is a line all the way across a circle and passing through its centre. It is the longest length across the circle.

A line from the centre of a circle to the circumference is called a **radius**. For any circle, the radius is always the same, that is, it is constant.

> **TIP**
>
> *Radius* is a Latin word. The plural is **radii**.

Here are some other terms relating to circles.

A **tangent** 'touches' a circle and is at right angles to the radius.
An **arc** is part of the circumference.
A **sector** is part of a circle between two radii, like a slice of cake.
A **chord** is a straight line dividing the circle in two parts.
A **segment** is the part cut off by a chord. The one shown is the **minor segment**. On the other side of the chord is the **major segment**.

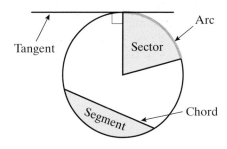

You should use these mathematical terms when talking about parts of a circle.

Circumference of a circle

Discovery 31.1

Find a number of circular or cylindrical items.

Measure the circumference and diameter of each item and complete a table like this.

Item name	Circumference	Diameter	Circumference ÷ Diameter

What do you notice?

For any circle, $\dfrac{\text{circumference}}{\text{diameter}} \approx 3$.

If it were possible to take very accurate measurements, you would find that $\dfrac{\text{circumference}}{\text{diameter}} = 3.141\ 592 \ldots$

This number is called **pi** and is represented by the symbol π.

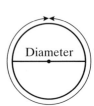

This means that you can write a formula for the circumference of any circle.

Circumference = π × diameter   or $C = \pi d$

π is a decimal number which does not terminate and does not recur: it goes on for ever. In calculations, you can either use the $\boxed{\pi}$ button on your calculator or use an approximation: 3.142 is suitable.

EXAMPLE 31.1

Find the circumference of a circle with a diameter of 45 cm.

Solution

Circumference = π × diameter
  = 3.142 × 45
  = 141.39   3.142 is an approximation for π so your answer is
  = 141.4 cm   not exact and should be rounded. You will often
              be told to what accuracy to give your answer.
              Here the answer is given correct to 1 decimal place.

TIP

You could do this calculation on your calculator, using the $\boxed{\pi}$ button.

Input $\boxed{\pi}\ \boxed{\times}\ \boxed{4}\ \boxed{5}\ \boxed{=}$. The answer on your display will be 141.371 669 4.

Use $C = \pi d$ to find the circumferences of circles with these diameters.

1 12 cm   **2** 25 cm   **3** 90 cm   **4** 37 mm   **5** 66 mm   **6** 27 cm

7 52 cm   **8** 4.7 cm   **9** 9.2 cm   **10** 7.3 m   **11** 2.9 m   **12** 1.23 m

Since a diameter is made up of two radii, $d = 2r$ so the formula can be expressed as shown below.

$$\text{Circumference} = \pi \times 2r = 2\pi r$$

Challenge 31.1

Find the circumferences of circles with these radii.

a) 8 cm   **b)** 30 cm   **c)** 65 cm   **d)** 59 mm   **e)** 0.7 m   **f)** 1.35 m

Challenge 31.2

The wheel of a bicycle has a diameter of 66 cm.

How many complete revolutions will the wheel complete in a journey of 1 kilometre?

Area of a circle

The area of a circle is the surface it covers.

Discovery 31.2

Take a disc of paper and cut it into 12 narrow sectors, all the same size. Cut one sector in half.

Arrange them, reversing every other piece, like this, with the halves at either end.

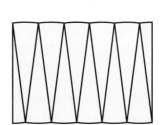

This is nearly a rectangle. If you had cut the disc into 100 sectors it would be more accurate.

a) What are the dimensions of the rectangle?

b) What is its area?

The height of the rectangle in Discovery 31.2 is the radius of the circle, r.

The width is half the circumference of the circle, $\frac{1}{2}\pi d$ or πr.

This gives a formula to calculate the area of a circle.

Area $= \pi r^2$ where r is the radius of the circle

The formula is Area $= \pi r^2$. This means $\pi \times r^2$: that is, square r first then multiply by π. Do not work out $(\pi r)^2$.

EXAMPLE 31.2

Find the area of a circle with a radius of 23 cm.

Solution

Area $= \pi r^2$
       $= 3.142 \times 23^2$
       $= 1662.118$
       $= 1662$ cm^2 (to the nearest whole number)

You could do this calculation on your calculator, using the $\boxed{\pi}$ button.
Input $\boxed{\pi}\,\boxed{\times}\,\boxed{2}\,\boxed{3}\,\boxed{x^2}\,\boxed{=}$. The answer on your display will be 1661.902 514.

EXERCISE 31.2

1 Use $A = \pi r^2$ to find the areas of circles with these radii.

a) 14 cm	**b)** 28 cm	**c)** 80 cm	**d)** 35 mm
e) 62 mm	**f)** 43 cm	**g)** 55 cm	**h)** 4.9 cm
i) 9.7 cm	**j)** 3.4 m	**k)** 2.6 m	**l)** 1.25 m

2 Find the areas of circles with these diameters.

a) 16 cm	**b)** 24 cm	**c)** 70 cm	**d)** 36 mm
e) 82 mm	**f)** 48 cm	**g)** 54 cm	**h)** 4.4 cm
i) 9.8 cm	**j)** 3.8 m	**k)** 2.8 m	**l)** 2.34 m

Challenge 31.3

A metal washer has the dimensions shown in the diagram.

Calculate the area of the washer.

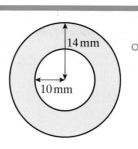

Area of complex shapes

The formula for the area of a rectangle is

$$\text{Area} = \text{length} \times \text{width} \qquad \text{or} \qquad A = l \times w$$

The formula for the area of a triangle is

$$\text{Area} = \tfrac{1}{2} \times \text{base} \times \text{height} \qquad \text{or} \qquad A = \tfrac{1}{2} \times b \times h$$

You can use these formulae to find the area of more complex shapes, which can be broken down into rectangles and right-angled triangles.

EXAMPLE 31.3

Find the area of this shape.

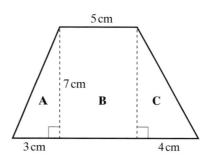

Solution

Work out the area of the rectangle and each of the triangles separately and then add them together to find the area of the whole shape.

Area of shape = Area of triangle **A** + Area of rectangle **B** + Area of triangle **C**

$$= \frac{3 \times 7}{2} \qquad + \qquad 5 \times 7 \qquad + \qquad \frac{4 \times 7}{2}$$

$$= \quad 10.5 \quad + \quad 35 \quad + \quad 14$$

$$= 59.5 \text{ cm}^2$$

Find the area of each of these shapes.

Break them down into rectangles and right-angled triangles first.

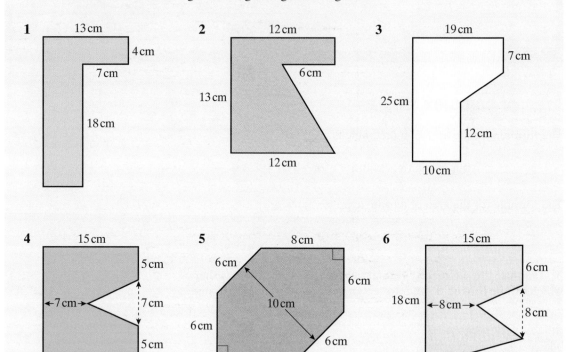

1   13 cm   4 cm   7 cm   18 cm

2   12 cm   6 cm   13 cm   12 cm

3   19 cm   7 cm   25 cm   12 cm   10 cm

4   15 cm   5 cm   ←7 cm→   7 cm   5 cm

5   8 cm   6 cm   6 cm   10 cm   6 cm   6 cm   8 cm

6   15 cm   6 cm   18 cm   ←8 cm→   8 cm

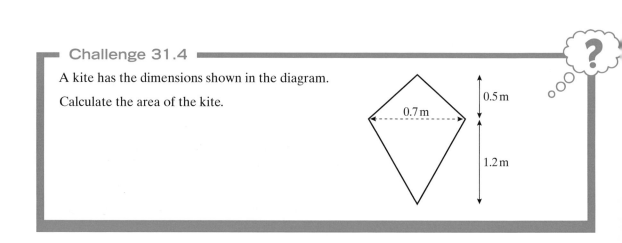

Challenge 31.4

A kite has the dimensions shown in the diagram.

Calculate the area of the kite.

0.5 m
0.7 m
1.2 m

A blade for a lawnmower has the dimensions shown in the diagram.

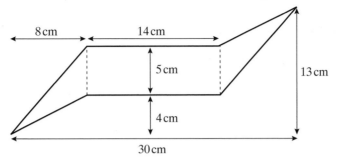

Calculate the area of the blade.

Volume of complex shapes

The formula for the volume of a cuboid is

$$\text{Volume} = \text{length} \times \text{width} \times \text{height} \quad \text{or} \quad V = l \times w \times h$$

It is possible to find the volume of shapes made from cuboids by breaking them down into smaller parts.

EXAMPLE 31.4

Find the volume of this shape.

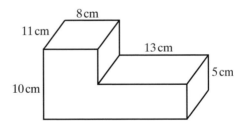

Solution

This shape can be broken down into two cuboids, **A** and **B**.
Work out the volumes of these two cuboids and add them together to find the volume of the whole shape.

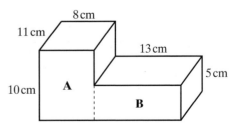

Volume of shape = volume of cuboid **A** + volume of cuboid **B**

$$
\begin{aligned}
&= \quad 8 \times 11 \times 10 \quad + \quad 13 \times 11 \times 5 \qquad \text{The width of cuboid } \textbf{B} \text{ is the}\\
&= \qquad\quad 880 \qquad\quad + \qquad 715 \qquad\qquad \text{same as the width of cuboid } \textbf{A}.\\
&= 1595 \text{ cm}^3
\end{aligned}
$$

1 Find the volume of each of these shapes.

a)

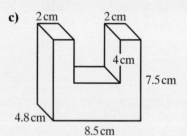

5 cm · 6 cm · 8 cm · 5 cm · 13 cm · 5 cm · 11 cm

b)

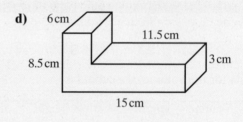

8 cm · 14 cm · 8 cm · 15 cm · 18 cm

c)

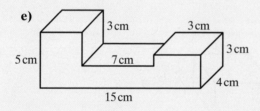

2 cm · 2 cm · 4 cm · 7.5 cm · 4.8 cm · 8.5 cm

d)

6 cm · 11.5 cm · 8.5 cm · 3 cm · 15 cm

e)

3 cm · 3 cm · 3 cm · 5 cm · 7 cm · 4 cm · 15 cm

f)

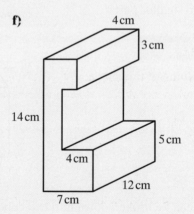

4 cm · 3 cm · 14 cm · 5 cm · 4 cm · 12 cm · 7 cm

2 The diagram shows a concrete lintel used by builders.
Calculate the volume of concrete needed to make
the lintel.

(Note: The diagram is *not* drawn to scale.)

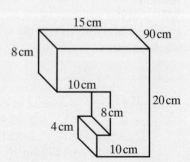

15 cm · 90 cm · 8 cm · 10 cm · 20 cm · 8 cm · 4 cm · 10 cm

Volume of a prism

A **prism** is a three-dimensional object that is the same 'shape' throughout. The correct definition is that the object has a **uniform cross-section**.

In this diagram the shaded area is the cross-section.

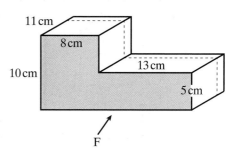

Looking at the shape from point F you see the cross-section as an L-shape. If you were to cut through the shape along the dotted line you would still see the same cross-section.

You could cut the shape into slices, each 1 cm thick. The volume of each slice, in cubic centimetres, would be the area of the cross-section × 1.
As the shape is 11 cm thick, you would have 11 identical slices. So the volume of the whole shape would be the area of the cross-section × 11.

This tells you that the formula for the volume of a prism is

Volume = area of cross-section × length

The area of the cross-section (shaded) = $(10 \times 8) + (13 \times 5)$
$$= 80 + 65$$
$$= 145 \text{ cm}^2$$

Volume = 145×11
$$= 1595 \text{ cm}^3$$

This is the same answer as in Example 31.4, when the volume of this shape was found by breaking it down into cuboids.

The formula works for any prism.

EXAMPLE 31.5

This prism has a cross-section of area 374 cm², and is 26 cm long.

Find its volume.

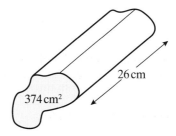

Solution

Volume = area of cross-section × length
$$= 374 \times 26$$
$$= 9724 \text{ cm}^3$$

1 Find the volume of each of these prisms.

a)

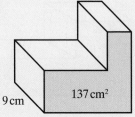

9 cm   137 cm²

b)

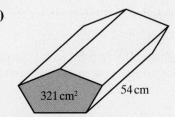

321 cm²   54 cm

c)

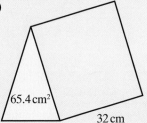

65.4 cm²   32 cm

d)

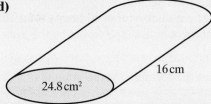

24.8 cm²   16 cm

e)

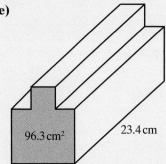

96.3 cm²   23.4 cm

f)

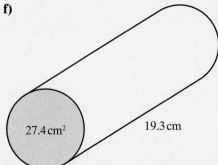

27.4 cm²   19.3 cm

2 The diagram shows a clay plant pot support with a uniform cross-section.

The area of the cross-section is 3400 mm² and the length of the support is 35 mm.
What is the volume of the clay in the support?

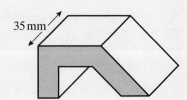

35 mm

Volume of a cylinder

A **cylinder** is a special kind of prism: the cross-section is always a circle.

Cylinder **A** and cylinder **B** are identical prisms.

You can find the volume of both cylinders using the formula for the volume of a prism.

Volume of cylinder **A**
  = area of cross-section × length
  = 77 × 18
  = 1386 cm^3

Volume of cylinder **B**
  = area of cross-section (area of circle) × length (height)
  = $\pi r^2 \times h$ cm^3

This gives you the formula for the volume of any cylinder.

Volume = $\pi r^2 h$ where r is the radius of the circle and h is the height of the cylinder

---

EXAMPLE 31.6

Find the volume of a cylinder with radius 13 cm and height 50 cm.

Solution

Volume = $\pi r^2 h$
  = 3.142 × 13^2 × 50
  = 26 549.9
  = 26 550 cm^3 (to the nearest whole number)

> **TIP**
>
> You could do this calculation on your calculator, using the $\boxed{\pi}$ button.
>
> Input $\boxed{\pi}$ $\boxed{\times}$ $\boxed{1}$ $\boxed{3}$ $\boxed{x^2}$ $\boxed{\times}$ $\boxed{5}$ $\boxed{0}$ $\boxed{=}$.
>
> The answer on your display will be 26 546.457 92.
>
> Rounded to the nearest whole number, this is 26 546 cm^3. This is different from the answer you get using 3.142 as an approximation for π because your calculator uses a more accurate value for π.

1 Use the formula to find the volumes of cylinders with these dimensions.

   a) Radius 8 cm and height 35 cm   **b)** Radius 14 cm and height 42 cm

   c) Radius 20 cm and height 90 cm   **d)** Radius 12 mm and height 55 mm

   e) Radius 25 mm and height 6 mm   **f)** Radius 0.7 mm and height 75 mm

   g) Radius 3 m and height 25 m   **h)** Radius 5.8 m and height 3.5 m

2 A capping stone for the top of a wall is in the shape of a half cylinder as shown in the diagram.

Calculate the volume of the capping stone.

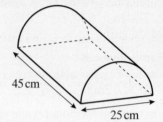

45 cm

25 cm

Challenge 31.6

A water tank in a factory is a cylinder, as shown in the diagram.

The tank is full and the water is pumped out at a rate of 600 litres per minute.

How long does it take to empty the tank?

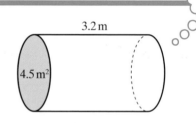

3.2 m

4.5 m²

Surface area of a cylinder

You can probably think of lots of examples of cylinders. Some of them, like the inner tube of a roll of kitchen paper, have no ends: these are called **open cylinders**. Others, like a can of beans, do have ends: these are called **closed cylinders**.

Curved surface area

If you took an open cylinder, cut straight down its length and opened it out, you would get a rectangle.
The **curved surface area** of the cylinder has become a flat shape.

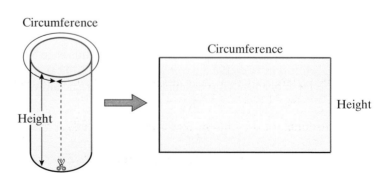

Circumference

Circumference

Height

Height

The area of the rectangle is circumference × height.

You know that the formula for the circumference of a circle is

$$\text{Circumference} = \pi \times \text{diameter} \quad \text{or} \quad C = \pi d$$
$$\text{or} \quad C = 2\pi r \text{ (for radius } r)$$

So the formula for the area of the curved surface of any cylinder is

$$\text{Curved surface area} = \pi \times \text{diameter} \times \text{height} \quad \text{or} \quad \pi d h$$

This formula is usually written in terms of the radius. You know that the radius is half the length of the diameter, or $d = 2r$.

So you can also write the formula for the area of the curved surface of any cylinder as

$$\text{Curved surface area} = 2 \times \pi \times \text{radius} \times \text{height} \quad \text{or} \quad 2\pi r h$$

EXAMPLE 31.7

Find the curved surface area of a cylinder with radius 4 cm and height 0.7 cm.

Solution

$$\begin{aligned}
\text{Curved surface area} &= 2\pi r h \\
&= 2 \times 3.142 \times 4 \times 0.7 \\
&= 17.5952 \\
&= 17.6 \text{ cm}^2 \text{ (correct to 1 decimal place)}
\end{aligned}$$

TIP

You could do this calculation on your calculator, using the $\boxed{\pi}$ button.

Input $\boxed{2}$ $\boxed{\times}$ $\boxed{\pi}$ $\boxed{\times}$ $\boxed{4}$ $\boxed{\times}$ $\boxed{0}$ $\boxed{\cdot}$ $\boxed{7}$ $\boxed{=}$. The answer on your display will be 17.592 918 86.

Total surface area

The total surface area of a closed cylinder is made up of the curved surface area and the area of the two circular ends.

So the formula for the total surface area of a (closed) cylinder is

$$\text{Total surface area} = 2\pi r h + 2\pi r^2$$

EXAMPLE 31.8

Find the total surface area of a closed cylinder with radius 13 cm and height 1.5 cm.

Solution

$$
\begin{aligned}
\text{Total surface area} &= 2\pi rh + 2\pi r^2 \\
&= (2 \times 3.142 \times 13 \times 1.5) + (2 \times 3.142 \times 13^2) \\
&= 122.538 + 1061.996 = 1184.534 \\
&= 1185 \text{ cm}^2 \text{ (to the nearest whole number)}
\end{aligned}
$$

> **TIP**
>
> You could do this calculation on your calculator, using the π button.
> Input (2 × π × 1 3 × 1 . 5) + (2 × π 1 3 × x^2) =.
> The answer on your display will be 1184.380 43.
>
> Rounded to the nearest whole number, this is 1184 cm². This is different from the answer you get using 3.142 as an approximation for π because your calculator uses a more accurate value for π.

⊚ EXERCISE 31.7

1 Find the curved surface areas of cylinders with these dimensions.
- **a)** Radius 12 cm and height 24 cm
- **b)** Radius 11 cm and height 33 cm
- **c)** Radius 30 cm and height 15 cm
- **d)** Radius 18 mm and height 35 mm
- **e)** Radius 15 mm and height 4 mm
- **f)** Radius 1.3 mm and height 57 mm
- **g)** Radius 2.1 m and height 10 m
- **h)** Radius 3.5 m and height 3.5 m

2 The cardboard tube in a toilet roll is in the shape of a cylinder.
The diameter of the tube is 4.4 cm and the tube is 11 cm long.
Find the area of card used to make the tube.

3 Find the total surface areas of cylinders with these dimensions.
- **a)** Radius 14 cm and height 10 cm
- **b)** Radius 21 cm and height 32 cm
- **c)** Radius 35 cm and height 12 cm
- **d)** Radius 18 mm and height 9 mm
- **e)** Radius 25 mm and height 6 mm
- **f)** Radius 3.5 mm and height 50 mm
- **g)** Radius 1.8 m and height 15 m
- **h)** Radius 2.5 m and height 1.3 m

Challenge 31.7

The curved surface area of an open cylinder is 550 cm², to the nearest whole number.
The height of the cylinder is 19.5 cm.
Find the diameter of the cylinder.

Challenge 31.8

The total surface area of a closed cylinder is 2163 cm², to the nearest whole number.

The radius of the cylinder is 8.5 cm.

Find the height of the cylinder.

Plans and elevations

This diagram is part of a builder's plan for a housing estate.

It shows the shapes of the houses as seen from above. You may have heard the term 'bird's eye view' for this sort of picture. The mathematical term is **plan view**.

From the plan you can tell only what shape the buildings are from above. You cannot tell whether they are bungalows, two-storey houses or even blocks of flats.

The view of the front of an object is called a **front elevation** and the view from the side is called a **side elevation**. An elevation shows you the height of an object.

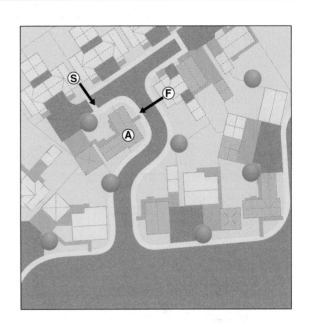

EXAMPLE 31.9

For house A, sketch
a) a possible view from F.

b) a possible view from S.

Solution

a) View from F

Roof

Wall

The dotted lines show hidden detail.

b) View from S

Roof

Wall

EXAMPLE 31.10

For this shape, draw
a) the plan.
b) the front elevation (view from F).
c) the side elevation from S.

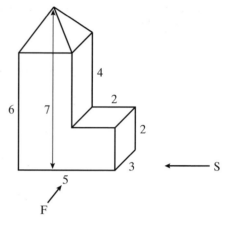

Solution

a) Plan

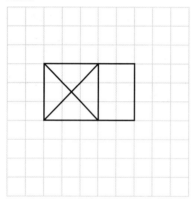

The cross shows the edges of the pyramid on top of the tower. The rectangle on the right is the flat top of the lower part of the shape.

b) Front elevation

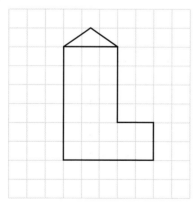

c) Side elevation

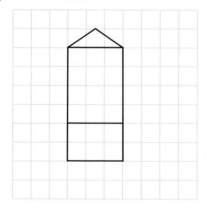

EXAMPLE 31.11

Draw the plan view, front elevation and side elevation of this child's building block.

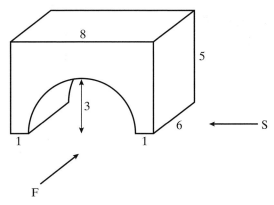

Solution

a) **Plan**

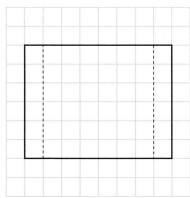

You can use dotted lines to show hidden detail.

In the plan, the dotted lines show the sides of the tunnel at floor level.

In the side elevation, the dotted line shows the top of the tunnel.

b) **Front elevation**

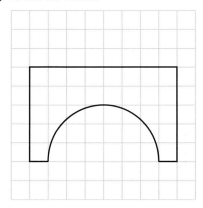

c) **Side elevation**

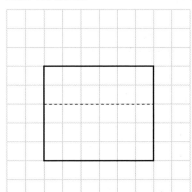

Draw the plan view, front elevation and side elevation of each of these objects.

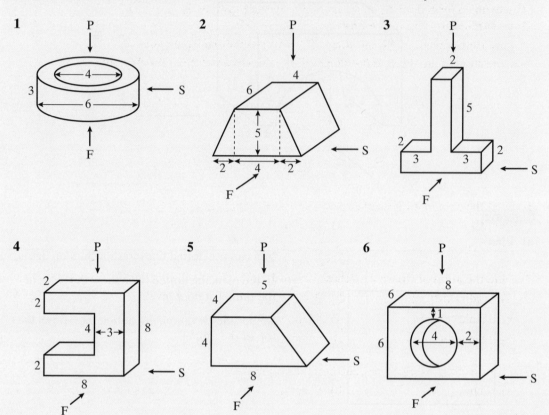

- The names of parts of a circle
- The formula for the circumference of a circle is Circumference $= \pi d$ or $2\pi r$
- The formula for the area of a circle is Area $= \pi r^2$
- The area of a complex shape can be found by breaking the shape down into rectangles and right-angled triangles
- The volume of a complex shape can be found by breaking the shape down into cuboids
- A prism is a three-dimensional object that has a uniform cross-section
- The formula for the volume of a prism is Volume $=$ area of cross-section $\times$ length
- A cylinder is a special kind of prism whose cross-section is a circle

- The formula for the volume of a cylinder is Volume $= \pi r^2 h$
- The formula for the curved surface area of a cylinder is
  Curved surface area $= 2\pi rh$ or πdh
- The formula for the total surface area of a (closed) cylinder is
  Total surface area $= 2\pi rh + 2\pi r^2$
- A plan view of an object is the shape of the object viewed from above
- An elevation of an object is the shape of the object viewed from the front or side

MIXED EXERCISE 31

1 Find the circumferences of circles with these diameters.

    a) 14.2 cm   **b)** 29.7 cm

    c) 65 cm   **d)** 32.1 mm

2 Find the areas of circles with these dimensions.

    a) Radius 6.36 cm   **b)** Radius 2.79 m

    c) Radius 8.7 mm   **d)** Diameter 9.4 mm

    e) Diameter 12.6 cm   **f)** Diameter 9.58 m

3 Draw a circle with radius 4 cm. On it draw and label

    a) a chord.   **b)** a sector.   **c)** a tangent.

4 Work out the area of each of these shapes.

    a)

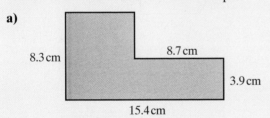

    b)

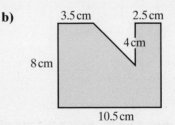

    c)

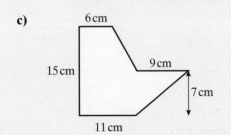

    d)

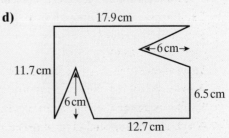

5 Find the volume of each of these shapes.

a)
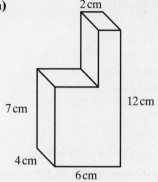

b)
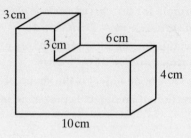

c)
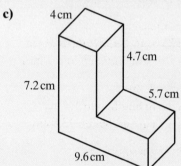

d)
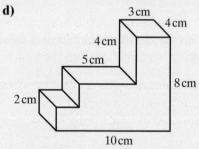

e)
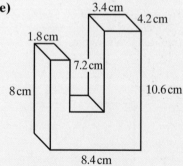

f)
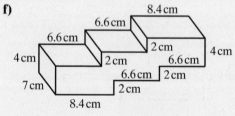

6 Find the volume of each of these prisms.

a)
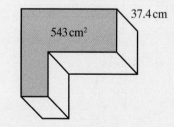

b)

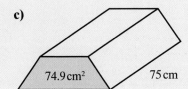

c) 74.9 cm²   75 cm

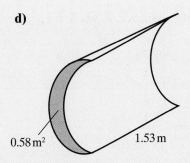

d) 0.58 m²   1.53 m

7 Find the volumes of cylinders with these dimensions.
  a) Radius 6 mm and height 23 mm   **b)** Radius 17 mm and height 3.6 mm
  c) Radius 22 cm and height 70 cm   **d)** Radius 12 cm and height 0.4 cm
  e) Radius 35 m and height 6 m   **f)** Radius 1.8 m and height 2.7 m

8 Find the curved surface areas of cylinders with these dimensions.
  a) Radius 9.6 m and height 27.5 m   **b)** Radius 23.6 cm and height 16.4 cm
  c) Radius 1.7 cm and height 1.5 cm   **d)** Radius 16.7 mm and height 6.4 mm

9 Find the total surface areas of cylinders with these dimensions.
  a) Radius 23 mm and height 13 mm   **b)** Radius 3.6 m and height 1.4 m
  c) Radius 2.65 cm and height 7.8 cm   **d)** Radius 4.7 cm and height 13.8 cm

10 Draw the plan view, front elevation and side elevation of these objects.

a)

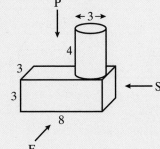

b)

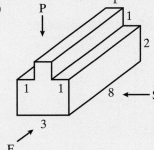

c)

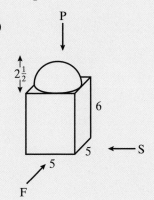

d)

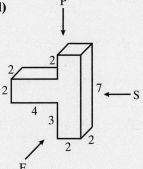

32 → TRIGONOMETRY 1

Labelling the sides

You already know that the longest side is called the **hypotenuse**.

The side opposite the angle you are using (θ) is called the **opposite**.

The remaining side is called the **adjacent**.

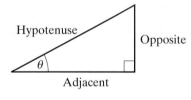
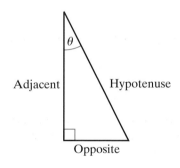

> **TIP**
>
> Label in the order hypotenuse, opposite, adjacent.
>
> To identify the opposite side, go straight out from the middle of the angle. The side you hit is the opposite.
>
> You can shorten the labels to 'H', 'O' and 'A'.

Label each of the three sides in these triangles.

a)
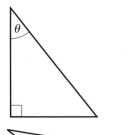

b)
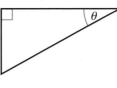

c)

d)
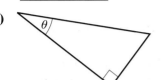

e)
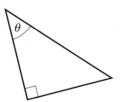

f)
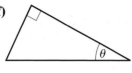

Discovery 32.1

- Make four drawings of this triangle.
  Draw the angles accurately but make each triangle a different size.
- Label the sides H, O, and A appropriately.
- Measure each of the sides of all your triangles.

- Use your calculator to work out $\dfrac{\text{opposite}}{\text{hypotenuse}}$ for each triangle.

What do you notice?

Repeat the task with four right-angled triangles each containing an angle of 70°

You should have found that your answers for each set of triangles were very close to each other. This is because all your first four triangles were similar to each other and all your second four triangles were similar to each other.

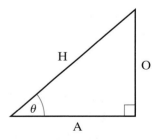

The ratio you worked out, $\dfrac{\text{opposite}}{\text{hypotenuse}}$, is called the **sine** of the angle.

This is often shortened to 'sin'.

$$\sin \theta = \frac{\text{opposite}}{\text{hypotenuse}}$$

Notice that the ratio of the lengths is written as a fraction, $\dfrac{\text{opposite}}{\text{hypotenuse}}$, rather than opposite : hypotenuse.

Find the $\boxed{\textbf{sin}}$ button on your calculator.

Find sin 40° on your calculator by pressing these keys.

$\boxed{\textbf{sin}}\ \boxed{4}\ \boxed{0}\ \boxed{=}$

Now find sin 70°.

Check that your answers in Discovery 32.1 are close to these answers.

Make sure that your calculator is set to degrees. This is the default setting but, if you see 'rad' or 'R' or 'grad' or 'G' in the window, change the setting using the $\boxed{\textbf{DRG}}$ button.

Discovery 32.2

Use again the first four triangles you drew in Discovery 32.1.

Use your calculator to work out the ratio $\dfrac{\text{adjacent}}{\text{hypotenuse}}$ for each triangle.

What do you notice?

Repeat this for the second group of four triangles you drew in Discovery 32.1.

The ratio you worked out, $\dfrac{\text{adjacent}}{\text{hypotenuse}}$, is called the **cosine** of the angle.

This is often shortened to 'cos'.

$$\cos\theta = \frac{\text{adjacent}}{\text{hypotenuse}}$$

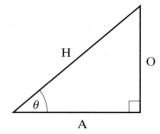

Find the $\boxed{\textbf{cos}}$ button on your calculator.

Find cos 40° on your calculator by pressing these keys.

$\boxed{\textbf{cos}}\ \boxed{4}\ \boxed{0}\ \boxed{=}$

Now find cos 70°.

Check that your answers in Discovery 32.2 are close to these answers.

Use again the first four triangles you drew in Discovery 32.1.

Use your calculator to work out the ratio $\dfrac{\text{opposite}}{\text{adjacent}}$ for each triangle.

What do you notice?

Repeat this for the second group of four triangles you drew in Discovery 32.1.

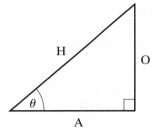

The ratio you worked out, $\dfrac{\text{opposite}}{\text{adjacent}}$, is called the **tangent** of the angle.

This is often shortened to 'tan'.

$$\tan \theta = \frac{\text{opposite}}{\text{adjacent}}$$

Find the tan button on your calculator.

Find tan 40° on your calculator by pressing these keys.

tan 4 0 =

Now find tan 70°.

Check that your answers in Discovery 32.3 are close to these answers.

TIP

You need to learn the three ratios

$$\sin \theta = \frac{O}{H}, \qquad \cos \theta = \frac{A}{H} \qquad \text{and} \qquad \tan \theta = \frac{O}{A}.$$

There are various ways of remembering these but one of the most popular is to learn the 'word' '**SOHCAHTOA**'.

This stands for

S	O	H	C	A	H	T	O	A
i	p	y	o	d	y	a	p	d
n	p	p	s	j	p	n	p	j
e	o	o	i	a	o	g	o	a
	s	t	n	c	t	e	s	c
	i	e	e	e	e	n	i	e
	t	n		n	n	t	t	n
	e	u		t	u		e	t
	s				s			
	e				e			

Using the ratios 1

When you need to solve a problem using one of the ratios, you should follow these steps.

- Draw a clearly labelled diagram.
- Label the sides H, O and A.
- Decide which ratio you need to use.
- Solve the equation.

In one type of problem you will encounter, you are required to find the numerator (top) of the fraction. This is demonstrated in the following examples.

EXAMPLE 32.1

Find the length marked x.

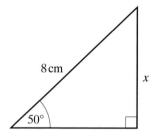

Solution

Draw a diagram and label the sides H, O and A.

Since you know the hypotenuse (H) and want to find the opposite (O), you use the sine ratio.

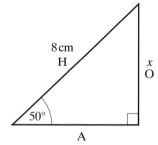

$$\sin 50° = \frac{O}{H} = \frac{x}{8}$$

$8 \times \sin 50° = x$ Multiply both sides by 8.

Press these keys on your calculator to find x.

$$\boxed{8}\ \boxed{\times}\ \boxed{\sin}\ \boxed{5}\ \boxed{0}\ \boxed{=}$$

$x = 6.128\ 35... = 6.13$ cm correct to 3 significant figures.

EXAMPLE 32.2

In triangle DEF, EF = 12 cm, angle $E = 90°$ and angle $F = 35°$.
Find the length DE.

Solution

Draw the triangle and label the sides.

Since you know the adjacent (A) and want to find the opposite (O),
you use the tangent ratio.

$$\tan 35° = \frac{O}{A} = \frac{x}{12}$$

$12 \times \tan 35° = x$ \qquad Multiply both sides by 12.

Press these keys on your calculator to find x.

$\boxed{1}\ \boxed{2}\ \boxed{\times}\ \boxed{\tan}\ \boxed{3}\ \boxed{5}\ \boxed{=}$

$x = 8.402\ 49... = 8.40$ cm correct to 3 significant figures.

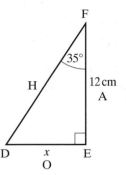

EXERCISE 32.1

1 In these diagrams find the lengths marked a, b, c, d, e, f, g and h.

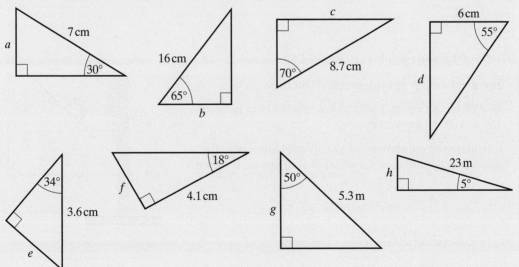

2 The ladder in the picture is 6 metres long.
The angle between the ladder and the ground is 70°.
How far from the wall is the foot of the ladder?

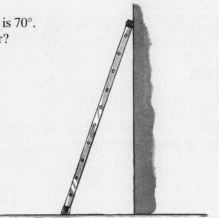

3 a) Find the height, h, of the triangle.
    b) Use the height you found in part **a)**
        to find the area of the triangle.

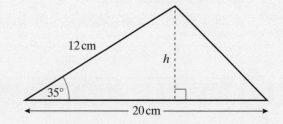

Challenge 32.1

The length of the crane's arm is 20 metres.

The crane can operate with the arm anywhere between
15° and 80° to the vertical.

Calculate the minimum and maximum values of x, the
distance from the crane at which a load can be lowered.

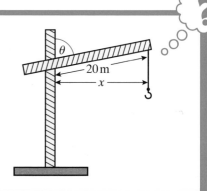

Using the ratios 2

Check up 32.2

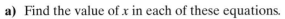

a) Find the value of x in each of these equations.

(i) $4 = \dfrac{8}{x}$ **(ii)** $4 = \dfrac{12}{x}$ **(iii)** $2 = \dfrac{20}{x}$

(iv) $3 = \dfrac{15}{x}$ **(v)** $2 = \dfrac{18}{x}$ **(vi)** $6 = \dfrac{24}{x}$

b) Copy and complete this general statement.

If $a = \dfrac{b}{x}$ then $x = \ldots\ldots\ldots$.

In the second type of problem you will encounter, you are required to find the denominator (bottom) of the fraction. This is demonstrated in the following examples.

EXAMPLE 32.3

Find the length marked x.

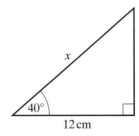

Solution

Draw the triangle and label the sides.

Since you know the A and want to find H, you use the cosine ratio.

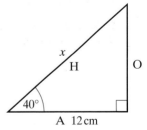

$$\cos 40° = \frac{A}{H} = \frac{12}{x}$$

$$x = \frac{12}{\cos 40°} \quad \text{Using the rule you found in Check up 32.2}$$

Press these keys on your calculator to find x.

$$\boxed{1}\,\boxed{2}\,\boxed{\div}\,\boxed{\cos}\,\boxed{4}\,\boxed{0}\,\boxed{=}$$

$x = 15.664\,88\ldots = 15.7$ cm correct to 3 significant figures.

EXAMPLE 32.4

Find the length marked *x*.

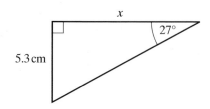

Solution

Draw the triangle and label the sides.

Since you know O and want to find A, you use the tangent ratio.

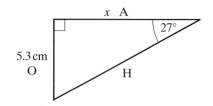

$$\tan 27° = \frac{O}{A} = \frac{5.3}{x}$$

$$x = \frac{5.3}{\tan 27°} \qquad \text{Using the rule you found in Check up 32.2.}$$

$$x = 10.401\ 83... \qquad \text{Using your calculator.}$$

$$x = 10.4 \text{ cm correct to 3 significant figures.}$$

TIP

Always look to see whether the length you are trying to find should be longer or shorter than the one you are given. If your answer is obviously wrong, you have probably multiplied instead of dividing.

EXERCISE 32.2

1 In these diagrams find the lengths marked *a*, *b*, *c*, *d*, *e*, *f*, *g* and *h*.

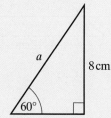

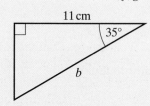

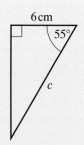

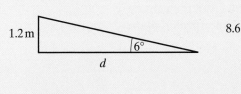

1.2 m

d

6°

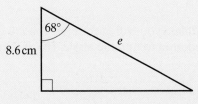

8.6 cm

68°

e

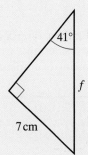

41°

f

7 cm

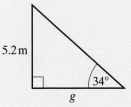

5.2 m

34°

g

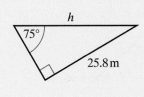

75°

h

25.8 m

2 The bearing of A from B is 040°.
A is 8 kilometres east of B.
Calculate how far A is north of B.

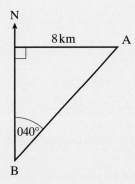

N

8 km

A

040°

B

3 The diagram shows a lean-to shed.

   a) Find the length *d*.

   b) The length of the shed is 2.5 m.
       Find the volume of the shed.

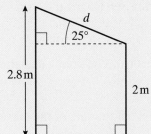

d

25°

2.8 m

2 m

Challenge 32.2

Mr Jones wants to buy a ladder.
His house is 5.3 metres high and he needs to reach the top.
The ladders are in two sections, each section being the same length.
When extended there must be an overlap of 1.5 metres between the two sections.

The safe operating angle between the ladder and the ground is 76°.

Calculate the length of each of the sections of ladder he needs to buy.

Using the ratios 3

In the third type of problem you will encounter, you are given the value
of two sides and are required to find the angle. This is demonstrated in the
following examples.

EXAMPLE 32.5

Find the angle θ.

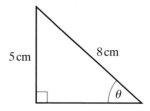

Solution

Draw the triangle and label the sides.

This time, look at the two sides you know.

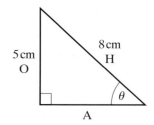

Since they are O and H, you use the sine ratio.

$$\sin \theta = \frac{O}{H} = \frac{5}{8}$$

Work out $5 \div 8 = 0.625$ on your calculator and leave this number in
your display.

You have worked out the sine of the angle and now need to work back
to the angle.

To do this, you use the $\sin^{-1}$ function (the inverse of sin).

You will find $\sin^{-1}$ above the $\boxed{\sin}$ button on your calculator.

To use this function you press the button labelled SHIFT, INV or
2nd F, followed by the $\boxed{\sin}$ button.

With 0.625 still in your display, press $\boxed{\text{SHIFT}}$ $\boxed{\sin}$ $\boxed{=}$, or the
equivalent on your calculator.

You should see 38.682 18… .

So θ = 38.7° correct to 3 significant figures or 39° correct to the nearest degree.

Check that you do get this answer using your calculator.

You can also do the calculation in one stage by pressing these buttons, or the equivalent on your calculator. Note that you *must* use the brackets.

$\boxed{\text{SHIFT}}\ \boxed{\text{sin}}\ \boxed{(}\ \boxed{5}\ \boxed{\div}\ \boxed{8}\ \boxed{)}\ \boxed{=}$

EXAMPLE 32.6

Find the angle θ.

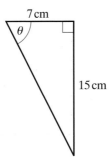

Solution

Draw the triangle and label the sides.

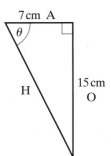

The two sides you know are O and A so you use the tangent ratio.

$$\tan\theta = \frac{O}{A} = \frac{15}{7} \qquad \text{so} \qquad \theta = \tan^{-1}\frac{15}{7}$$

This is the sequence of keys to press on your calculator.

$\boxed{\text{SHIFT}}\ \boxed{\text{tan}}\ \boxed{(}\ \boxed{1}\ \boxed{5}\ \boxed{\div}\ \boxed{7}\ \boxed{)}\ \boxed{=}$

This gives the answer θ = 64.983… = 65.0° correct to 3 significant figures.

EXAMPLE 32.7

Find the angle θ.

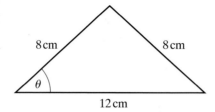

Solution

Since this is an isosceles triangle, not a right-angled triangle, you need to draw in the line of symmetry. This splits the triangle into two equal right-angled triangles.

The sides you know are A and H so you use the cosine ratio.

$$\cos \theta = \frac{A}{H} = \frac{6}{8} \qquad \text{so} \qquad \theta = \cos^{-1} \frac{6}{8}$$

This is the sequence of keys to press on your calculator.

[SHIFT] [cos] [(] [6] [÷] [8] [)] [=]

This gives the answer $\theta = 41.4°$ correct to 3 significant figures.

TIP

Example 32.7 shows how to deal with isosceles triangles. You use the line of symmetry to split the triangle into two equal right-angled triangles. This works only with isosceles triangles because they have a line of symmetry.

EXERCISE 32.3

1 In these diagrams find the angles marked a, b, c, d, e, f, g and h.

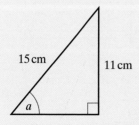

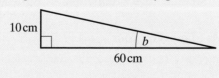

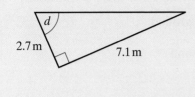

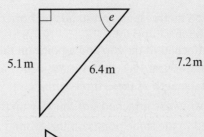

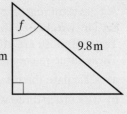

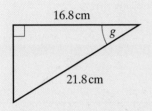

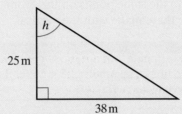

2 The diagram represents a ladder leaning against a wall.
Find the angle the ladder makes with the horizontal.

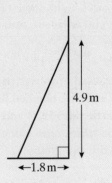

3 In the picture the kite is 15 metres above the girl.
The string is 25 metres long.

Find the angle the string makes with the horizontal.

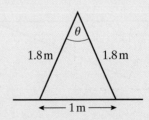

4 The diagram represents a pair of step ladders
standing on a horizontal floor.

Find the angle, θ, between the two parts of the
step ladder.

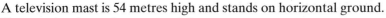

Challenge 32.3

A television mast is 54 metres high and stands on horizontal ground.

Six guy wires keep the mast upright.

Three of these are attached to the top and a point on the ground.

These three make an angle of 16.5° with the vertical.

a) Calculate the total length of these three wires.

b) The other three wires are attached $\frac{2}{3}$ of the way up the mast.

They are attached to the same points on the ground as the previous three.
Calculate the angle these make with the vertical.

WHAT YOU HAVE LEARNED

- **The sides of a right-angled triangle are labelled hypotenuse, opposite and adjacent, usually abbreviated to H, O and A**

- $\sin \theta = \dfrac{O}{H}, \quad \cos \theta = \dfrac{A}{H}, \quad \tan \theta = \dfrac{O}{A}$

- **The steps used to solve a trigonometry problem are as follows**
  1 **Draw a clearly labelled diagram**
  2 **Label the sides H, O and A**
  3 **Decide which ratio you need to use**
  4 **Solve the equation**

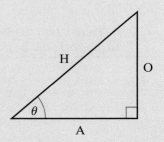

MIXED EXERCISE 32

1 In these diagrams find the lengths marked a, b, c, d, e and f.

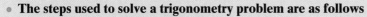

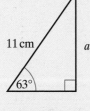

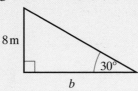

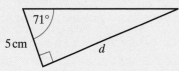

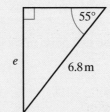

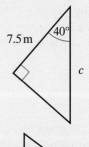

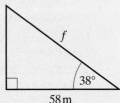

2 In these diagrams find the angles marked *a* and *b*.

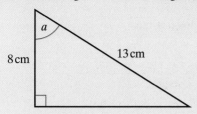

3 **a)** Find the length *a* in this diagram.

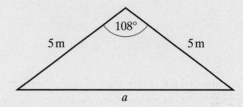

b) Find the angle *b* in this diagram.

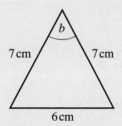

4 The picture shows a ladder leaning against a wall.

a) For safety, the angle the ladder makes with the horizontal should be between 75° and 77°.
Is this ladder safe? Show your working.

b) Find the length of the ladder.

c) A man is standing on the ladder.
His feet are 3.5 metres from the bottom of the ladder.
How far above the ground are his feet?

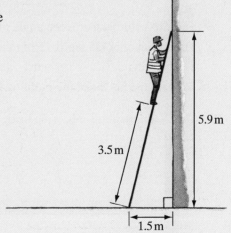

5 A ship sails on a bearing of 070° for 120 kilometres.

a) Draw a diagram to show this.

b) Calculate how far the ship is
   (i) east of its starting point.
   (ii) north of its starting point.

6 Find the acute angle between the diagonals of a rectangle with sides 7 cm and 10 cm.

7 Calculate the length of the equal sides in this isosceles triangle.

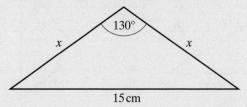

8 a) Find angle θ in this trapezium.
  b) Find the area of the trapezium.

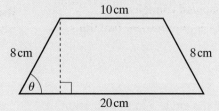

9 A survey is being made of a tall building.

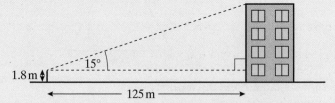

The surveyor measures the angle of elevation of the top of the building as 15°.
The surveyor is 125 metres from the foot of the building.
The sighting device is 1.8 metres above the ground.
Calculate the height of the building.

10 The diagrams show a bascule bridge in the closed and open positions.

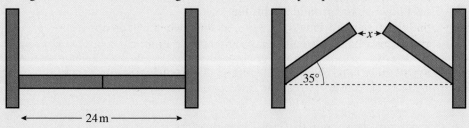

Calculate the distance, x, between the two parts of the bridge in the open position.

33 → TRIGONOMETRY 2

THIS CHAPTER IS ABOUT

- Finding areas, angles and lengths in non-right-angled triangles
- Graphs of trigonometrical functions

YOU SHOULD ALREADY KNOW

- How to find the area of a right-angled triangle
- Basic trigonometry within right-angled triangles

The area of a triangle

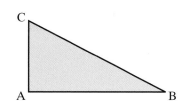

For any right-angled triangle, you can find the area of the triangle using the formula

$$\text{Area} = \tfrac{1}{2} \times \text{base} \times \text{height} = \tfrac{1}{2} \times AB \times AC.$$

For a non-right-angled triangle, you can use the formula to find the area of the triangle if you know the perpendicular height of the triangle. This is because the triangle can be divided into two right-angled triangles.

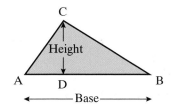

$$\text{Area} = \left(\tfrac{1}{2} \times AD \times CD\right) + \left(\tfrac{1}{2} \times DB \times CD\right)$$

$$= \left(\tfrac{1}{2} \times CD\right) \times (AD + DB)$$

$$= \tfrac{1}{2} \times CD \times AB$$

$$= \tfrac{1}{2} \times \text{base} \times \text{height}$$

You cannot find the area of a triangle using the formula when you do not know the perpendicular height. An alternative method is required.

There is a convention to label the sides and angles of a triangle. The angles are labelled with capital letters. The side opposite an angle is labelled with the same letter but in lower case.

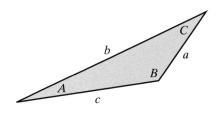

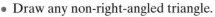

Discovery 33.1

- Draw any non-right-angled triangle.
- Label it according to convention.
- Measure accurately each of the sides and each of the angles of your triangle.
- For your triangle calculate $ab \sin C$, $bc \sin A$ and $ac \sin B$.

a) What do you notice?

b) Compare your answers with those of other students.

$ab \sin C$ means $a \times b \times \sin C$.

General formula for the area of a triangle

ACD is a right-angled triangle.

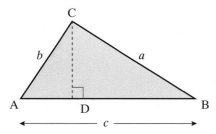

As you learned in Chapter 32, you can find the length of CD using simple trigonometry.

$$\sin A = \frac{O}{H} = \frac{CD}{b}$$

so $\qquad CD = b \sin A$

Now you can work out a formula to find the area of any triangle.

$$\begin{aligned} \text{Area} &= \tfrac{1}{2} \times \text{base} \times \text{height} \\ &= \tfrac{1}{2} \times \text{AB} \times \text{CD} \\ &= \tfrac{1}{2} \times c \times b \sin A \\ &= \tfrac{1}{2} bc \sin A \end{aligned}$$

Using different sides as the base you get three alternative versions of the formula.

$$\text{Area} = \tfrac{1}{2} bc \sin A$$
$$\text{Area} = \tfrac{1}{2} ac \sin B$$
$$\text{Area} = \tfrac{1}{2} ab \sin C$$

EXAMPLE 33.1

Find the area of this triangle.

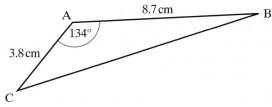

Solution

Angle A is given so the version of the formula to use is

$$\text{Area} = \frac{1}{2} \times b \times c \sin A$$
$$= \frac{1}{2} \times 3.8 \times 8.7 \times \sin 134°$$
$$= 11.9 \text{ cm}^2$$

EXERCISE 33.1

1 Find the area of each of these triangles.

a)
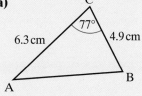

b)
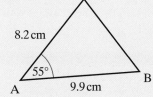

c)
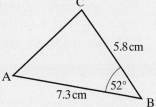

d)
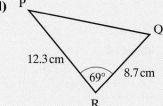

e)
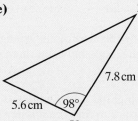

f)
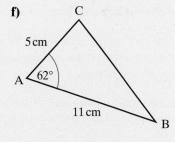

g)
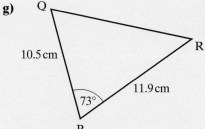

h)
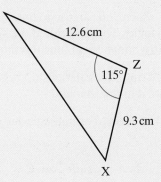

2 a) In triangle ABC, $a = 5$ cm, $b = 6$ cm and the area is 11 cm².
Find the size of angle C.

b) In triangle PQR, $p = 6.4$ cm, $r = 7.8$ cm and the area is 23.4 cm².
Find the size of angle Q.

c) In triangle XYZ, $y = 10.7$ cm, $z = 7.6$ cm and the area is 16.9 cm².
Find the size of angle X.

3 a) In triangle ABC, $a = 27$ cm, angle B is 54° and the area is 345 cm².
Find the length of side c.

b) In triangle PQR, $r = 9.3$ cm, angle P is 123° and the area is 74.1 cm².
Find the length of side q.

Challenge 33.1

a) A triangle ABC has side AC of length 13.7 cm, side BC of length 8.7 cm and an area of 50 cm².
Calculate the size of angle ACB.

b) A triangle ABC has side AB of length 8.9 cm, angle ABC of 78° and an area of 29.2 cm².
Calculate the length of side BC.

The sine rule

The general formula works for finding the area of any triangle where the lengths of two sides and the size of the angle between them (the **included angle**) are known.

If the angle given is a non-included angle or if two angles are given and the length of just one side, it is necessary to find the missing information first.

There are two rules which you can use to find missing information in non-right-angled triangles. They are called the **sine rule** and the **cosine rule** because of the trigonometrical function they each rely upon.

The sine rule is based upon the relationship between an angle and the side opposite it in the triangle.

The sine rule takes two forms. For finding sides you use this form.

$$\frac{a}{\sin A} = \frac{b}{\sin B} = \frac{c}{\sin C}$$

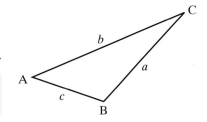

For finding angles you use this form.

$$\frac{\sin A}{a} = \frac{\sin B}{b} = \frac{\sin C}{c}$$

You use any two of the fractions at the same time.

It is useful to learn and remember the sine rule but it is not essential that you do so. The first version will be provided in an examination and the second version is just the first version inverted.

The sine rule works for any triangle in which you know
- the lengths of two sides and the size of the non-included angle.
- the length of one side and the sizes of any two angles (because this means you in fact know the size of all the angles).

Proof 33.1

It is easy to prove the sine rule, although you will not be required to do so in an examination.

In triangle ADC,   AD = $b \sin C$.
In triangle ADB,   AD = $c \sin B$.

Therefore $b \sin C = c \sin B$

or $\dfrac{b}{\sin B} = \dfrac{c}{\sin C}$

Using another perpendicular will involve a and A.

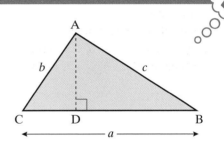

EXAMPLE 33.2

Find the size of each of the missing angles and sides in these diagrams.

a)

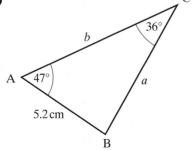

b)

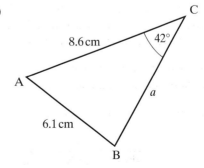

Solution

a) **Side *a***

The fractions to use are $\dfrac{a}{\sin A} = \dfrac{c}{\sin C}$.

$$\frac{a}{\sin 47} = \frac{5.2}{\sin 36}$$

$$a = \frac{5.2 \times \sin 47}{\sin 36}$$

$$a = 6.47 \text{ cm}$$

Angle *B*

Angle $B = 97°$, because the angles of a triangle add up to $180°$.

Side *b*

$$\frac{b}{\sin B} = \frac{c}{\sin C}$$

$$\frac{b}{\sin 97} = \frac{5.2}{\sin 36}$$

$$b = \frac{5.2 \times \sin 97}{\sin 36}$$

$$b = 8.78 \text{ cm}$$

b) **Angle *B***

Use the alternative form of the sine rule since you want to find an angle.

$$\frac{\sin B}{b} = \frac{\sin C}{c}$$

$$\frac{\sin B}{8.6} = \frac{\sin 42}{6.1}$$

$$\sin B = \frac{8.6 \times \sin 42}{6.1}$$

$$B = 70.6°$$

Angle *A*

Angle $A = 67.4°$, because the angles of a triangle add up to $180°$.

Side *a*

$$\frac{a}{\sin A} = \frac{c}{\sin C}$$

$$\frac{a}{\sin 67.4} = \frac{6.1}{\sin 42}$$

$$a = \frac{6.1 \times \sin 67.4}{\sin 42}$$

$$a = 8.42 \text{ cm}$$

> **TIP**
> In your calculations use the sides and angles that you are given, rather than ones you have calculated, where possible.

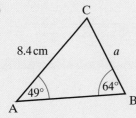

EXERCISE 33.2

1 Find the size of each of the angles and sides marked.

a)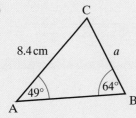

b)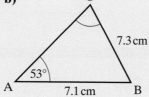

c)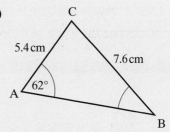

d)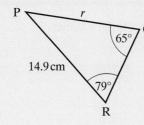

e)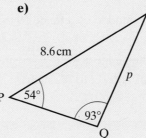

f)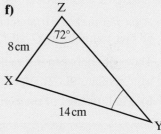

2 Find the size of each of the sides and angles not given in these diagrams.

a)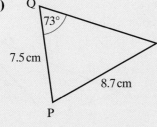

b)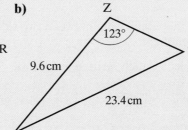

c)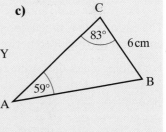

d)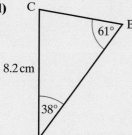

e)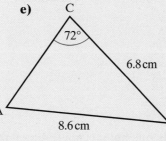

f)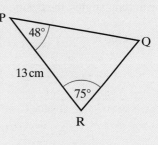

Challenge 33.2

The diagram represents a river.

The bearing of a tree on one bank is measured from two points, R and S, that are 50 metres apart on the opposite bank.

Calculate the width of the river.

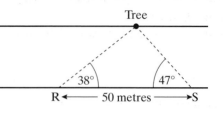

Tree

38°   47°

R ◄——— 50 metres ———► S

Challenge 33.3

A farmer makes a triangular pen bounded by a fence 59 metres long, a hedge 68 metres long and a wall.

The angle between the wall and the hedge is 49°.

a) Calculate the length of the wall, correct to the nearest metre.

b) Find the area of the pen.

The cosine rule

The cosine rule is based upon the relationship between two sides of a triangle and the angle included between them.

For finding sides, the cosine rule is usually written as follows.

$$a^2 = b^2 + c^2 - 2bc \cos A$$
$$b^2 = a^2 + c^2 - 2ac \cos B$$
$$c^2 = a^2 + b^2 - 2ab \cos C$$

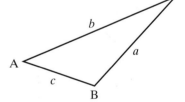

For finding angles, the cosine rule is usually written as follows.

$$\cos A = \frac{b^2 + c^2 - a^2}{2bc}$$
$$\cos B = \frac{a^2 + c^2 - b^2}{2ac}$$
$$\cos C = \frac{a^2 + b^2 - c^2}{2ab}$$

Again, it is useful to learn the cosine rule but it is not essential that you do so. One version of each of the forms of the cosine rule will be provided in an examination. You will, however, need to remember how to manipulate them.

The cosine rule works for any triangle in which you know
- the lengths of all three sides.
- the lengths of two sides and the size of the included angle.

Proof 33.2

Again, you do not need to be able to prove the cosine
rule in an examination. It is, however, quite easy to
prove using Pythagoras' theorem.

In triangle ADC, $\qquad b^2 = AD^2 + DC^2$

$$DC = a - BD$$

Therefore $\qquad DC^2 = a^2 - 2aBD + BD^2$

In triangle ADB: $\qquad AD^2 = c^2 - BD^2$

and $\qquad BD = c \cos B$

Therefore $\qquad DC^2 = a^2 - 2ac \cos B + BD^2$

so $\qquad b^2 = c^2 - BD^2 + a^2 - 2ac \cos B + BD^2$

so $\qquad b^2 = c^2 + a^2 - 2ac \cos B$

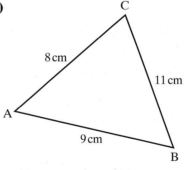

EXAMPLE 33.3

Find the size of each of the missing angles and sides in these diagrams.

a)

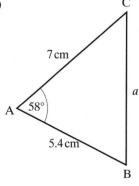

b)

Solution

a) Side a

$$a^2 = b^2 + c^2 - 2bc \cos A$$
$$a^2 = 7^2 + 5.4^2 - 2 \times 7 \times 5.4 \times \cos 58°$$
$$a^2 = 49 + 29.16 - 40.06$$
$$a^2 = 38.1$$
$$a = 6.2 \text{ cm}$$

Angle B

$$\cos B = \frac{a^2 + c^2 - b^2}{2ac}$$

$$\cos B = \frac{38.1 + 5.4^2 - 7^2}{2 \times 6.2 \times 5.4}$$

$$\cos B = \frac{18.26}{66.96}$$

$$B = 74.2°$$

Angle C

Angle $C = 47.8°$, because the angles of a triangle add up to $180°$.

b) Angle A

$$\cos A = \frac{b^2 + c^2 - a^2}{2bc}$$

$$\cos A = \frac{8^2 + 9^2 - 11^2}{2 \times 8 \times 9}$$

$$\cos A = \frac{24}{144}$$

$$A = 80.4°$$

Angle B

$$\cos B = \frac{a^2 + c^2 - b^2}{2ac}$$

$$\cos B = \frac{11^2 + 9^2 - 8^2}{2 \times 11 \times 9}$$

$$\cos B = \frac{138}{198}$$

$$B = 45.8°$$

Angle C

Angle $C = 47.8°$, because the angles of a triangle add up to $180°$.

EXERCISE 33.3

1 Find the size of each of the sides and angles marked in these diagrams.

a)

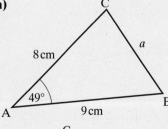

b)

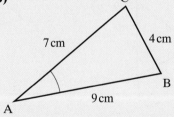

c)

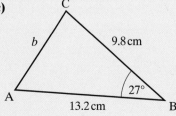

d)

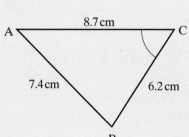

2 Find the size of each of the sides and angles not given in these diagrams.

a)
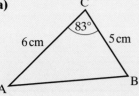

b)
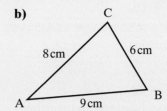

c)
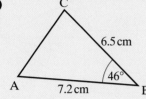

d)
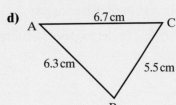

e)
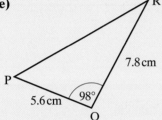

f)
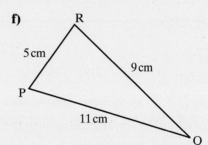

g)
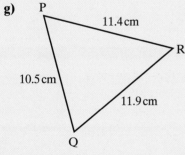

h)
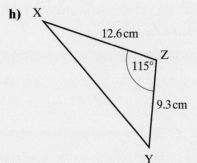

i)
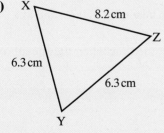

j)

3 a) In triangle ABC, a = 23 cm, b = 19.4 cm and angle C is 54°.
   Find the length of AB.

b) In triangle PQR, p = 12 cm, q = 13.4 cm and r = 15.6 cm.
   Find the size of angle Q.

4 ABCDEFGH is a cuboid.
   ACH is a triangle contained within the cuboid.
   Calculate the size of each of these angles.

   a) Angle ACH

   b) Angle AHC

   c) Angle CAH

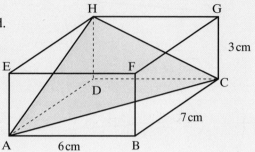

Challenge 33.4

A runner takes part in a race. The course is in the shape of a triangle.

Three posts, X, Y and Z, mark the corners of the triangle.

Angle XYZ is 54°, XY is 100 metres and YZ is 120 metres.

What is the length of the race?

Challenge 33.5

A harbour (H) is 1500 metres due south of a lighthouse (L).

A speedboat is travelling towards H.

When the speedboat is first seen from the lighthouse it is on a bearing of 060° and at a distance of 2800 m from L.

Calculate the distance of the speedboat from the harbour at the time when the speedboat is first seen from the lighthouse.

Graphs of trigonometrical functions

When using the sine and cosine rules in the previous sections you will, from time to time, have been dealing with angles greater than 90° and your calculator will have given you the sine or cosine value of these angles. This is possible because the values of sine and cosine follow a repeating pattern. Consider the following situation.

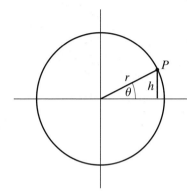

$$\sin \theta = \frac{h}{r}$$

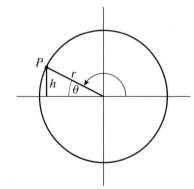

Rotating the radius anticlockwise until the height of P is again h gives rise to the diagram on the left.

$$\sin (180 - \theta) = \frac{h}{r}$$

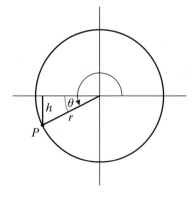

Rotating the radius anticlockwise until the height of P is again h gives rise to the diagram on the left.

$$\sin (180 + \theta) = \frac{-h}{r}$$

A further rotation to $(360 - \theta)$

would produce $\sin (360 - \theta) = \frac{-h}{r}$.

This pattern can be repeated for any height h (different values of θ). Plotting the value of h against the value of θ, for all angles θ, produces the graph shown below. It is called the **sine curve**.

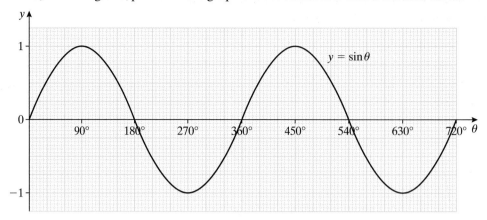

The height (or depth) of the wave from zero is known as the **amplitude**.
The distance over which the wave repeats itself is known as the **period**.

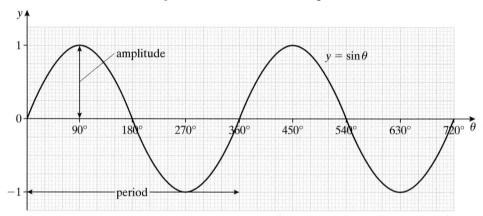

It is a useful skill to be able to sketch the sine curve and, for this reason, you should try to remember the basic features of the curve.

The cosine curve takes a similar form, as can be seen in the diagram below.

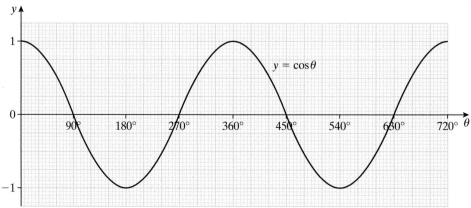

Challenge 33.6

The curve of cos θ is simply the curve of sin x translated by 90°.
Why do you think that this is the case?

A similar diagram can be drawn for tan θ but this does not produce a continuous curve, as can be seen in the diagram below.

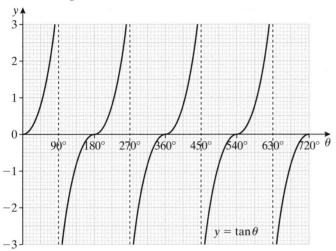

The diagram for the tan θ function does have a repeating pattern, like that for sin θ and cos θ. Its period is 180°, as shown by the dashed vertical lines known as **asymptotes**.

EXAMPLE 33.4

Draw accurately the graph of $y = \sin \theta$ for values of θ from −180° to 180°.
From your graph read off all the values of θ in this range for which $\sin \theta = 0.4$.

Solution

Draw the curve and the line $y = 0.4$.

Read off the values of θ where the curve and the line intersect.
The approximate possible solutions from the graph are 24° and 156°.

Note that, if you were to find $\sin^{-1} 0.4$ using a scientific calculator you would only obtain the one solution of $23.6°$.

Drawing the curve and the line on a graphic calculator would enable you to see both solutions. Check that you know how to use the ZOOM, TRACE and INTERSECT functions on your graphic calculator.

EXERCISE 33.4

1 **a)** Sketch the graph of $y = \sin \theta$ for values of θ from $0°$ to $540°$.
   b) For what values of θ in this range does $\sin \theta = 0.5$?

2 **a)** Sketch the graph of $y = \cos \theta$ for values of θ from $-180°$ to $360°$.
   b) For what values of θ in this range does $\cos \theta = -0.5$?

3 **a)** Draw accurately the graph of $y = \sin \theta$ for values of θ from $-180°$ to $180°$.
   b) From your graph read off all the values of θ for which $\sin \theta = 0.7$.

4 **a)** Draw accurately the graph of $y = \cos \theta$ for values of θ from $0°$ to $540°$.
   b) From your graph read off all the values of θ for which $\cos \theta = -0.4$.

5 One solution to $\sin \theta = 0.8$ is approximately $53°$.
   Using only the symmetry of the sine curve, find the other angles between $0°$ and $720°$ that also satisfy the equation $\sin \theta = 0.8$.

6 **a)** Draw a sketch graph of $y = \tan \theta$ for values of θ from $0°$ to $360°$.
   b) Find, from your graph, the angles for which $\tan \theta = 1.2$.

Transforming graphs of trigonometrical functions

The graph below shows the curves $y = \sin \theta$ and $y = 4 \sin \theta$, i.e. $4 \times (\sin \theta)$.

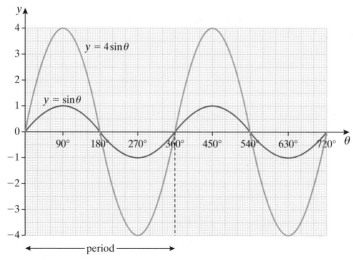

You can see that the amplitude of the green curve, $y = 4\sin\theta$, is 4 times the amplitude of the curve $y = \sin\theta$. This is because you are plotting 4 times $\sin\theta$ and this has the effect of stretching the curve in the y direction by a scale factor of 4.

Fractional scale factors such as $y = \frac{1}{2}\sin\theta$ have the effect of compressing the curve as the amplitude is halved and all the y values are consequently reduced.

In both cases the period of the graph remains unchanged.

The graph below shows the curves $y = \cos\theta$ and $y = \cos 4\theta$, i.e. $\cos(4 \times \theta)$. Note the difference between this graph and the graph above.

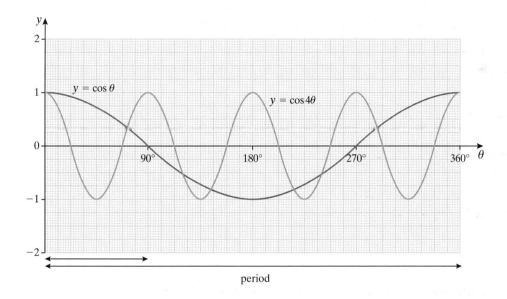

period

You will see that the period of the green curve, $y = \cos 4\theta$, is one quarter the period of the curve $y = \cos\theta$. This is because you are plotting $\cos(4 \times \theta)$ and this has the effect of compressing the curve in the θ direction by a scale factor of 4.

Fractional scale factors, such as $y = \cos\frac{1}{2}\theta$ have the effect of stretching the curve in the θ direction. This is because the angle is effectively halved and all the θ values are consequently reduced.

In both cases the amplitude of the graph remains unchanged.

For all graphs of the type $y = A\sin B\theta$ and $y = A\cos B\theta$, the amplitude of the curve is A and the period of the curve is $\dfrac{360}{B}$.

EXAMPLE 33.5

Draw the graph of $y = 2\sin 3\theta$ between $0°$ and $360°$.
Find the solutions of $2\sin 3\theta = 1.5$ between $150°$ and $270°$

a) from your graph.

b) from your calculator, giving your answers to 1 decimal place.

Solution

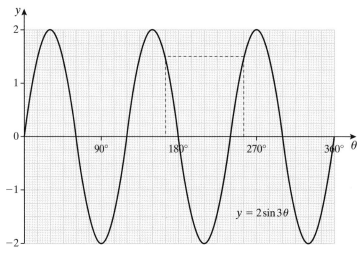

a) Approximately $165°$ and $255°$

b) $166.2°$ and $253.8°$

EXERCISE 33.5

1 Find the amplitude and the period of each of these curves.

a) $y = 3\sin\theta$ b) $y = \cos 5\theta$ c) $y = 3\sin 4\theta$

d) $y = 4\sin\frac{1}{2}\theta$ e) $y = 5\cos 3\theta$ f) $y = 2\cos 0.8\theta$

2 Draw the graph of $y = 3\cos\theta$ for values of θ from $0°$ to $180°$.

3 Draw the graph of $y = \sin 2\theta$ for values of θ from $0°$ to $180°$.

4 Sketch the graph of $y = \cos 2\theta$ for values of θ from $0°$ to $360°$.

5 Sketch the graph of $y = 2.5\sin\theta$ for values of θ from $0°$ to $360°$.

6 Find the solutions of $\cos 3\theta = -0.5$ between $0°$ and $180°$.

Challenge 33.7

Using a scale of 1 cm to represent 30°, draw the curve of $y = \sin 2x$ for values of x from 0° to 360°.

The flow of tides on any given day can be modelled by the curve $y = \sin 2x$.

The first high tide of a day occurs at 3 a.m.

What is the time of

a) the next high tide?

b) the low tides that day?

Challenge 33.8

a) On one set of axes, draw sketch graphs of $y = 1.2 \sin \theta$ and $y = \cos 0.8\theta$ for values of θ from 0° to 360°.

b) Find, from your graph, the solutions to the equation $1.2 \sin \theta = \cos 0.8\theta$.

WHAT YOU HAVE LEARNED

- **The convention for labelling angles and sides in a triangle**

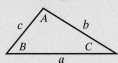

- **Area of any triangle** $= \frac{1}{2}bc \sin A$
- **The sine rule is used for calculating sides or angles in non-right-angled triangles when you know either**
  - **the lengths of two sides and the size of the non-included angle**
  - **the length of one side and the sizes of any two angles**
- **The sine rule can be written** $\dfrac{a}{\sin A} = \dfrac{b}{\sin B} = \dfrac{c}{\sin C}$ **or** $\dfrac{\sin A}{a} = \dfrac{\sin B}{b} = \dfrac{\sin C}{c}$
- **The cosine rule is used for calculating sides or angles in non-right-angled triangles when you know either**
  - **the lengths of all three sides**
  - **the length of two sides and the size of the included angle**
- **The cosine rule can be written** $\cos A = \dfrac{b^2 + c^2 - a^2}{2bc}$

- **The amplitude of a trigonometrical graph is the maximum height (or depth) of the curve**
- **The period of a trigonometrical graph is the distance over which the curve repeats**

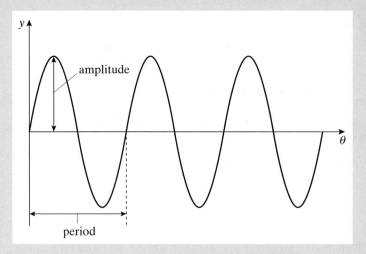

- **For all graphs of the type $y = A \sin B\theta$ and $y = A \cos B\theta$, the amplitude is A and the period, in degrees, is $\dfrac{360}{B}$**

MIXED EXERCISE 33

1 Work out the area of each of these triangles.

a)

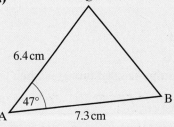

b)

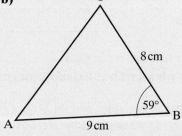

c)

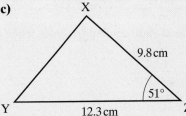

d)

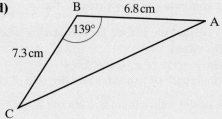

2 **a)** In triangle ABC, $a = 9$ cm, $b = 11$ cm and the area is 20 cm^2.
Find the size of angle C.

b) In triangle PQR, $p = 10.4$ cm, $r = 12.4$ cm and the area is 33.2 cm^2.
Find the size of angle Q.

c) In triangle XYZ, $y = 15.3$ cm, $z = 9.4$ cm and the area is 68.8 cm^2.
Find the size of angle X.

3 **a)** In triangle ABC, $a = 15$ cm, angle B is $49°$ and the area is 45 cm^2.
Find the length AB.

b) In triangle XYZ, $x = 18.4$ cm, angle Y is $96°$ and the area is 148.2 cm^2.
Find the length XY.

4 Find the size of each of the marked angles and sides in these diagrams.

a)
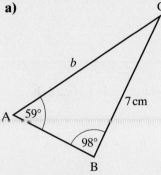

b)
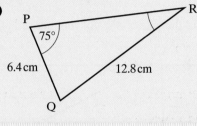

c)
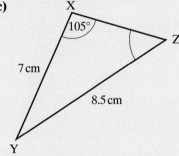

d)
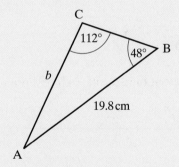

5 Find the size of each of the sides and angles not given in these diagrams.

a)
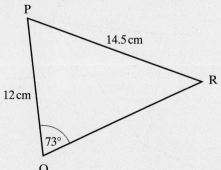

b)
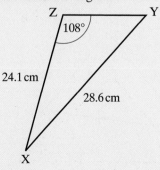

c)

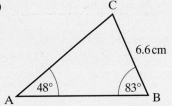

d)

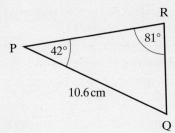

6 In the triangle XYZ, angle XYZ is 43°, angle ZXY is 65° and side YZ is 8.3 cm. Calculate the length of the longest side of the triangle.

7 Find the size of each of the marked angles and sides in these diagrams.

a)

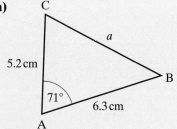

b)

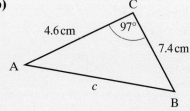

c)

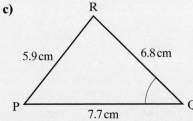

d)

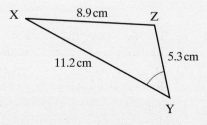

8 Find the size of each of the sides and angles not given in these diagrams.

a)

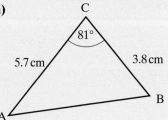

b)

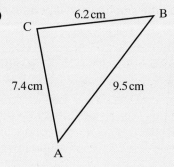

c)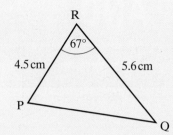

d)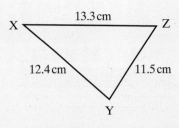

9 A scalene triangle has sides of 53 metres, 74 metres and 85 metres.
Calculate the sizes of the angles of the triangle.

10 ABCDEFGH is a cuboid. M is the midpoint
of CD. AHM is a triangle contained within
the cuboid.
Calculate the size of each of these angles.

    a) Angle AHM

    b) Angle HMA

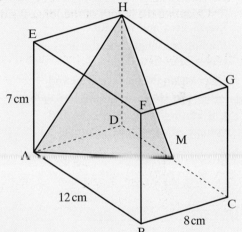

11 a) (i) Sketch the graph of $y = \sin\theta$ for values of θ from $0°$ to $360°$.
    (ii) For what values of θ in this range does $\sin\theta = 0.7$?

    b) (i) Draw accurately the graph of $y = \cos\theta$ for values of θ from $-180°$ to $180°$.
    (ii) For what values of θ in this range does $\cos\theta = -0.3$?

    c) One solution to $\cos\theta = -0.6$ is approximately $127°$.
    Using only the symmetry of the cosine curve, find the other angles between $0°$ and
    $720°$ that also satisfy the equation $\cos\theta = -0.6$.

    d) (i) Sketch the graph of $y = \tan\theta$ for values of θ from $-180°$ to $180°$.
    (ii) Find, from your graph, the angles for which $\tan\theta = 2$.

    e) Find the amplitude and period of each of these curves.
    (i) $y = 3\sin\theta$ **(ii)** $y = \cos 5\theta$ **(iii)** $y = 3\sin 4\theta$
    (iv) $y = 4\sin\frac{1}{2}\theta$ **(v)** $y = 5\cos\frac{2}{3}\theta$ **(vi)** $y = 2\cos 0.8\theta$
    (vii) $y = 5\cos 3\theta$ **(viii)** $y = 2\cos 0.6\theta$

    f) Sketch these graphs.
    (i) $y = 3\sin\theta$ for values of θ from $0°$ to $180°$
    (ii) $y = \cos 4\theta$ for values of θ from $0°$ to $180°$
    (iii) $y = 3\cos 2\theta$ for values of θ from $0°$ to $360°$
    (iv) $y = 0.5\sin 3\theta$ for values of θ from $0°$ to $360°$

    g) Find the solutions of $\sin 5\theta = 0.4$ between $-180°$ and $180°$.

34 → LENGTH, AREA AND VOLUME

Dimension analysis

Discovery 34.1

Write down all the formulae you can remember for length, area and volume, keeping each of these categories in a separate list.

Include the formulae given above.

Look at the formulae in each group, to see what they have in common.

Looking at the dimensions of the formulae:

number $\times$ length = length (1 dimension)
length + length = length (1 dimension)
length $\times$ length = area (2 dimensions)
length $\times$ length $\times$ length = volume (3 dimensions)
area $\times$ length = volume (3 dimensions)

Check that these rules fit all the formulae you remembered in Discovery 34.1.

TIP

Looking at dimensions can help you to remember formulae. For example, you can find out whether the area of a circle is πr^2 or $2\pi r$ if you have forgotten which it is. Remember that π is a number and has no dimensions.

Looking at the number of dimensions can also help you to remember which units to use. For example

cm for length (1 dimension)
cm^2 for area (2 dimensions)
cm^3 for volume (3 dimensions).

EXAMPLE 34.1

a, b, h and r are lengths.
Decide whether each of these formulae represents a length, an area, a volume or none of these.

a) $r + 2h$ **b)** $\frac{1}{2}(a + b)h$ **c)** $r^2 + 2\pi r$ **d)** $\frac{4}{3}\pi r^3$

Solution

a) length + length = length
b) (length + length) × length = length × length = area
c) area + length = none of these
d) length^3 = volume

EXERCISE 34.1

Throughout this exercise, letters in algebraic expressions represent lengths and numbers have no dimensions.

1 Which of the following expressions could be a length?
     a) $3r$ **b)** $a + 2b$ **c)** rh

2 Which of the following expressions could be an area?
     a) $bc - a^2$ **b)** $\pi h(a^2 + b^2)$ **c)** $\frac{1}{3}\pi r^2$

3 Which of the following expressions could be a volume?
     a) $2\pi r^2$ **b)** $3ab(c + d)$ **c)** $\frac{1}{3}\pi r^2 h$

4 State whether each of these expressions represents a length, an area or a volume.
     a) $a + p$ **b)** $\pi r^2 h$ **c)** $\pi r l$

5 State whether each of these expressions represents a length, an area, a volume or is none of these.
     a) $8c^3$ **b)** $6c^2$ **c)** $4c + \pi r^2$ **d)** $12c$

Arcs and sectors

An **arc** is part of the circumference of a circle.

The shape in the diagram is a **sector** of a circle.
It is formed by an arc and two radii.

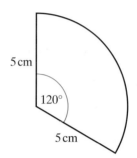

Any sector is a fraction $\dfrac{\theta}{360}$ of a circle, where θ is the sector angle, as shown in the diagram.

The area of the sector is this fraction of the area of the circle.

The length of the arc is this fraction of the circumference of the circle.

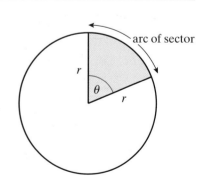

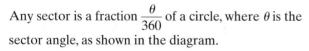

$$\text{Area of a sector} = \frac{\theta}{360} \times \pi r^2$$

$$\text{Length of the arc of a sector} = \frac{\theta}{360} \times 2\pi r$$

EXAMPLE 34.2

Calculate the area and the perimeter of this sector.

4.6 cm

75°

4.6 cm

Solution

Sector area $= \dfrac{\theta}{360} \times \pi r^2$

$\qquad = \dfrac{75}{360} \times \pi \times 4.6^2$

$\qquad = 13.8$ cm^2 to 1 decimal place

Perimeter $=$ Arc length $+ 2 \times 4.6$

$\qquad = \dfrac{\theta}{360} \times 2\pi r + 2 \times 4.6$

$\qquad = \dfrac{75}{360} \times 2\pi \times 4.6 + 2 \times 4.6$

$\qquad = 15.2$ cm to 1 decimal place

EXAMPLE 34.3

A sector has area 92 cm^2. The sector angle is 210°.
Find the radius of the circle.

Solution

Area of the sector $= \dfrac{\theta}{360} \times \pi r^2$

$\qquad\qquad 92 = \dfrac{210}{360} \times \pi r^2$

$\qquad \dfrac{92 \times 360}{210 \times \pi} = r^2$

$\qquad\qquad r = \sqrt{\dfrac{92 \times 360}{210 \times \pi}}$

$\qquad\qquad\quad = 7.1$ cm to 1 decimal place

> **TIP**
>
> Solving an equation like this involves rearranging the equation. If you prefer, you can rearrange the formula before substituting.

1   Find the arc length of each of these sectors. Give your answers to the nearest millimetre.

   a)   **b)**   **c)**

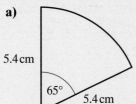   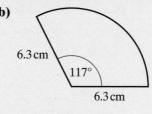   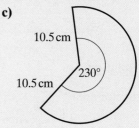

2   Find the area of each of the sectors in question **1**.

3   Find the sector angle of each of these sectors. Give your answers to the nearest degree.

   a)   **b)**

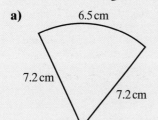   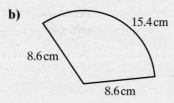

   c)   **d)**

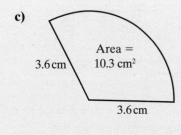   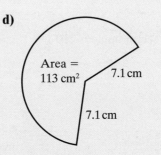

4   Find the radius of each of these sectors.

   a)   **b)**   **c)**

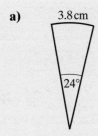   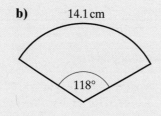   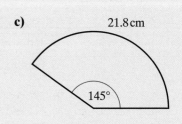

5 Find the radius of each of these sectors.

a)

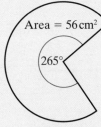

Area = 56 cm²

265°

b)

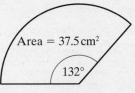

Area = 37.5 cm²

132°

c)

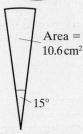

Area = 10.6 cm²

15°

d)

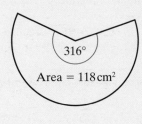

316°

Area = 118 cm²

6 A sector of a circle of radius 6.2 cm has an arc length of 20.2 cm.
Calculate the angle of the sector and hence find the area of the sector.

7 A cushion is in the shape of a sector with a sector angle of 150°. The radius is 45 cm.
Edging is sewn all round the cushion. What length of edging is required?

Pyramids, cones and spheres

Discovery 34.2

a) Cut out a sector of a circle of radius 12 cm.
Each class/group member should choose a different sector angle.

b) Stick the straight edges together to form a cone.

c) Calculate the arc length of your sector.
This has become the circumference of the base of your cone.

d) Work out the radius of the base of your cone.
Check your answer by measuring your cone.

e) The sector area has become the curved surface area of the cone. Calculate this too.

Challenge 34.2

Replace 12 cm in Discovery 34.2 with the slant height, ℓ, of the cone.
Using a sector angle of θ, find the radius, r, of the base of the cone in terms of ℓ and θ.
Then find the area of the sector in terms of r and ℓ.

Here are some formulae for pyramids and cones.

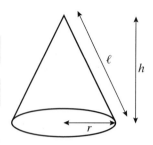

$$\text{Curved surface area of a cone} = \pi r \ell$$

$$\text{Volume of a pyramid} = \tfrac{1}{3} \times \text{area of base} \times \text{height}$$

A cone is a special sort of pyramid with a circular base.

$$\text{Volume of a cone} = \tfrac{1}{3}\pi r^2 h$$

EXAMPLE 34.4

A pyramid has a rectangular base 8 cm by 5 cm and a height of 9 cm.
Calculate its volume.

Solution

$$\begin{aligned} \text{Volume} &= \tfrac{1}{3} \times \text{area of base} \times \text{height} \\ &= \tfrac{1}{3} \times (8 \times 5) \times 9 \\ &= 120 \text{ cm}^3 \end{aligned}$$

EXAMPLE 34.5

A cone has a slant height of 6.8 cm and its curved surface area is 91 cm^2.
Find the radius, r, of its base.

Solution

$$\begin{aligned} \text{Curved surface area} &= \pi r \ell \\ 91 &= \pi r \times 6.8 \\ r &= \frac{91}{\pi \times 6.8} \\ &= 4.3 \text{ cm to the nearest millimetre} \end{aligned}$$

The other shape you meet in this section is a **sphere**. Proving the formulae
for a sphere is beyond the scope of this course, but here are the results
you will need.

$$\text{Volume of sphere} = \tfrac{4}{3}\pi r^3$$

$$\text{Surface area of sphere} = 4\pi r^2$$

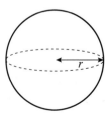

EXAMPLE 34.6

A bowl is in the shape of a hemisphere of diameter 25 cm.
Calculate its volume, in litres.

Solution

Volume of hemisphere $= \frac{1}{2} \times$ volume of sphere

$$= \frac{2}{3}\pi r^3$$

$$= \frac{2}{3}\pi \times 12.5^3$$

$$= 4090.6\ldots \text{ cm}^3 \qquad (1 \text{ litre} = 1000 \text{ cm}^3)$$

$$= 4.09 \text{ litres to 3 significant figures}$$

EXERCISE 34.3

1 Find the curved surface area of each of these cones.
Give your answers to 3 significant figures.

a)

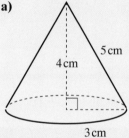

5 cm

4 cm

3 cm

b)

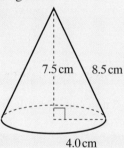

7.5 cm   8.5 cm

4.0 cm

c)

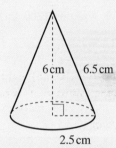

6 cm   6.5 cm

2.5 cm

2 Calculate the volume of each of the cones in question **1**.

3 Calculate the volume of each of these pyramids.

a)

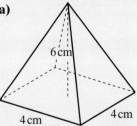

6 cm

4 cm   4 cm

b)

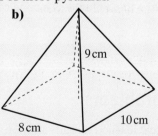

9 cm

8 cm   10 cm

c)

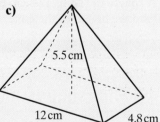

5.5 cm

12 cm   4.8 cm

4 A solid cone has a base of radius 4.2 cm and a slant height of 7.8 cm.
Calculate its total surface area.

5 A pyramid has a square base with sides of 8 cm. Its volume is 256 cm³.
Find its height.

6 Find the base radius of cones with these measurements.
   a) Volume 114 cm³, height 8.2 cm   **b)** Volume 52.9 cm³, height 5.4 cm
   c) Volume 500 cm³, height 12.5 cm

7 A sector of a circle is joined to form a cone (as in Discovery 34.2).
Find the base radius of the cones made with these sectors.
   a) Radius 5.8 cm, angle 162°   **b)** Radius 9.7 cm, angle 210°
   c) Radius 12.1 cm, angle 295°

8 Find the surface area of spheres with these measurements.
   a) Radius 5.1 cm   **b)** Radius 8.2 cm
   c) Diameter 20 cm

9 Find the volume of each of the spheres in question **8**.

10 Find the radius of spheres with these measurements.
   a) Surface area 900 cm²   **b)** Surface area 665 cm²
   c) Volume 1200 cm³   **d)** Volume 8000 cm³

Challenge 34.3

How many glass marbles of radius 8 mm can be made from 500 cm³ of molten glass?

Challenge 34.4

A sphere of radius 12.8 cm has the same volume as a cone of base radius 8.0 cm.
Find the height of the cone.

Compound shapes and problems

In this section you apply what you have learned already to more complex
shapes and to problems involving shapes.

Two important examples of compound shapes are a segment of a circle
and a frustum of a cone.

A line drawn between any two points on the circumference of a circle is called a **chord**.

A chord divides a circle into two **segments**.

The larger segment (white) is called the **major** segment; the smaller segment (green) is called the **minor** segment.

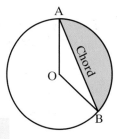

Area of the shaded segment = area of sector AOB − area of triangle AOB

EXAMPLE 34.7

Calculate the area of the shaded segment shown in the diagram.

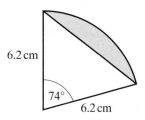

6.2 cm

74°

6.2 cm

Solution

Area of the sector $= \dfrac{\theta}{360} \times \pi r^2$

$\qquad\qquad\quad = \dfrac{74}{360} \times \pi \times 6.2^2$

$\qquad\qquad\quad = 24.823\ldots \text{ cm}^2$

Area of triangle $= \frac{1}{2}ab \sin C$

$\qquad\qquad\quad = \frac{1}{2} \times 6.2^2 \times \sin 74°$

$\qquad\qquad\quad = 18.475\ldots \text{ cm}^2$

Area of segment = area of sector − area of triangle

$\qquad\qquad\quad\quad = 24.823\ldots - 18.475\ldots$

$\qquad\qquad\quad\quad = 6.35 \text{ cm}^2$ to 3 significant figures

The **frustum** of a cone is the shape remaining when the top of a cone has been removed.

The circle on the top of the frustum is in a plane parallel to the base.

Volume of frustum = volume of whole cone − volume of missing cone

EXAMPLE 34.8

Find the volume of the frustum remaining when a cone of height 8 cm is removed from a cone of height 12 cm and base radius 6 cm.

Solution

First, use similar triangles to find the base radius, r cm, of the cone which has been removed.

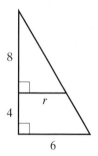

$$\frac{r}{8} = \frac{6}{12}$$

$$r = 4$$

Then find the volume of the frustum.

Volume of frustum = volume of whole cone − volume of missing cone

$$= \tfrac{1}{3}\pi r_1^2 h_1 - \tfrac{1}{3}\pi r_2^2 h_2$$

$$= \tfrac{1}{3}\pi \times 6^2 \times 12 - \tfrac{1}{3}\pi \times 4^2 \times 8$$

$$= 318 \text{ cm}^3 \quad \text{to 3 significant figures}$$

Other shape problems may use simpler shapes but require several steps to solve the problem, as in the next example.

EXAMPLE 34.9

A cone has a base of radius 5.6 cm and a volume of 82 cm³.

a) Calculate its height.

b) Calculate also its slant height and hence find its curved surface area.

Solution

Volume of cone $= \tfrac{1}{3}\pi r^2 h$

$$82 = \tfrac{1}{3}\pi \times 5.6^2 \times h$$

$$\frac{3 \times 82}{\pi \times 5.6^2} = h$$

$$h = 2.496\ldots$$

$$= 2.5 \text{ cm to 1 decimal place}$$

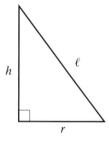

Using Pythagoras' theorem,

$$\ell^2 = r^2 + h^2 = 5.6^2 + 2.496\ldots^2$$

$$= 37.59\ldots$$

$$\ell = \sqrt{37.59\ldots} = 6.13\ldots$$

$$= 6.1 \text{ cm to 1 decimal place}$$

Curved surface area $= \pi r \ell = \pi \times 5.6 \times 6.13\ldots$

$$= 108 \text{ cm}^2 \text{ to 3 significant figures}$$

> **TIP**
> When using a result in further working, make sure you use the unrounded value. Do not clear your calculator until you are sure that you do not need the result again.

Some problems may use algebra, as in the next challenge.

Challenge 34.5

The curved surface area of a particular cylinder of radius r cm and height h cm has the same surface area as a cube of side r cm.

Find h in terms of r.

EXERCISE 34.4

1   Calculate the area of each of the orange segments.

a)
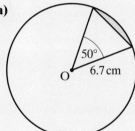

b)
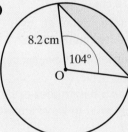

c)
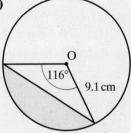

2   A grain silo is in the shape of a cylinder of height 25 m and radius 4 m topped by a hemisphere.
    Calculate the volume of the silo.

3   Calculate the area of each of the green major segments.

a)
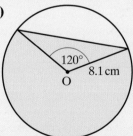

b)
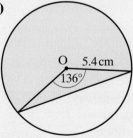

c)
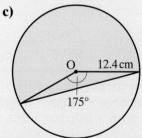

4   A piece of cheese is a prism of height 7 cm.
    Its cross-section is a sector of a circle with radius 15 cm and sector angle 72°.
    Calculate the volume of the piece of cheese.

5 Calculate the perpendicular height of each of these cones and hence find their volumes.

a)

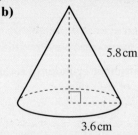

9.1 cm

5.4 cm

b)

5.8 cm

3.6 cm

c)

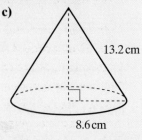

13.2 cm

8.6 cm

6 A solid cone has a base of radius 6.9 cm and a height of 8.2 cm.

   a) Calculate its volume.

   b) Find its slant height and hence its total surface area.

7 The top of a flowerpot is a circle of radius 10 cm.
Its base is a circle of radius 8 cm.
The height of the flowerpot is 10 cm.

   a) Show that the flowerpot is a frustum of an inverted cone of complete height 50 cm and base radius 10 cm.

   b) Calculate how many litres of soil the flowerpot can contain.

8 A cone has a height of 20 cm and a base radius of 12 cm.
The top 15 cm of the cone is removed to leave a frustum of height 5 cm.
Find the volume of this frustum.

9 A pyramid has a square base of side 10.4 cm and its sloping edges are all 8.8 cm long.
Calculate the perpendicular height of the pyramid and its volume.

10 A sector with angle 300° and radius 12.5 cm is joined to form a hollow cone.
Showing your method clearly, calculate the volume of the cone.

WHAT YOU HAVE LEARNED

- **Dimensions of formulae:**
  - **number × length = length (1 dimension)**
  - **length + length = length (1 dimension)**
  - **length × length = area (2 dimensions)**
  - **length × length × length = volume (3 dimensions)**
  - **area × length = volume (3 dimensions)**
- **Area of a sector = $\dfrac{\theta}{360} \times \pi r^2$**

- **Arc length of a sector** $= \dfrac{\theta}{360} \times 2\pi r$
- **Volume of a pyramid** $= \frac{1}{3} \times$ **area of base** $\times$ **height**
- **Volume of a cone** $= \frac{1}{3}\pi r^2 h$
- **Curved surface area of a cone** $= \pi r \ell$
- **Volume of a sphere** $= \frac{4}{3}\pi r^3$
- **Surface area of a sphere** $= 4\pi r^2$
- **Area of a minor segment = area of sector − area of triangle**
- **Volume of a frustum = volume of whole cone − volume of missing cone**

◎ MIXED EXERCISE 34

1 The letters in these formulae all represent lengths.
   Which of the formulae represent a length?

   a) $\sqrt{6a^2}$ b) $\dfrac{4bc}{d}$ c) $2a(c + d)$ d) $\dfrac{2\pi a}{5}$

2 Classify these formulae into those representing length, area, volume or none of these.
   The letters in the formulae all represent lengths.

   a) $\sqrt{a^2 + b^2}$ b) $a^2(2a + b)$ c) $\pi(4a + 3bc)$ d) $\pi a^2 + 3ac$

3 Calculate the arc length and the area of sectors with these measurements.

   a) Radius 6.2 cm, sector angle 62°

   b) Radius 7.8 cm, sector angle 256°

4 Find the sector angle of sectors with these measurements.

   a) Radius 4.6 cm, arc length 9.2 cm

   b) Radius 5.7 cm, area 85.6 cm²

5 A spinning top is a solid inverted cone of height 12.5 cm and base radius 6.7 cm.
   Find its total surface area.

6 A sphere has a volume of 127 cm³.
   Find its radius.

7 A wastepaper bin is the frustum of a cone.
   It has a top radius of 20 cm, a bottom radius of 15 cm and a height of 40 cm.
   Find the volume of rubbish that it can contain.

8 A sphere has a radius of 3.5 cm. A cone with a volume equal to that of the sphere has a
   base radius of 4.5 cm. Calculate the height of the cone.

35 ➡ SIMILARITY AND ENLARGEMENT

<table>
<tr><td>

THIS CHAPTER IS ABOUT

- **Recognising and working with similar triangles**
- **Enlarging triangles and other plane shapes using negative scale factors**
- **Area and volume of similar figures**

</td><td>

YOU SHOULD ALREADY KNOW

- **The angle facts for parallel lines: alternate angles, *a* and *b*, are equal and corresponding angles, *a* and *c*, are equal**

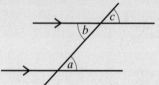

- **The angles in a triangle add up to 180°**
- **The angles on a straight line add up to 180°**
- **The exterior angle of a triangle equals the sum of the opposite, interior angles**

</td></tr>
</table>

Similarity of triangles and other plane shapes

Two triangles are **similar** if the angles in one triangle are equal to the angles in the other.

For other polygons, the test for similarity is that the angles in the two shapes are equal *and* the corresponding sides are in proportion.

In this diagram, triangle ABC is similar to triangle PQR because angles A, B and C equal angles P, Q and R.

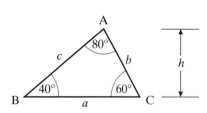
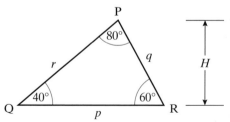

For similar triangles the ratios of the corresponding sides are equal.
As you did in Chapter 32, you can write the ratios as fractions.

$$\frac{a}{p} = \frac{b}{q} = \frac{c}{r} = k$$

The ratio k is the scale factor.

The heights are also in the same ratio.

$$\frac{h}{H} = k$$

Two triangles are also similar if their corresponding sides are in the same proportion.

EXAMPLE 35.1

The diagram shows two triangles, triangle ABC and triangle XYZ.

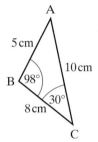
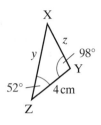

a) Explain why the two triangles are similar.

b) Calculate the values of y and z.

Solution

a) Angle BAC = 52°   Angles in a triangle add up to 180°.
Angle ZXY = 30°   Angles in a triangle add up to 180°.

Therefore the corresponding angles in each triangle are equal.
Therefore the triangles are similar.

b) AB and ZY are corresponding sides, both are opposite the angles of 30°.
BC and YX are corresponding sides, both are opposite the angles of 52°.

$$\frac{ZY}{AB} = \frac{4}{5} \quad \text{so} \quad \frac{YX}{BC} = \frac{4}{5}$$

$$YX = \tfrac{4}{5} \times BC$$

$$= \tfrac{4}{5} \times 8$$

$$= 0.8 \times 8 = 6.4$$

$$z = 6.4 \text{ cm}$$

Similarly,

$$\frac{ZX}{AC} = \frac{4}{5}$$

$$ZX = \tfrac{4}{5} \times AC$$

$$= \tfrac{4}{5} \times 10$$

$$= 0.8 \times 10 = 8$$

$$y = 8 \text{ cm}$$

EXAMPLE 35.2

Explain why triangles ABC and PQR are *not* similar.

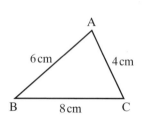 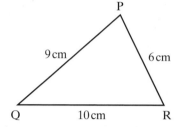

Solution

The shortest side in triangle ABC is AC and this corresponds with the shortest side in triangle PQR, namely PR.

They are in the ratio (or proportion) $\dfrac{PR}{AC} = \dfrac{6}{4} = 1.5$

The longest sides in both triangles also correspond. These are BC and QR.

They are in the ratio (or proportion) $\dfrac{QR}{BC} = \dfrac{10}{8} = 1.25$

Clearly the corresponding sides are not in the same proportion.
Therefore the triangles ABC and PQR are not similar.

⦿ EXERCISE 35.1

1 Which of the triangles in this diagram are similar?

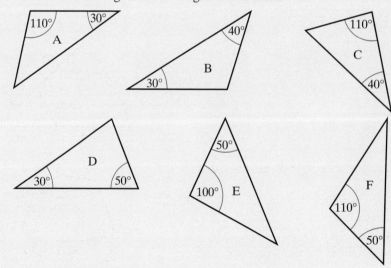

2 These two triangles are similar.

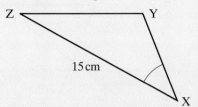

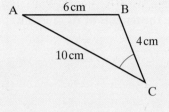

Calculate the lengths of XY and YZ.

3 In triangle ABC, XY is parallel to AC,
AB = 12 cm, YC = 3 cm, YB = 5 cm
and XY = 3 cm.
Prove that triangles BXY and BAC
are similar.
Calculate the lengths of AC, AX and XB.

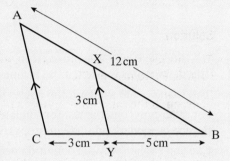

4 At noon a radio mast of height 12 m has a shadow length of 16 m.
Calculate the height of a tower with a shadow length of 56 m.

5 Calculate the lengths x and y in this diagram.

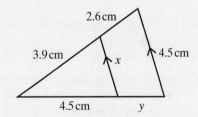

6 Calculate the lengths x and y in this diagram.

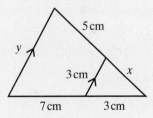

7 A stand for a small statue is made from part of a cone
(called a frustum).
The circular top of the stand has a diameter of 12 cm; the base
has a diameter of 18 cm.
The slant height of the stand is 9 cm.
Calculate the slant height of the complete cone.

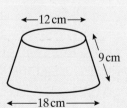

8 **a)** Explain how you know that the two triangles in the diagram are similar.

   b) Calculate the lengths of AC and CB.

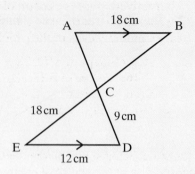

Challenge 35.1

This definition was given at the beginning of this chapter.

Two triangles are **similar** if the angles in one triangle are equal to the angles in the other.

For other polygons, the test for similarity is that the angles in the two shapes are equal *and* the corresponding sides are in proportion.

Can you explain why the definition needs extending for other polygons?

Enlargement

To carry out an enlargement you need two pieces of information
- the scale factor
- the centre of enlargement.

In Chapter 28 you learned about enlargements with fractional scale factors. The next example reminds you about this.

EXAMPLE 35.3

Plot the coordinates $A(4, 4)$, $B(16, 4)$, $C(16, 8)$ and $D(4, 8)$ and join them to form a rectangle.
Enlarge rectangle ABCD by a scale factor of $\frac{1}{2}$ using the origin as the centre of enlargement.

Solution

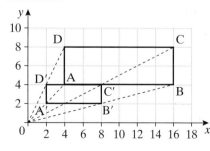

You draw the rectangle ABCD and join each of the vertices to the centre of enlargement.

Then you mark the position of the image of each of the vertices.

$OA' = \frac{1}{2} \times OA$ measured along OA.

$OB' = \frac{1}{2} \times OB$ measured along OB.

$OC' = \frac{1}{2} \times OC$ measured along OC.

$OD' = \frac{1}{2} \times OD$ measured along OD.

Alternatively, since the centre of enlargement is the origin, you can work out the coordinates of the image without drawing lines from the vertices. You simply multiply each of the coordinates of the object by the scale factor.

A(4, 4) maps to A′(2, 2).
B(16, 4) maps to B′(8, 2).
C(16, 8) maps to C′(8, 4).
D(4, 8) maps to D′(2, 4).

If the scale factor of an enlargement is negative, the image is on the opposite side of the centre of enlargement from the object, and the image is inverted. This is shown in the next example.

EXAMPLE 35.4

Plot the coordinates A(2, 2), B(4, 2) and C(2, 4) and join them to form a triangle.

Enlarge triangle ABC by a scale factor of −3 using the point O(1, 1) as the centre of enlargement.

Solution

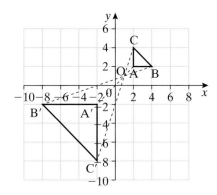

You plot the triangle.

Then you draw a line from each of the vertices through the centre of enlargement O and extend it on the other side.

You measure the distance from the vertex A, to the centre of enlargement O, and multiply it by 3.

You then mark the point A′ at a distance of 3 × OA along OA extended on the other side of the centre of enlargement.

> **TIP**
>
> You can find the position of the image points by counting squares from the centre of enlargement as you have done previously, so long as you remember that the image will be on the opposite side from the object. For example, from the centre of enlargement to point B is 3 units in the positive x direction and 1 unit in the positive y direction. Multiplying by the scale factor, point B′ is 9 units in the negative x direction and 3 unit in the negative y direction.

If you are given both the original shape and the enlarged shape you can find the centre of enlargement as you have previously. You join corresponding points on the two shapes with straight lines. The centre of enlargement is where the lines cross.

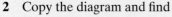
1   Draw a set of axes with the x-axis from -16 to 6 and the y-axis from -6 to 4.

Plot the points $A(2, 2)$, $B(5, 0)$, $C(5, -1)$ and $D(2, -1)$ and join them to form a quadrilateral.

Enlarge the quadrilateral by a scale factor of -3 using the origin as the centre of enlargement.

2   Copy the diagram and find
    a) the centre of enlargement.
    b) the scale factor.

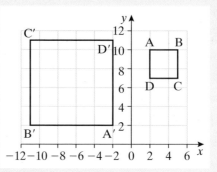

3   Draw a set of axes with the x-axis from 0 to 16 and the y-axis from 0 to 12.

Plot the points $A(2, 8)$, $B(6, 8)$, $C(6, 4)$ and $D(2, 4)$ and join them to form a square.

Enlarge the square by a scale factor of -2 using the point $O(5, 6)$ as the centre of enlargement.

4   Copy the diagram and find
    a) the centre of enlargement.
    b) the scale factor.

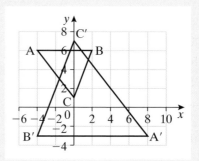

5   The diagram shows a quadrilateral, ABCD, and its image $A'B'C'D'$.

Copy the diagram and find
    a) the centre of enlargement.
    b) the scale factor.

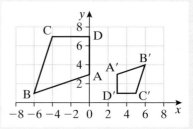

Challenge 35.2

a) Draw a scalene triangle.
Choose a centre and enlarge the triangle by a scale factor of −1.
Can the same result be achieved with other transformations?

b) Can you find another transformation that gives the same result if the triangle has symmetry?

The area and volume of similar figures

'Figure' is a mathematical term for a shape.

Discovery 35.1

Here are two similar figures. The sides of the larger figure are three times the length of those of the smaller figure. (The linear scale factor is 3.)

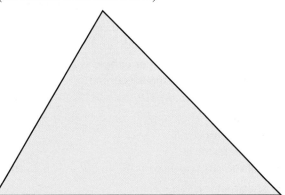

Tessellate the small triangle into the larger. How many will fit?

What is the area scale factor?

What is the connection between the area scale factor and the linear scale factor?

Discovery 35.2

Use some centimetre cubes to make a larger cube with each side three times as long. (The linear scale factor is 3.)

Are the small cube and the large cube similar figures?

What is the volume scale factor?

What is the connection between the volume scale factor and the linear scale factor?

The results of Discoveries 35.1 and 35.2 can be generalised.

> For similar shapes, the area scale factor is
> the square of the linear scale factor.
>
> For similar solids, the volume scale factor is
> the cube of the linear scale factor.

EXAMPLE 35.5

Two cuboids are similar, with a linear scale factor of 2.5.

a) The volume of the smaller cuboid is 10 cm³.
   Find the volume of the larger cuboid.

b) The total surface area of the larger cuboid is 212.5 cm².
   Find the surface area of the smaller cuboid.

Solution

a) The linear scale factor is 2.5.
   The volume scale factor is 2.5^3.
   The volume of the larger cuboid $= 10 \times 2.5^3 = 156.25$ cm³

b) The linear scale factor is 2.5.
   The area scale factor is 2.5^2.
   The surface area of the smaller cuboid $= 212.5 \div 2.5^2 = 34$ cm²

EXERCISE 35.3

1 Triangles PQR and XYZ are similar.

a) What is the linear scale factor of the enlargement?

b) Find the height of triangle XYZ.

c) Calculate the area of triangle PQR.

d) Calculate the area of triangle XYZ.

e) What do you notice about the ratio of the areas?

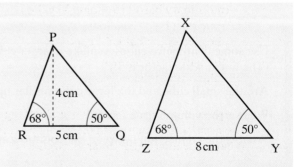

2 State the area scale factor and the volume scale factor for each of these linear scale factors.

    a) 2   **b)** 5   **c)** 10   **d)** 0.5

3 Find the linear scale factor and the volume scale factor for each of these area scale factors.

    a) 36   **b)** 64   **c)** 50   **d)** 0.1

4 A carton of cream holds 125 ml.
A similar carton is 1.5 times as tall.
How much cream does it hold?

5 A model car is built to a scale of 1 : 24.
It is 15 cm long.

    a) How long is the real car?

    b) 10 ml of paint are required to paint the model.
How many litres of paint will be needed for the real car?

6 An artist makes a model for a large sculpture. It is 24 cm high.
The finished sculpture will be 3.6 m high.

    a) Find the linear scale factor.

    b) Find the area scale factor.

    c) The model has a volume of 1340 cm^3.
What is the volume of the sculpture?

7 A tumbler is 12 cm high.
How tall is a similar glass which holds twice as much?

8 A model aircraft is made to a scale of 1 : 48.
The area of the wing of the real aircraft is 52 m^2.
What is the area of the wing of the model?

9 Barrels to hold liquid come in various sizes with different names.
    • An ordinary barrel (b) holds 36 gallons.
    • A firkin (f) holds 9 gallons.
    • A hogshead (h) holds 54 gallons.

    All the barrels are similar.
Find the ratio of their heights.

10 A square is enlarged by increasing the length of its sides by 10%.
The area of the square was originally 64 cm^2.
What is the area of the enlarged square?

a) A simple model for the heat lost by birds is that it is proportional to their surface area.

Also, the energy a bird can produce, to replace the heat loss, is proportional to its volume.

Investigate whether similar birds (meaning birds of the same species) are larger or smaller in colder climates.

b) Investigate the relationship for similar aircraft between wing area (which gives the lift) and volume (which is closely associated with the mass).

WHAT YOU HAVE LEARNED

- **Triangles are similar if the corresponding angles in each are equal**
- **A negative enlargement moves the image to the opposite side of the centre of enlargement from the original and inverts it**
- **The area scale factor for similar figures is the square of the linear scale factor**
- **The volume scale factor for similar figures is the cube of the linear scale factor**

MIXED EXERCISE 35

1 Calculate the lengths a, b, c, d and e in these diagrams.

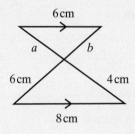

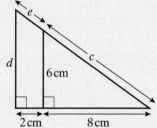

2 Triangle ADE has an area of 24 cm².

a) Calculate the area of triangle ABC.

b) Calculate the length of BC.

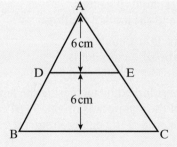

3 Draw a set of axes with the x-axis from 0 to 16 and the y-axis from 0 to 8.

Plot the points A(4, 5), B(9, 5), C(9, 2) and D(4, 2) and join them to form a rectangle.

Enlarge the rectangle by a scale factor of -1.5 using the point O(8, 4) as the centre of enlargement.

Write down the coordinates of the enlarged rectangle.

4 Copy the diagram and find
  a) the centre of enlargement.
  b) the scale factor.

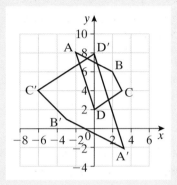

5 P and Q are two regular octagons. Q is an enlargement of P with a scale factor of 3.

The area of octagon Q is 90 cm². What is the area of octagon P?

6 An architect has made a model of a new hotel complex using a scale of 1 : 100.
  a) The area of the garden on the model is 650 cm².
    Find the area of the real garden.
  b) The real swimming pool will hold 500 000 litres of water.
    How much water is needed in the model?

7 Two bottles of lemonade are similar.
  One holds 2 litres and the other holds 1.5 litres.
  a) Find the ratio of their heights.
  b) The label on the larger bottle has an area of 75 cm².
    What is the area of the label on the smaller bottle?

36 → THREE-DIMENSIONAL GEOMETRY

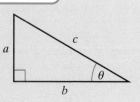
Finding lengths and angles in three dimensions

You can find lengths and angles of **3-D** objects by identifying **right-angled triangles** within the object and using **Pythagoras' theorem** or **trigonometry**.

EXAMPLE 36.1

A vertical flagpole, VO, which is 12 m high, is attached to the ground by three ropes of equal length.

The ropes reach the ground at A, B and C, 2 m from the foot of the flagpole.

The ground is horizontal.

Calculate

a) the length of a rope, VA.

b) the angle which it makes with the ground, VÂO.

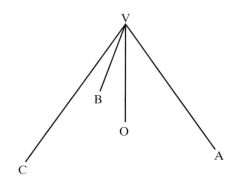

Solution

First draw a diagram.

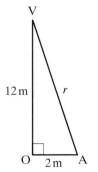

> **TIP**
>
> In a 3-D diagram, right angles often don't look their true size.
>
> Work out where the right angles are and draw a 2-D diagram of a relevant triangle.

a) $r^2 = 12^2 + 2^2$
$r^2 = 148$
$r = \sqrt{148}$
$r = 12.2$ m (to 1 d.p.)

Use Pythagoras' theorem to find the length of the rope.

b) $\tan V\hat{A}O = \dfrac{12}{2}$
$= 6$

Use trigonometry to find $V\hat{A}O$.

Angle VAO $= \tan^{-1} 6$
$= 80.5°$ (to 1 d.p.)

Use the $\boxed{\text{SHIFT}}$ $\boxed{\text{tan}}$ keys on your calculator to obtain the inverse function. It may be labelled arctan, invtan or $\tan^{-1}$.

EXAMPLE 36.2

Calculate the length of the diagonal of this cuboid.

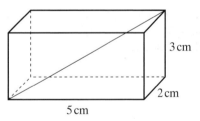

Solution

First identify a relevant right-angled triangle.

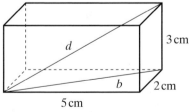
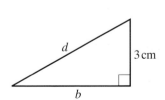

$d^2 = b^2 + 3^2$ (1)   Use Pythagoras' theorem to link the sides of the triangle. You cannot solve this equation yet because you don't know the value of b, so label the equation (1).

b is the length of the diagonal of the base of the cuboid.

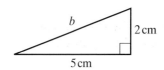

$b^2 = 5^2 + 2^2$ (2)   Use Pythagoras' theorem to form a second equation linking
$b^2 = 29$ the sides of this triangle, and use it to find b^2.
$d^2 = b^2 + 3^2$ (1)   Substitute for b^2 in equation (1).
$d^2 = 29 + 9$
$d^2 = 38$
$d = \sqrt{38}$
$d = 6.2$ cm (to 1 d.p.)

> **TIP** Don't forget to include the units in your answer.

Proof 36.1

Use a similar method to that used in Example 36.2 to show that the length of the diagonal of a cuboid measuring a cm by b cm by c cm is $\sqrt{a^2 + b^2 + c^2}$.

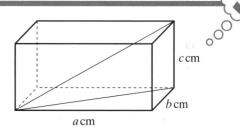

You can use the formula:

Diagonal of a cuboid measuring a cm by b cm by c cm $= \sqrt{a^2 + b^2 + c^2}$

when working with coordinates in three dimensions.

EXERCISE 36.1

1   Calculate the length of the diagonal of a cuboid measuring 5 cm by 8 cm by 3 cm.

2   Calculate the length of the diagonal of a cube with sides 5.6 cm.

3   A box is a cuboid with base 6 cm by 15 cm.
     A pencil 17 cm long just fits in the box.
     Calculate the height of the box.

4   A square-based pyramid has sloping edges of length 9.5 cm.
     Its base has diagonals of length 8.4 cm.
     Calculate the vertical height of the pyramid.

5 A pyramid has a rectangular base of sides 8.2 cm and 7.6 cm. Its height is 6.5 cm. All its sloping edges are the same length. Calculate the length of a sloping edge.

6 ABCDEF is a triangular wedge.
The faces ABFE, BCDF and ACDE are rectangles.

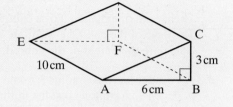

   a) Calculate the length CE.

   b) Calculate each of these angles.

     (i) CÂB

     (ii) CÊB

7 The pyramid OABCD has a horizontal rectangular base ABCD as shown.

O is vertically above A.

Calculate

   a) the length of OC.

   b) OĈA.

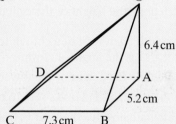

8 A vertical mast, MT, has its foot, M, on horizontal ground. It is supported by a wire, AT, which makes an angle of 65° with the horizontal and is of length 12 m, and two more wires, BT and CT, where A, B and C are on the ground.

BM = 4.2 m.

Calculate

   a) the height of the mast.

   b) the angle which BT makes with the ground.

   c) the length of wire BT.

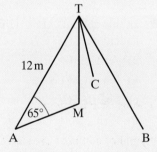

WHAT YOU HAVE LEARNED

- **How to find lengths and angles in 3-D shapes by finding right-angled triangles and using Pythagoras' theorem or trigonometry as appropriate**
- **That the length of the diagonal of a cuboid measuring a cm by b cm by c cm is $\sqrt{a^2 + b^2 + c^2}$**

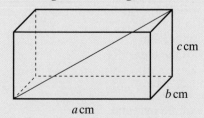

1 Calculate the length of the diagonal of a cuboid measuring 5.8 cm by 6.7 cm by 3.8 cm.

2 The length of a diagonal of a cuboid is 12.4 cm.
The base of the cuboid is a square of side 5.3 cm.
Calculate the height of the cuboid.

3 A cylinder has base radius 3.7 cm.
A pencil 15.6 cm long just fits in the cylinder.
  a) Calculate the angle between the pencil and the base of the cylinder.
  b) Calculate the height of the cylinder.

4 Amy stands 30 m west of a church tower.
She measures the angle of elevation of the top of the tower from ground level as 52°.
  a) Calculate the height of the tower.
  Amy then walks 25 m due south to point B.
  b) Calculate how far she is, at B, from the tower.
  c) Calculate the angle of elevation of the top of the tower from B.

5 In this cuboid, AB = 12 cm, BC = 5 cm and CG = 7 cm.
Calculate
  a) $A\hat{B}E$.
  b) length EG.
  c) length EC.
  d) $G\hat{E}C$.

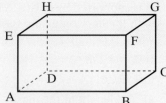

6 This 'lean-to' conservatory is a prism with a trapezium as its cross-section.
  a) Calculate the length of the sloping edge of the roof.
  b) Calculate the angle between the roof and the wall against which the conservatory is built.

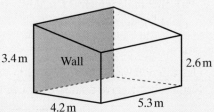

7 A pyramid has height 7.5 cm and a square base of side 6.3 cm.
  a) Calculate the length of a sloping edge of the pyramid.
  b) Calculate the angle that a sloping edge makes with the vertical.

8 A pyramid has a rectangular base of sides 3.6 cm and 5.2 cm.
The faces with base 5.2 cm make an angle of 62° with the base.
Calculate
  a) the height of the pyramid.
  b) the angle made with the base by a sloping edge of the pyramid.

- Solving problems in geometry and giving reasons/proof for your solutions
- Recognising and working with congruent triangles

- The angle facts for parallel lines: alternate angles, a and b, are equal and corresponding angles, a and c, are equal

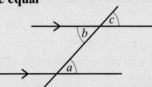

- The angles in a triangle add up to 180°
- The sum of the angles on a straight line is 180°
- The exterior angle of a triangle is equal to the sum of the opposite, interior angles

Angle properties of triangles

In Chapter 26 you learned how to use the properties of parallel lines to prove that the angles in a triangle add up to 180°.

Check up 37.1

Use the diagram to write a proof that the angles in a triangle add up to 180°. Remember to give a reason for each step.

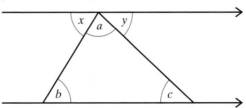

In Chapter 26 you saw one proof that the exterior angle of a triangle is equal to the sum of the opposite, interior angles. There is an alternative proof of this fact using the properties of parallel lines.

Proof 37.1

Using the properties of parallel lines, prove that the exterior angle of a triangle is equal to the sum of the opposite, interior angles.

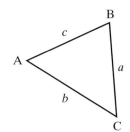

Solution

A line parallel to AB from C is constructed, as shown in the diagram.

$a = y$ Corresponding angles
$b = x$ Alternate angles

So

$a + b = x + y$ Since angles in a triangle add up to 180° and angles on a straight line add up to 180°.

This proves that the exterior angle of a triangle is equal to the sum of the opposite, interior angles.

Congruent triangles

In Chapter 35 you learned that two triangles are **similar** if all their angles are the same. If the angles of two triangles are the same, all their sides are in the same ratio. They are the same shape but not necessarily the same size.

Two triangles are **congruent** if they are the same shape and the same size. This means that they will fit exactly on to each other when one of them is rotated, reflected or translated.

Discovery 37.1

For each of the following triangles

(i) Construct the triangle.
You will need to use compasses for some of them.

(ii) Compare each of your triangles with your neighbours'.
Are they all congruent?

a) $a = 4$ cm, $b = 5$ cm, $c = 6$ cm
b) $A = 50°$, $b = 5$ cm, $c = 6$ cm
c) $A = 50°$, $B = 60°$, $c = 6$ cm
d) $A = 90°$, $a = 10$ cm, $b = 6$ cm
e) $A = 20°$, $a = 6$ cm, $b = 10$ cm

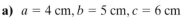

Two triangles are congruent if any of these conditions are satisfied.

- The three sides of one triangle are equal to the corresponding three sides of the other triangle (side, side, side or SSS).

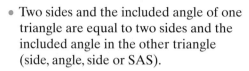 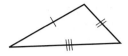 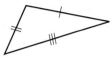

- Two sides and the included angle of one triangle are equal to two sides and the included angle in the other triangle (side, angle, side or SAS).

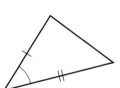 

- Two angles and the side between them of one triangle are equal to two angles and the side between them in the other triangle (angle, side, angle or ASA).

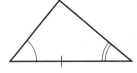 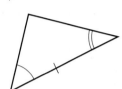

- Each triangle is right-angled and the hypotenuse and one other side of one triangle are equal to the hypotenuse and one side in the other triangle (right angle, hypotenuse, side or RHS).

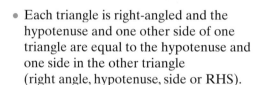 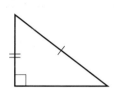 

EXAMPLE 37.1

State whether or not each of these pairs of triangles are congruent. If the triangles are congruent, give a reason for your answer.

a)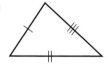

b)

c)

d)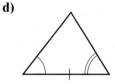

e)

f)

Solution

a) Congruent: SSS
b) Congruent: SAS
c) Not congruent
d) Congruent: ASA
e) Not congruent
f) Congruent: the missing third angles are also equal, ASA

Proof 37.2

Triangle XYZ is isosceles with XY = XZ.

The bisector of angle Y meets XZ at M; the bisector of angle Z meets XY at N.

Prove that YM = ZN.

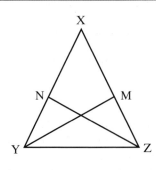

Solution

$$X\hat{Y}Z = X\hat{Z}Y$$ Base angles of an isosceles triangle are equal.

so

$$M\hat{Y}Z = N\hat{Z}Y.$$ Since YM bisects $X\hat{Y}Z$ and ZN bisects $X\hat{Z}Y$.

In triangles YMZ and ZNY, YZ is common.

So triangles YMZ and ZNY are congruent.   ASA

So YM = ZN.

⊙ EXERCISE 37.1

1 ABC is an isosceles triangle with AC = BC.
X and Y are points on AC and BC so that CX = CY.

Prove that triangle CXB is congruent to triangle CYA.

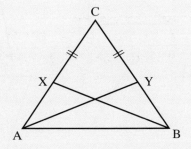

2 In this diagram, DX = XC, DV = ZC and AB is parallel to DC.

Prove that triangle DBZ is congruent to triangle CAV.

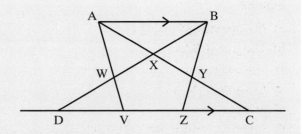

3 These two triangles are congruent.

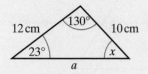

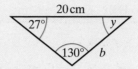

a) Write down the length of sides a and b.

b) Write down the size of angles x and y.

4 PQRS is a rhombus. PQ = QR = RS = SP.
PQ is parallel to RS, QR is parallel to PS.

a) Prove that
   (i) triangles PQX and RSX are congruent.
   (ii) triangles PSX and QRX are congruent.

b) Prove that X is the midpoint of SQ and of PR, and hence that the diagonals of a rhombus intersect at right angles.

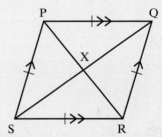

5 Which two of these triangles are congruent?

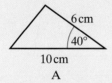

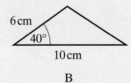

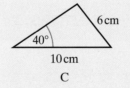

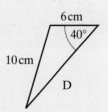

6 ABCD is a square.
P, Q, R and S are the midpoints of the sides.

Prove that the quadrilateral PQRS is a square.

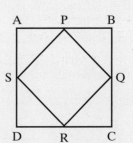

7 In this diagram, CD is the bisector of $A\hat{C}B$ and AE is parallel to DC.

Prove that AC = CE.

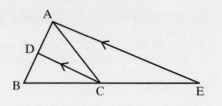

8 In this diagram, AB = AC, CD is parallel to BA, CD = CB and $B\hat{A}C = 40°$.

Calculate the size of angle DBC, giving reasons for your answer.

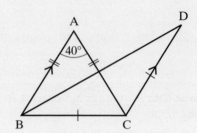

9 In this diagram, $C\hat{B}X = 122°$, $A\hat{E}C = 33°$ and AX is parallel to DY.

Calculate the size of $B\hat{A}E$ and $B\hat{C}Y$, giving reasons for your answers.

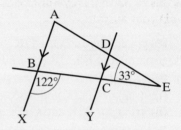

10 In triangle ABC, BD bisects angle $A\hat{B}C$, $A\hat{C}B = 80°$ and BD = BC.

Prove that $B\hat{A}C = 60°$.

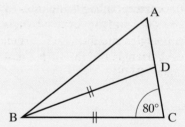

11 ABC is a right-angled triangle. $A\hat{C}B = 55°$ and angle PQC = 90°.

Prove that angle $a = 145°$.

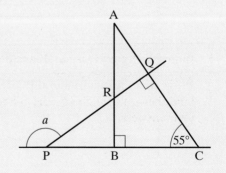

Challenge 37.1

Draw a triangle with one side 6 cm and angles 35°, 66° and 79°.

Compare your triangle with your neighbours'. Are they all congruent?
If the answer is 'No', how many different triangles are there?

Are all the triangles similar?

Challenge 37.2

Plane figures other than triangles, for example, rectangles and octagons, can be congruent.

Two rectangles have the same area and the same perimeter.
Does this mean that they must be congruent?
Explain your answer.

WHAT YOU HAVE LEARNED

* **Two triangles are congruent if one of these four statements is true:**
  * **the corresponding three sides of each triangle are equal (SSS)**
  * **two sides and the included angle in each triangle are equal (SAS)**
  * **two angles and the side between them in each triangle are equal (ASA)**
  * **each triangle is right-angled and the hypotenuse and one other side in each triangle are equal (RHS)**

MIXED EXERCISE 37

1 In this diagram PQ is parallel to SR, SP = SR, $\hat{SPR} = 66°$ and $\hat{PQS} = 22°$.

Find the size of angles x, y, z, giving reasons for your answers.

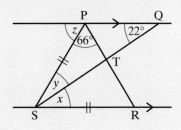

2 Calculate the size of angles x, y and z, giving reasons for your answers.

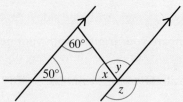

3 ABCD is a quadrilateral.
$\hat{CBA} = 70°$, $\hat{BAD} = 80°$ and $\hat{ADC} = 130°$.
CA bisects $\hat{DCB}$.

Prove that triangle CAB is isosceles.

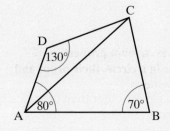

4 Find the size of angles a and b, giving reasons for your answers.

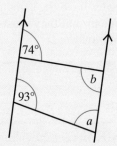

5 Find the size of x, giving reasons for your answer.

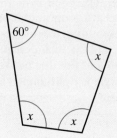

6 Use congruent triangles to prove that the diagonals of a kite cross at right angles.

7 Two straight lines, AB and CD, bisect each other at right angles at E.
Prove that the straight lines AC, CB, BD and DA are equal in length.

38 → CIRCLE THEOREMS

The perpendicular from the centre to a chord bisects the chord

In Chapter 37 you learned how to prove that triangles are **congruent**. The proof that the perpendicular from the centre of a circle to a chord bisects the chord uses congruent triangles. Remember, **bisect** means to divide exactly in two.

The letter O is generally used to represent the centre of the circle.

Proof 38.1

Look at triangles OAM and OBM.

OM is common to both triangles.

OA = OB   (Both are radii of the circle.)

$O\hat{M}A = O\hat{M}B = 90°$ (Given)

So triangles OAM and OBM are congruent.   (RHS)

Therefore AM = BM.

So the chord is bisected.

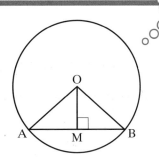

You can use the properties of circles to solve geometry problems. The next example shows how the fact that the perpendicular from the centre of a circle to a chord bisects the chord can be used.

EXAMPLE 38.1

Find the length of the chord PQ.
Give a reason for each step of your work.

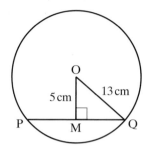

Solution

$$MQ = \sqrt{13^2 - 5^2} \qquad \text{(By Pythagoras' theorem)}$$
$$= \sqrt{144}$$
$$= 12 \text{ cm}$$

$$PQ = 2MQ \qquad \text{(The perpendicular from the centre bisects the chord.)}$$
$$= 24 \text{ cm}$$

The angle subtended by an arc at the centre of a circle is twice the angle that it subtends at any point on the circumference

The proof that the angle subtended by an arc at the centre of a circle is twice the angle that it subtends at any point on the circumference uses the facts about triangles that you met in Chapter 26. **Subtended by** is a mathematical expression meaning 'opposite to'.

Proof 38.2

Let $\hat{CAO} = x$ and $\hat{BAO} = y$.

Then $\hat{CAB} = x + y$.

Triangles OAB and OAC are isosceles
(OA, OB and OC are radii of the circle.)

$\hat{ACO} = x$ and $\hat{ABO} = y$
(Base angles of isosceles triangles)

$\hat{DOC} = \hat{OAC} + \hat{OCA} = 2x$ (The exterior angle of a triangle
$\hat{DOB} = \hat{OAB} + \hat{OBA} = 2y$ equals the sum of opposite interior angles.)

$\hat{COB} = \hat{DOC} + \hat{DOB}$
$\qquad = 2x + 2y = 2(x + y)$
$\qquad = 2 \times \hat{CAB}$

So the angle at the centre is equal to twice the angle at the circumference.

Example 38.2 shows a typical application of the fact that the angle subtended by an arc at the centre of a circle is twice the angle that it subtends at any point on the circumference.

EXAMPLE 38.2

Find $A\hat{C}B$.
Give a reason for each step of your work.

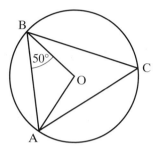

Solution

Triangle OAB is isosceles.	(OA and OB are radii.)
$B\hat{O}A = 180 - (50 + 50) = 80°$	(The angle sum of a triangle is 180°.)
$B\hat{C}A = 40°$	(The angle at the centre is twice the angle at the circumference.)

The angle subtended at the circumference in a semicircle is a right angle

This proof is a special case of the proof that the angle subtended by an arc at the centre of a circle is twice the angle that it subtends at any point on the circumference.

Proof 38.3

$A\hat{O}B = 2 \times A\hat{P}B$ (The angle at the centre is twice the angle at the circumference.)

$A\hat{O}B = 180°$ (AB is a diameter.)

$A\hat{P}B = 90°$ (Half of $A\hat{O}B$)

So the angle in a semicircle is 90°.

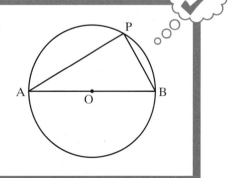

This theorem is often used in conjunction with the angle sum of a triangle, as is shown in the next example.

EXAMPLE 38.3

Work out the size of angle x.
Give a reason for each step of your work.

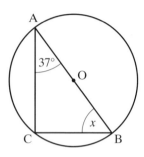

Solution

$\hat{ACB} = 90°$	(The angle in a semicircle is a right angle.)
$x = 180 - (90 + 37)$	(The angle sum of a triangle is 180°.)
$x = 53°$	

Angles in the same segment of a circle are equal

The proof that angles in the same segment of a circle are equal also uses the proof that the angle at the centre is twice the angle at the circumference. In Chapter 31 you learned that a **segment** is formed when a chord divides a circle into two pieces. In the diagram in Proof 38.4, the chord is between A and B.

Proof 38.4

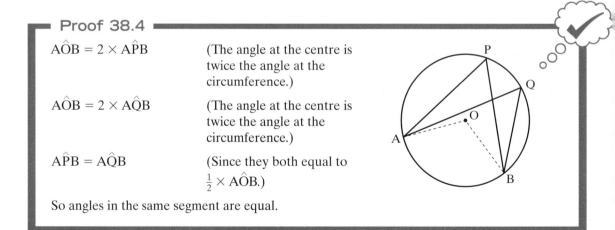

$\hat{AOB} = 2 \times \hat{APB}$	(The angle at the centre is twice the angle at the circumference.)
$\hat{AOB} = 2 \times \hat{AQB}$	(The angle at the centre is twice the angle at the circumference.)
$\hat{APB} = \hat{AQB}$	(Since they both equal to $\frac{1}{2} \times \hat{AOB}$.)

So angles in the same segment are equal.

To use this property, you have to identify angles subtended by the same arc since these are in the same segment. This is shown in the next example.

EXAMPLE 38.4

Find the sizes of angles a, b and c.
Give a reason for each step of your work.

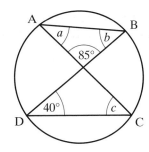

Solution

$a = \hat{BDC} = 40°$ (Angles in the same segment are equal.)

$b = 180 − (40 + 85) = 55°$ (The angle sum of a triangle is 180°.)

$c = b = 55°$ (Angles in the same segment are equal.)

Challenge 38.1

Alan (A), Ben (B), Clive (C) and Dave (D) are fishing on the edge of a circular lake.

There is a small island (I) in the centre of the lake.

Alan, the island and Dave lie in a line running due south to due north.

Ben, the island and Clive lie in a straight line.

Clive is on a bearing of 038° from Alan.

a) What is the bearing of Clive from Ben?

b) Find the angles DBC and CDA.

Give a reason for each step of your work.

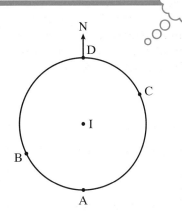

The opposite angles of a cyclic quadrilateral add up to 180°

A **cyclic quadrilateral** is one with each of its vertices on the circumference of a circle. The proof that the opposite angles of a cyclic quadrilateral add up to 180° relies, yet again, on the proof that the angle at the centre is twice the angle at the circumference.

Proof 38.5

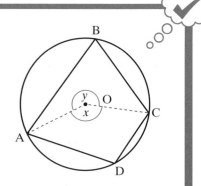

$$\hat{ABC} = \tfrac{1}{2}x$$ 　　(The angle at the centre is twice the angle at the circumference.)

$$\hat{ADC} = \tfrac{1}{2}y$$ 　　(The angle at the centre is twice the angle at the circumference.)

$$x + y = 360°$$ 　　(The angles around a point add up to 360°.)

$$\hat{ABC} + \hat{ADC} = \tfrac{1}{2}(x + y)$$

$$\hat{ABC} + \hat{ADC} = 180°$$

So opposite angles of a cyclic quadrilateral add up to 180°.

The next example shows a typical application of the fact that the opposite angles of a cyclic quadrilateral add up to 180°.

EXAMPLE 38.5

Find the sizes of angles c and d.
Give a reason for each step of your work.

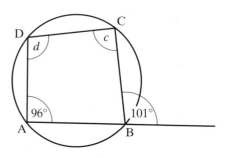

Solution

$c = 180 - 96 = 84°$ 　　(Opposite angles of a cyclic quadrilateral add up to 180°.)

$\hat{ABC} = 180 - 101 = 79°$ 　　(Angles on a straight line add up to 180°.)

$d = 180 - 79 = 101°$ 　　(Opposite angles of a cyclic quadrilateral add up to 180°.)

When you are presented with a geometry problem you have to decide which of the theorems you need to use. The next exercise gives you practice choosing and using the theorems you have met so far.

In each of the questions, find the size of the angle or length marked with a lower-case letter. Set out your answers as in the examples, giving a reason for each step of your work.

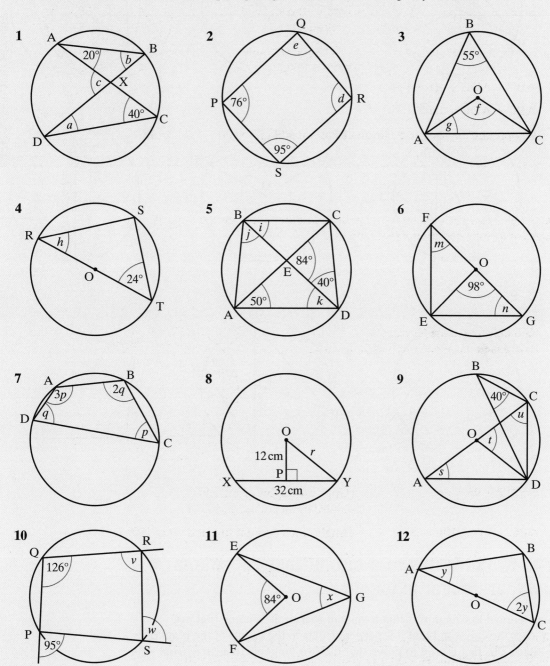

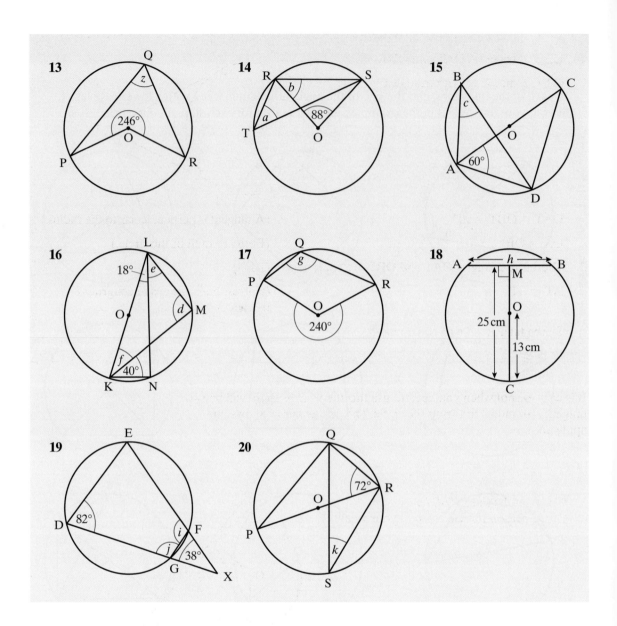

The two tangents to a circle from an external point are equal in length

You learned in Chapter 31 that a tangent to a circle is a line that just touches the circle and that it is perpendicular to the radius at the point of contact. This fact is used to prove that the two tangents to a circle from an external point are equal in length.

Proof 38.6

Look at triangles OAT and OBT.

OT is common to both triangles.

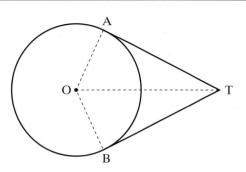

$\hat{OAT} = \hat{OBT} = 90°$ (A tangent is perpendicular to the radius.)

OA = OB (Both are radii of the circle.)

Therefore triangles OAT and OBT are congruent. (RHS)

AT = BT (Corresponding sides of congruent triangles)

So the tangents are of equal length.

It is easy to spot when you need to use this theorem as there will be two tangents (or more) drawn to the circle. The next example shows one application.

EXAMPLE 38.6

Work out the size of $\hat{OBA}$.
Give a reason for each step of your work.

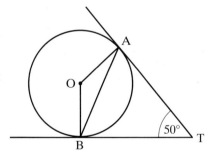

Solution

TA = TB (Equal length tangents)

$\hat{TBA} = \hat{TAB}$ (Base angles of an isosceles triangle)

$\hat{TBA} = \frac{1}{2}(180 - 50) = 65°$ (The angle sum of a triangle is 180°.)

$\hat{OBT} = 90°$ (A tangent is perpendicular to the radius.)

$\hat{OBA} = 90 - 65 = 25°$

A sphere of ice-cream is put into a hollow cone.

The diagram shows a section through the centre of the ice-cream and the cone.

The sides of the cone are tangents to the sphere.

a) Find angle x.

b) Calculate the slant height, y cm, of the cone.

Give a reason for each step of your work.

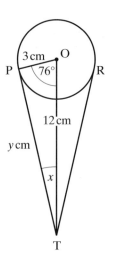

The angle between a tangent and a chord is equal to the angle subtended by the chord in the alternate segment

Look at the diagram in Proof 38.7 and identify first the angles and segments that are being referred to. The point of contact of the tangent is A. The chord is drawn from the point of contact, A, to another point on the circle, B. The angle between the tangent and the chord is labelled x. The chord divides the circle into two segments. One segment lies between the chord and the tangent, in this case the minor or smaller segment. The other segment is called the **alternate segment**.

Draw the diameter from A to meet the circle again at C.

$$A\hat{B}C = 90° \quad \text{(The angle in a semicircle is 90°.)}$$
$$y + r = 90° \quad \text{(The angle sum of a triangle is 180°.)}$$
$$y + x = 90° \quad \text{(A tangent is perpendicular to the radius.)}$$

So
$$x = r$$
$$r = p \quad \text{(Angles in the same segment)}$$

So
$$x = p$$

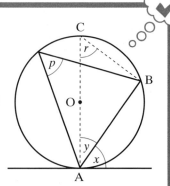

So the angle between a tangent and a chord is equal to the angle in the alternate segment.

EXAMPLE 38.7

Find angles x and y.
Give a reason for each step of your work.

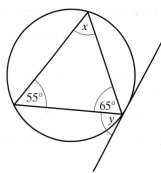

Solution

$x = 180 - (55 + 65) = 60°$ (Angles in a triangle add up to 180°.)

$y = 60°$ (Angle in the alternate segment)

EXERCISE 38.2

In each of the questions, find the size of the angles marked with a lower-case letter.
Set out your answers as in the examples, giving a reason for each step of your work.

1

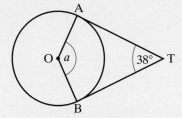

2

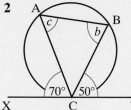

3

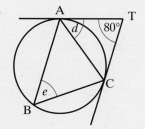

4

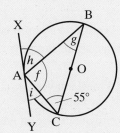

5

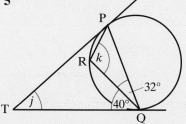

6

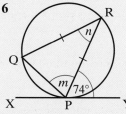

7

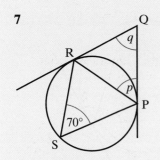

8

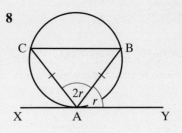

9

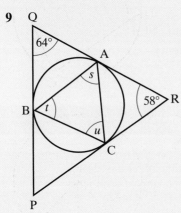

10

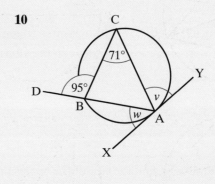

In a circle
- The perpendicular from the centre to a chord bisects the chord
- The angle subtended by an arc at the centre is twice the angle subtended at the circumference
- The angle in a semicircle is a right angle
- Angles in the same segment are equal
- The opposite angles of a cyclic quadrilateral add up to 180°
- The two tangents from a point are equal
- The angle between a tangent and a chord is equal to the angle in the alternate segment

In each of the questions, find the size of the angle or length marked with a lower-case letter.
Set out your answers as in the examples, giving a reason for each step of your work.

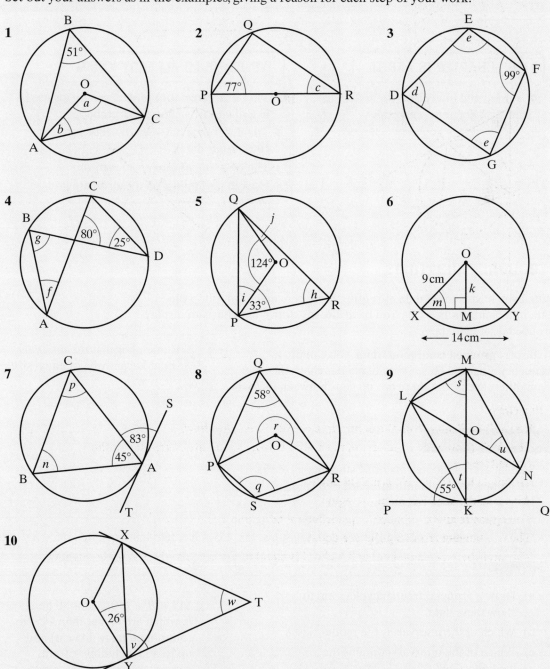

1

2

3

4

5

6

7

8

9

10

39 → STATISTICAL DIAGRAMS

THIS CHAPTER IS ABOUT

- **Drawing and interpreting a variety of statistical diagrams and tables**

YOU SHOULD ALREADY KNOW

- That a large amount of data can be displayed in a diagram, such as a bar graph
- The difference between discrete and continuous data
- The meaning of *mode* and *median*
- How to plot points on a coordinate grid

Frequency diagrams

When you have a lot of data, it is often more convenient to group the data into bands or intervals. You have already drawn bar charts to display grouped discrete data.

To display **grouped continuous data**, you can use a **frequency diagram**. This is very like a bar chart: the main difference is that there are no gaps between the bars.

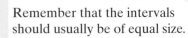

TIP Remember that the intervals should usually be of equal size.

EXAMPLE 39.1

Saul measured the heights of 34 students.
He grouped the data into intervals of 5 cm.
Here is his table of values.

Height (h cm)	$140 < h \leqslant 145$	$145 < h \leqslant 150$	$150 < h \leqslant 155$	$155 < h \leqslant 160$	$160 < h \leqslant 165$	$165 < h \leqslant 170$
Frequency	3	8	8	9	2	4

a) Draw a grouped frequency diagram to show these data.

b) Which of the intervals is the modal class?

c) Which of the intervals contains the median value?

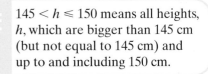

TIP $145 < h \leqslant 150$ means all heights, h, which are bigger than 145 cm (but not equal to 145 cm) and up to and including 150 cm.

Solution

a)

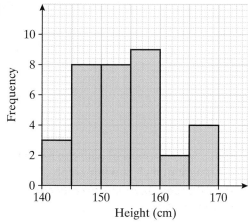

Don't forget to label the axes.

The horizontal axis shows the type of data being collected.

The vertical axis shows the **frequency**, or how many data items there are in each of the intervals.

b) $155 < h \leqslant 160$

The modal class is the one with the highest frequency. It has the highest number in the 'frequency' row of the table, and the highest bar in the grouped frequency diagram.

c) The median value is the value halfway along the ordered list.
As there are 34 values, the median will lie between the 17th and 18th values.
Add on the frequency for each interval until you find the interval containing the 17th and 18th values:

	3 is smaller than 17. The 17th and 18th values do not lie in interval $140 < h \leqslant 145$.
$3 + 8 = 11$	11 is smaller than 17. The 17th and 18th values do not lie in interval $145 < h \leqslant 150$.
$11 + 8 = 19$	19 is larger than 18. The 17th and 18th values must lie in interval $150 < h \leqslant 155$.

Interval $150 < h \leqslant 155$ contains the median value.

EXERCISE 39.1

1 The manager of a leisure centre recorded the ages of the women who used the swimming pool one morning. Here are his results.

Age (a years)	$15 \leqslant a < 20$	$20 \leqslant a < 25$	$25 \leqslant a < 30$	$30 \leqslant a < 35$	$35 \leqslant a < 40$	$40 \leqslant a < 45$	$45 \leqslant a < 50$
Frequency	4	12	17	6	8	3	12

Draw a grouped frequency diagram to show these data.

2 In a survey, the annual rainfall was measured in 100 different towns.
Here are the results of the survey.

Rainfall (r cm)	$50 \leqslant r < 70$	$70 \leqslant r < 90$	$90 \leqslant r < 110$	$110 \leqslant r < 130$	$130 \leqslant r < 150$	$150 \leqslant r < 170$
Frequency	14	33	27	8	16	2

   a) Draw a grouped frequency diagram to show these data.

   b) Which of the intervals is the modal class?

   c) Which of the intervals contains the median value?

3 As part of a fitness campaign, a business measured the weight of all of its workers.
Here are the results.

Weight (w kg)	$60 \leqslant w < 70$	$70 \leqslant w < 80$	$80 \leqslant w < 90$	$90 \leqslant w < 100$	$100 \leqslant w < 110$
Frequency	3	18	23	7	2

   a) Draw a grouped frequency diagram to show these data.

   b) Which of the intervals is the modal class?

   c) Which of the intervals contains the median value?

4 Here is a frequency diagram.

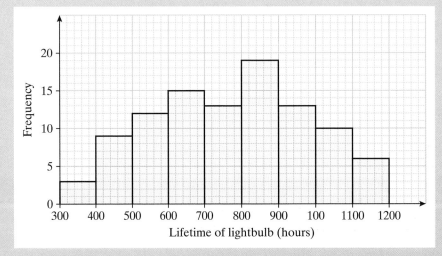

Use the grouped frequency diagram to make a grouped frequency table like those in
questions **1** to **3**.
The first interval includes 300 hours, but excludes 400 hours.

When you choose the size of the interval, make sure you don't end up with too many, or too few, groups. Between five and ten intervals is usually about right. Remember that the intervals should usually be equal.

Discovery 39.1

a) Measure the height of everyone in your class and record the data in two lists, one for boys and one for girls.

b) Choose suitable intervals for the data and organise the data in a two-way table.

c) Draw two frequency diagrams, one for the boys' data and one for the girls'. Use the same scales for both diagrams so that they can be compared easily.

d) Compare the two diagrams.
What do the shapes of the graphs tell you, in general, about the heights of the boys and girls in your class?

e) Compare your frequency diagrams with others in your class.
Have they used the same intervals for the data as you?
If they haven't, has this made a difference to their answers to part **d)**?
Which of the diagrams looks the best? Why?

Challenge 39.1

Lisa checked the price of kettles on the internet.

Here are the prices of the first 30 she saw.

£9.60	£6.54	£8.90	£12.95	£13.90	£13.95
£14.25	£16.75	£16.90	£17.75	£17.90	£19.50
£19.50	£21.75	£22.40	£23.25	£24.50	£24.95
£26.00	£26.75	£27.00	£27.50	£29.50	£29.50
£29.50	£29.50	£32.25	£34.50	£35.45	£36.95

Complete a tally chart and draw a grouped frequency diagram to show these data.

Use appropriate intervals for your groups.

Frequency polygons

A **frequency polygon** is another way of representing grouped continuous data.

A frequency polygon is formed by joining, with straight lines, the midpoints of the tops of the bars in a frequency diagram. The bars are not drawn. This means that several frequency polygons can be drawn on one grid, which makes them easier to compare.

To find the midpoint of each interval, add the bounds of each interval and divide the sum by 2.

EXAMPLE 39.2

The grouped frequency table shows the number of days that students in a tutor group were absent one term.

Days absent (d)	$0 \leqslant d < 5$	$5 \leqslant d < 10$	$10 \leqslant d < 15$	$15 \leqslant d < 20$	$20 \leqslant d < 25$
Frequency	11	8	6	0	5

Draw a frequency polygon to show these data.

Solution

First find the midpoint of each class.

$$\frac{0 + 5}{2} = 2.5 \qquad \frac{5 + 10}{2} = 7.5 \qquad \frac{10 + 15}{2} = 12.5 \qquad \frac{15 + 20}{2} = 17.5 \qquad \frac{20 + 25}{2} = 22.5$$

 Notice that the midpoints go up in fives: this is because the interval size is five.

Now you can draw your frequency polygon.

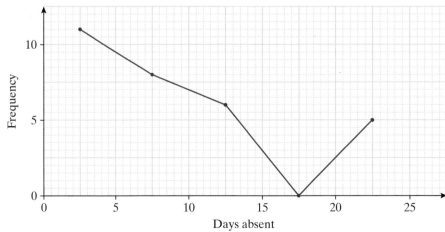

1 The table shows the weight loss of people in a slimming club over 6 months.

Weight (w kg)	$0 \leqslant w < 6$	$6 \leqslant w < 12$	$12 \leqslant w < 18$	$18 \leqslant w < 24$	$24 \leqslant w < 30$
Frequency	8	14	19	15	10

Draw a frequency polygon to show these data.

2 The table shows the length of time that cars stayed in a car park one day.

Time (t mins)	$15 \leqslant t < 30$	$30 \leqslant t < 45$	$45 \leqslant t < 60$	$60 \leqslant t < 75$	$75 \leqslant t < 90$	$90 \leqslant t < 105$
Frequency	56	63	87	123	67	2

Draw a frequency polygon to show these data.

3 The table shows the heights of 60 students.

Height (h cm)	$168 \leqslant h < 172$	$172 \leqslant h < 176$	$176 \leqslant h < 180$	$180 \leqslant h < 184$	$184 \leqslant h < 188$	$188 \leqslant h < 192$
Frequency	2	6	17	22	10	3

Draw a frequency polygon to show these data.

4 The table shows the number of words per sentence in the first 50 sentences of two books.

Number of words (w)	$0 < w \leqslant 10$	$10 < w \leqslant 20$	$20 < w \leqslant 30$	$30 < w \leqslant 40$	$40 < w \leqslant 50$	$50 < w \leqslant 60$	$60 < w \leqslant 70$
Frequency Book 1	2	9	14	7	4	8	6
Frequency Book 2	27	11	9	0	3	0	0

a) On the same grid, draw a frequency polygon for each book.

b) Use the frequency polygons to compare the number of words per sentence in the two books.

Scatter diagrams

A scatter diagram is used to find out whether there is a **correlation**, or relationship, between two sets of data.

Data are presented as pairs of values, each of which is plotted as a coordinate point on a graph.

Here are some examples of what a scatter diagram could look like and how we might interpret them.

Strong positive correlation

Here, one quantity increases as the other increases.
This is called **positive correlation**.

The trend is from bottom left to top right.

When the points are closely in line, we say that the correlation is **strong**.

Weak positive correlation

Here the points again display positive correlation.
The points are more scattered so we say that the correlation is **weak**.

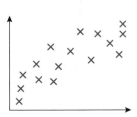

Strong negative correlation

Here, one quantity decreases as the other increases.
This is called **negative correlation**.

The trend is from top left to bottom right.

Again, the points are closely in line, so we say that the correlation is **strong**.

Weak negative correlation

Here the points again display negative correlation.
The points are more scattered so the correlation is **weak**.

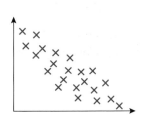

No correlation

When the points are totally scattered and there is no clear pattern we say that there is **no correlation** between the two quantities.

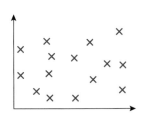

If a scatter diagram shows correlation, you can draw a **line of best fit** on it. Try putting your ruler in various positions on the scatter diagram until you have a slope which matches the general slope of the points. There should be roughly the same number of points on each side of the line.

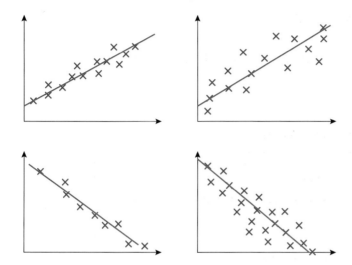

You cannot draw a line of best fit on a scatter diagram with no correlation.

You can use the line of best fit to predict a value when only one of the pair of quantities is known.

EXAMPLE 39.3

The table shows the weights and heights of 12 people.

Height (cm)	150	152	155	158	158	160	163	165	170	175	178	180
Weight (kg)	56	62	63	64	57	62	65	66	65	70	66	67

a) Draw a scatter diagram to show these data.

b) Comment on the strength and type of correlation between these heights and weights.

c) Draw a line of best fit on your scatter diagram.

d) Tom is 162 cm tall. Use your line of best fit to estimate his weight.

Solution

a), c)

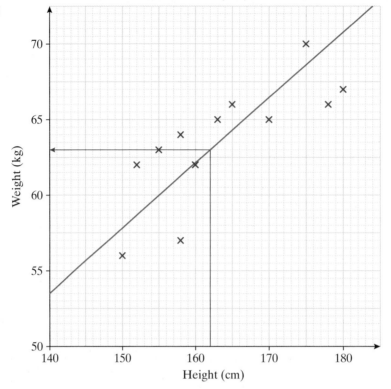

b) Weak positive correlation.

d) Draw a line from 162 cm on the Height axis, to meet your line of best fit.
Now draw a horizontal line and read off the value where it meets the
Weight axis.
Tom's probable weight is about 63 kg.

When drawing a line of best fit there is a particular point you can plot
which will help make your line a good one.

Look again at the data in Example 39.3.
The mean of the heights of the 12 people is:

$$\frac{(150 + 152 + 155 + 158 + 158 + 160 + 163 + 165 + 170 + 175 + 178\ 180)}{12}$$

$= 163.67 = 163.7$ (correct to 1 decimal place).

The mean of the weights of the people is 63.6 (correct to 1 decimal place).
Plot the point (163.7, 63.6) on the scatter diagram in Example 39.3.
You should find that it is on the line of best fit.

It can be shown that the line of best fit between two variables x and y should pass through the point $(\bar{x}, \bar{y})$, where $\bar{x}$, said as 'x bar', is the mean of the values of x and $\bar{y}$, said as 'y bar', is the mean of the values of y.

EXERCISE 39.3

1 The table shows the number of bad peaches per box after different delivery times.

Delivery time (hours)	10	4	14	18	6
Number of bad peaches	2	0	4	5	2

    a) Draw a scatter diagram to show this information.
    b) Describe the correlation shown in the scatter diagram.
    c) The mean number of bad peaches is 2.6. Calculate the mean delivery time.
    d) Plot the point that has these means as coordinates.
    e) Draw a line of best fit on your scatter diagram.
    f) Use your line of best fit to estimate the number of bad peaches expected after a 12-hour delivery time.

2 The table shows the marks of 15 students taking Paper 1 and Paper 2 of a maths exam. Both papers were marked out of 40.

Paper 1	36	34	23	24	30	40	25	35	20	15	35	34	23	35	27
Paper 2	39	36	27	20	33	35	27	32	28	20	37	35	25	33	30

    a) Draw a scatter diagram to show this information.
    b) Describe the correlation shown in the scatter diagram.
    c) Calculate the mean mark for Paper 1. The mean mark for Paper 2 is 30.5.
    d) Plot the point that has these means as coordinates.
    e) Draw a line of best fit on your scatter diagram.
    f) Joe scored 32 on Paper 1 but was absent for Paper 2.
       Use your line of best fit to estimate his score on Paper 2.

3 The table shows the engine size and petrol consumption of nine cars.

Engine size (litres)	1.9	1.1	4.0	3.2	5.0	1.4	3.9	1.1	2.4
Petrol consumption (mpg)	34	42	23	28	18	42	27	48	34

    a) Draw a scatter diagram to show this information.
    b) Describe the correlation shown in the scatter diagram.
    c) The mean of the engine sizes is 2.7. Calculate the mean petrol consumption.
    d) Plot the point that has these means as coordinates.
    e) Draw a line of best fit on your scatter diagram.
    f) Another car has an engine size of 2.8 litres.
       Use your line of best fit to estimate the petrol consumption of this car.

4 Tracy thinks that the larger your head, the cleverer you are.
The table shows the number of marks scored in a test by ten students, and the
circumference of their heads.

Circumference of head (mm)	600	500	480	570	450	550	600	460	540	430
Mark	43	33	45	31	25	42	23	36	24	39

   a) Draw a scatter diagram to show this information.

   b) Describe the correlation shown in the scatter diagram.

   c) Is Tracy correct?

   d) Can you think of any reasons why the comparison may not be valid?

Challenge 39.2

   a) What type of correlation, if any, would you expect to find if you drew a scatter
diagram for each of the sets of following data?
     (i) Sales of CDs and amount spent on advertising
     (ii) Number of accidents and number of speed cameras
     (iii) Height of adults and age of adults

   b) Write down your own examples of data which would show
     (i) positive correlation.
     (ii) negative correlation.
     (iii) no correlation.

WHAT YOU HAVE LEARNED

- **How to construct and interpret frequency diagrams, frequency polygons and scatter diagrams**
- **About the different types of correlation**
- **How to draw and use lines of best fit**

1   Emma kept a record of the time, in minutes, that she had to wait for the school bus each
     morning for 4 weeks.

         11   5   7   4   2   18   3   10   8   1
         13   4   9   10   14   4   5   17   6   7

     a) Make a frequency table for these values using the groups
          $0 \leqslant t < 5$, $5 \leqslant t < 10$, $10 \leqslant t < 15$ and $15 \leqslant t < 20$.

     b) Draw a grouped frequency diagram for these data.

     c) Which of the intervals is the modal class?

     d) Which of the intervals contains the median value?

2   The table shows the marks gained by students in an examination.

Mark	$30 \leqslant m < 40$	$40 \leqslant m < 50$	$50 \leqslant m < 60$	$60 \leqslant m < 70$	$70 \leqslant m < 80$	$80 \leqslant m < 90$
Frequency	8	11	18	13	8	12

     a) Draw a grouped frequency polygon to show these data.

     b) Describe the distribution of the marks.

     c) Which is the modal class?

     d) How many students took the examination?

     e) What fraction of students scored 70 or more in the examination?
          Give your answer in its simplest form.

3   A pet shop owner carried out a survey to investigate the average weight of a breed of
     rabbit at various ages. The table shows his results.

Age of rabbit (months)	1	2	3	4	5	6	7	8
Average weight (g)	90	230	490	610	1050	1090	1280	1560

     a) Draw a scatter diagram to show this information.

     b) Describe the correlation shown in the scatter diagram.

     c) Draw a line of best fit on your scatter diagram.

     d) Use your line of best fit to estimate
          (i)   the weight of a rabbit of this breed which is $4\frac{1}{2}$ months old.
          (ii) the weight of a rabbit of this breed which is 9 months old.

     e) If the line of best fit was extended you could estimate the weight of a rabbit of this
          breed which is 20 months old.
          Would this be sensible? Give a reason for your answer.

The mean from a frequency table

The **mean** of a set of data is found by adding the values together and dividing the total by the number of values used.

For example, the following set of data shows the number of pets owned by nine Year 10 students.

	8	4	4	6	3	7	3	2	8

The mean is $45 \div 9 = 5$.

What you are working out is (the total number of pets) $\div$ (the total number of students surveyed).

If you surveyed 150 people you would have a list of 150 numbers. You could find the mean by adding them all up and dividing by 150, but this would take a long time.

Instead, you can put the data in a frequency table and work out the mean using a different method.

EXAMPLE 40.1

Skye asked all the students in Year 10 at her local girls' school how many brothers they had. The table shows her results.

Work out the mean number of brothers for these students.

Number of brothers	Frequency (number of girls)
0	24
1	60
2	47
3	11
4	5
5	2
6	0
7	0
8	1
Total	150

Solution

The mean of this data is (the total number of brothers) ÷ (the total number of girls surveyed).

First you need to work out the total number of brothers.

You can see from the table that

- 24 girls do not have any brothers. They have 24 × 0 = 0 brothers between them.
- 60 girls have one brother each. They have 60 × 1 = 60 brothers between them.
- 47 girls have two brothers each. They have 47 × 2 = 94 brothers between them.

and so on.

If you add the results for each row of the table together, you will get the total number of brothers.

You can add some more columns to the table to show this.

Number of brothers (x)	Number of girls (f)	Number of brothers × frequency	Total number of brothers (fx)
0	24	0 × 24	0
1	60	1 × 60	60
2	47	2 × 47	94
3	11	3 × 11	33
4	5	4 × 5	20
5	2	5 × 2	10
6	0	6 × 0	0
7	0	7 × 0	0
8	1	8 × 1	8
Total	150		225

The 'Number of brothers' column is the variable and is usually labelled x.
The 'Number of girls' column is the frequency and is usually labelled f.
The 'Total number of brothers' column is usually labelled fx because it represents (Number of brothers) $\times$ (Number or girls) $= x \times f$.

The total number of brothers = 225
The total number of girls surveyed = 150
So the mean = 225 $\div$ 150 = 1.5 brothers.

You can enter the calculations into your calculator as a chain of numbers and then press the $=$ key to find the total before dividing by 150.

Input ⓪ ✕ ② ④ ＋ ① ✕ ⑥ ⓪ ＋ ② ✕ ④ ⑦ ＋ ③ ✕ ① ① ＋ ④ ✕ ⑤
＋ ⑤ ✕ ② ＋ ⑥ ✕ ⓪ ＋ ⑦ ✕ ⓪ ＋ ⑧ ✕ ① ＝ ＋ ① ⑤ ⓪ ＝

You can also work out the **mode**, **median** and **range** from the table.

The mode of the number of brothers is 1.
This is the number of brothers with the highest frequency (60).

The median number of brothers is 1.
As there are 150 values, the median will lie between the 75th and 76th values.
Add on the frequency for each number of brothers (row) until you find the interval containing the 75th and 76th values:

24 is smaller than 75. The 75th and 76th values do not lie in row 0.
24 + 60 = 84   84 is larger than 76. The 75th and 76th values must lie in row 1.

The range of the number of brothers is 8.
This is (the largest number of brothers) $-$ (the smallest number of brothers) $= 8 - 0 = 8$.

Using a spreadsheet to find the mean

You can also calculate the mean using a computer spreadsheet. Follow these steps to work out the mean for the data in Example 40.1.

Type the bold text carefully: do not put in any spaces.

1 Open a new spreadsheet.

2 In cell A1 type the title 'Number of brothers (x)'.
   In cell B1 type the title 'Number of girls (f)'.
   In cell C1 type the title 'Total number of brothers (fx)'.

3 In cell A2 type the number 0. Then type the numbers 1 to 8 in cells A3 to A10.

4 In cell B2 type the number 24. Then type the other frequencies in cells B3 to B10.

5 In cell C2 type **=A2*B2** and press the enter key.
Click on cell C2, click on Edit in the toolbar and select Copy.
Click on cell C3, and hold down the mouse key and drag down to cell C10. Then click on Edit in the toolbar and select Paste.

6 In cell A11 type the word 'Total'.

7 In cell B11 type **=SUM(B2:B10)** and press the enter key.
In cell C11 type **=SUM(C2:C10)** and press the enter key.

8 In cell A12 type the word 'Mean'.

9 In cell B12 type **=C11/B11** and press the enter key.

Your spreadsheet should look like this.

	A	B	C
1	Number of brothers (x)	Number of girls (f)	Total number of brothers (fx)
2	0	24	0
3	1	60	60
4	2	47	94
5	3	11	33
6	4	5	20
7	5	2	10
8	6	0	0
9	7	0	0
10	8	1	8
11	Total	150	225
12	Mean	1.5	

Answer one of the questions in the next exercise using a computer spreadsheet.

1 For each of these sets of data
   (i) find the mode.   **(ii)** find the median.
   (iii) find the range.   **(iv)** calculate the mean.

a)

Score on dice	Number of times thrown
1	89
2	77
3	91
4	85
5	76
6	82
Total	500

b)

Number of matches	Number of boxes
47	78
48	82
49	62
50	97
51	86
52	95
Total	500

c)

Number of accidents	Number of drivers
0	65
1	103
2	86
3	29
4	14
5	3
Total	300

d)

Number of cars per house	Number of students
0	15
1	87
2	105
3	37
4	6
Total	250

2 Calculate the mean for each of these sets of data.

a)

Number of passengers in taxi	Frequency
1	84
2	63
3	34
4	15
5	4
Total	200

b)

Number of pets owned	Frequency
0	53
1	83
2	23
3	11
4	5
Total	175

c)

Number of books read in a month	Frequency
0	4
1	19
2	33
3	42
4	29
5	17
6	6
Total	150

d)

Number of drinks in a day	Frequency
3	81
4	66
5	47
6	29
7	18
8	9
Total	250

3 Calculate the mean for each of these sets of data.

a)

x	Frequency
1	47
2	36
3	28
4	57
5	64
6	37
7	43
8	38

b)

x	Frequency
23	5
24	9
25	12
26	15
27	13
28	17
29	14
30	15

c)

x	Frequency
10	5
11	8
12	6
13	7
14	3
15	9
16	2

d)

x	Frequency
0	12
1	59
2	93
3	81
4	43
5	67
6	45

4 In Barnsfield, bus tickets cost 50p, £1.00, £1.50 or £2.00 depending on the length of the journey. The frequency table shows the numbers of tickets sold on one Friday. Calculate the mean fare paid on that Friday.

Price of ticket (£)	0.50	1.00	1.50	2.00
Number of tickets	140	207	96	57

5 800 people were asked how many newspapers they had bought one week. The table shows the data. Calculate the mean number of newspapers bought.

Number of newspapers	Frequency
0	20
1	24
2	35
3	26
4	28
5	49
6	97
7	126
8	106
9	54
10	83
11	38
12	67
13	21
14	26

Challenge 40.1

a) Design a data collection sheet for the number of pairs of trainers owned by each of the students in your class.

b) Collect the data for your class.

c) (i) Find the mode of your data.
(ii) Find the range of your data.
(iii) Calculate the mean number of pairs of trainers owned by the students in your class.

Grouped data

The table shows the number of CDs bought in January by a group of 75 people.

Number of CDs purchased	Number of people
0–4	35
5–9	21
10–14	12
15–19	5
20–24	2

Grouping data makes it easier to work with, but it also causes some problems when calculating the mode, median, mean or range.

For example, the modal class of these data is 0–4, because that is the class with the highest frequency.

However, it is impossible to say which number of CDs was the mode because we do not know exactly how many people in this class bought what number of CDs.

It is possible (though not very likely) that seven people bought no CDs, seven people bought one CD, seven people bought two CDs, seven people bought three CDs and seven people bought four CDs. If eight or more people bought nine CDs, then the mode would actually be 9, even though the modal class is 0–4!

The median presents a similar problem: you can see which class contains the median value, but you cannot work out what the median value actually is.

It is also impossible to work out the exact mean from a grouped frequency table. You can, however, calculate an estimate using a single value to represent each class: it is usual to use the middle value.

These middle values can also be used to calculate an estimate for the range. You cannot find the exact range because it is impossible to say for certain what the highest and lowest numbers of CDs purchased are. The maximum possible purchase is 24 but you cannot tell whether anyone did actually buy 24. The minimum possible purchase is 0 but, again, you cannot tell whether anyone did actually buy no CDs.

EXAMPLE 40.2

Use the data in the table above to calculate
a) an estimate of the mean number of CDs purchased.
b) an estimate of the range of the number of CDs purchased.
c) which of the classes contains the median value.

Solution

a)

Number of CDs purchased (x)	Number of people (f)	Middle value (x)	f × x	fx
0–4	35	2	35 × 2	70
5–9	21	7	21 × 7	147
10–14	12	12	12 × 12	144
15–19	5	17	5 × 17	85
20–24	2	22	2 × 22	44
Total	75			490

The estimate of the mean number of CDs purchased is
$490 \div 75 = 6.5$ (to 1 d.p.).

> **TIP**
>
> There are five groups in the table but the total number of people is 75.
>
> Do not be tempted to divide by 5!

b) The estimate of the range of the number of CDs purchased is
$22 - 2 = 20$ but it could be as high as 24 or as low as 16.

c) As there are 75 values, the median will be the 38th value.
Add on the frequency for each class until you find the class containing the 38th value.
   35 is smaller than 38. The 38th value does not lie in class 0–4.
$35 + 21 = 56$ 56 is larger than 38. The 38th value must lie in class 5–9.
So the 5–9 class contains the median value.

Using a spreadsheet to find the mean of grouped data

An estimate of the mean of grouped data can also be calculated using a computer spreadsheet. The method is the same as before, except for the addition of a 'Middle value (x)' column. This column can then also be used to calculate an estimate of the range.

Follow these steps to calculate estimates of the mean and range of the data in Example 40.2.

> **TIP**
>
> Type the bold text carefully: do not put in any spaces.

1. Open a new spreadsheet.

2. In cell A1 type the title 'Number of CDs purchased (x)'.
   In cell B1 type the title 'Number of people (f)'.
   In cell C1 type the title 'Middle value (x)'.
   In cell D1 type the title 'Total number of CDs purchased (fx)'.

3. In cell A2 type 0–4. Then type the other classes in cells A3 to A6.

4. In cell B2 type the number 35. Then type the other frequencies in cells B3 to B6.

5. In cell C2 type =(0+4)/2 and press the enter key.
   In cell C3 type =(5+9)/2 and press the enter key.
   In cell C4 type =(10+14)/2 and press the enter key.
   In cell C5 type =(15+19)/2 and press the enter key.
   In cell C6 type =(20+24)/2 and press the enter key.

6. In cell D2 type =B2*C2 and press the enter key.
   Click on cell D2, click on Edit in the toolbar and select Copy.
   Click on cell D3, and hold down the mouse key and drag down to cell D6. Then click on Edit in the toolbar and select Paste.

7. In cell A7 type the word 'Total'.

8. In cell B7 type =SUM(B2:B6) and press the enter key.
   In cell D7 type =SUM(D2:D6) and press the enter key.

9. In cell A8 type the word 'Mean'.

10. In cell B8 type =D7/B7 and press the enter key.

11. In cell A9 type the word 'Range'.

12. In cell B9 type =C6−C2 and press the enter key.

Your spreadsheet should look like this.

	A	B	C	D
1	Number of CDs purchased (x)	Number of people (f)	Middle value (x)	Total number of CDs purchased (fx)
2	0-4	35	2	70
3	5-9	21	7	147
4	10-14	12	12	144
5	15-19	5	17	85
6	20-24	2	22	44
7	Total	75		490
8	Mean	6.533333333		
9	Range	20		

Answer one of the questions in the next exercise using a computer spreadsheet.

1 For each of these sets of data calculate an estimate of
(i) the range. (ii) the mean.

a)

Number of texts received	Number of people	Middle value
0–9	99	4.5
10–19	51	14.5
20–29	28	24.5
30–39	14	34.5
40–49	7	44.5
50–59	1	54.5
Total	200	

b)

Number of telephone calls made	Number of people	Middle value
0–4	118	2
5–9	54	7
10–14	39	12
15–19	27	17
20–24	12	22
Total	250	

c)

Number of texts sent	Number of people	Middle value
0–9	79	4.5
10–19	52	14.5
20–29	31	24.5
30–39	13	34.5
40–49	5	44.5
Total	180	

d)

Number of calls received	Frequency	Middle value
0–4	45	2
5–9	29	7
10–14	17	12
15–19	8	17
20–24	1	22
Total	100	

2 For each of these sets of data
(i) find the modal class.
(ii) calculate an estimate of the range.
(iii) calculate an estimate of the mean.

a)

Number of DVDs owned	Number of people
0–4	143
5–9	95
10–14	54
15–19	26
20–24	12
Total	330

b)

Number of books owned	Number of people
0–9	54
10–19	27
20–29	19
30–39	13
40–49	7
Total	120

c)

Number of train journeys in a year	Number of people
0–49	118
50–99	27
100–149	53
150–199	75
200–249	91
250–299	136

d)

Number of flowers on a plant	Frequency
0–14	25
15–29	52
30–44	67
45–59	36

3 For each of these sets of data
  (i) find the modal class.
  (ii) calculate an estimate of the mean.

a)

Number of eggs in a nest	Frequency
0–2	97
3–5	121
6–8	43
9–11	7
12–14	2

b)

Number of peas in a pod	Frequency
0–3	15
4–7	71
8–11	63
12–15	9
16–19	2

c)

Number of leaves on a branch	Frequency
0–9	6
10–19	17
20–29	27
30–39	34
40–49	23
50–59	10
60–69	3

d)

Number of bananas in a bunch	Frequency
0–24	1
25–49	29
50–74	41
75–99	52
100–124	24
125–149	3

4 A company records the number of complaints they receive each week about their products.
The table shows the data for one year.

Calculate an estimate of the mean number of complaints each week.

Number of complaints	Frequency
1–10	12
11–20	5
21–30	10
31–40	8
41–50	9
51–60	5
61–70	2
71–80	1

5 An office manager records the number of photocopies made by his staff each day in September.
The data is shown in the table.

Calculate an estimate of the mean number of copies each day.

Number of photocopies	Frequency
0–99	13
100–199	8
200–299	3
300–399	0
400–499	5
500–599	1

Challenge 40.2

a) Design a data collection sheet, using appropriate groupings, for the number of CDs owned by the students in your class.

b) Collect the data for your class.

c) (i) Estimate the range of your data.
(ii) Calculate an estimate of the mean number of CDs owned by the students in your class.

Continuous data

So far all the data in this chapter has been **discrete data** (the result of objects beings counted).

When dealing with **continuous data** (the result of measurement) you estimate the mean in the same way as for grouped discrete data.

EXAMPLE 40.3

A manager records the lengths of telephone calls made by her employees.
The table shows the results for one week.

Duration of telephone call in minutes (x)	Frequency (f)
$0 \leqslant x < 5$	86
$5 \leqslant x < 10$	109
$10 \leqslant x < 15$	54
$15 \leqslant x < 20$	27
$20 \leqslant x < 25$	16
$25 \leqslant x < 30$	8
Total	300

Remember that $15 \leqslant x < 20$ means all times, x, which are greater than or equal to 15 minutes but less than 20 minutes.

Solution

Duration of telephone call in minutes (x)	Frequency (f)	Middle value (x)	$f \times x$
$0 \leqslant x < 5$	86	2.5	215
$5 \leqslant x < 10$	109	7.5	817.5
$10 \leqslant x < 15$	54	12.5	675
$15 \leqslant x < 20$	27	17.5	472.5
$20 \leqslant x < 25$	16	22.5	360
$25 \leqslant x < 30$	8	27.5	220
Total	300		2760

The estimate of the mean is $2760 \div 300 = 9.2$ minutes or 9 minutes and 12 seconds.

Remember that there are 60 seconds in 1 minute.
$60 \times 0.2 = 12$ seconds.

Use a spreadsheet to answer one of the questions in this exercise.

1 For each of these sets of data calculate an estimate of
   (i) the range.   **(ii)** the mean.

a)

Height of plant in centimetres (x)	Number of plants (f)
$0 \leqslant x < 10$	5
$10 \leqslant x < 20$	11
$20 \leqslant x < 30$	29
$30 \leqslant x < 40$	26
$40 \leqslant x < 50$	18
$50 \leqslant x < 60$	7
Total	96

b)

Weight of egg in grams (x)	Number of eggs (f)
$0 \leqslant x < 8$	3
$8 \leqslant x < 16$	18
$16 \leqslant x < 24$	43
$24 \leqslant x < 32$	49
$32 \leqslant x < 40$	26
$40 \leqslant x < 48$	5
Total	144

c)

Length of string in centimetres (x)	Frequency (f)
$60 \leqslant x < 64$	16
$64 \leqslant x < 68$	28
$68 \leqslant x < 72$	37
$72 \leqslant x < 76$	14
$76 \leqslant x < 80$	5
Total	100

d)

Height of plant in centimetres (x)	Number of plants (f)
$0 \leqslant x < 10$	151
$10 \leqslant x < 20$	114
$20 \leqslant x < 30$	46
$30 \leqslant x < 40$	28
$40 \leqslant x < 50$	17
$50 \leqslant x < 60$	9
Total	365

2 For each of these sets of data
   (i) write down the modal class.
   (ii) calculate an estimate of the mean.

a)

Age of chick in days (x)	Number of chicks (f)
$0 \leqslant x < 3$	61
$3 \leqslant x < 6$	57
$6 \leqslant x < 9$	51
$9 \leqslant x < 12$	46
$12 \leqslant x < 15$	44
$15 \leqslant x < 18$	45
$18 \leqslant x < 21$	46

b)

Weight of apple in grams (x)	Number of apples (f)
$90 \leqslant x < 100$	5
$100 \leqslant x < 110$	24
$110 \leqslant x < 120$	72
$120 \leqslant x < 130$	81
$130 \leqslant x < 140$	33
$140 \leqslant x < 150$	10

c)

Length of runner bean in centimetres (x)	Frequency (f)
$10 \leqslant x < 14$	16
$14 \leqslant x < 18$	24
$18 \leqslant x < 22$	25
$22 \leqslant x < 26$	28
$26 \leqslant x < 30$	17
$30 \leqslant x < 34$	10

d)

Time to complete race in minutes (x)	Frequency (f)
$40 \leqslant x < 45$	1
$45 \leqslant x < 50$	8
$50 \leqslant x < 55$	32
$55 \leqslant x < 60$	26
$60 \leqslant x < 65$	5
$65 \leqslant x < 70$	3

3 The table shows the weekly wages of the manual workers in a factory.

Wage in £ (x)	$150 \leqslant x < 200$	$200 \leqslant x < 250$	$250 \leqslant x < 300$	$300 \leqslant x < 350$
Frequency (f)	4	14	37	15

a) What is the modal class?   **b)** In which class is the median wage?
c) Calculate an estimate of the mean wage.

4 The table shows the masses, in grams, of the first 100 letters posted one day.

Mass in grams (x)	$0 \leqslant x < 15$	$15 \leqslant x < 30$	$30 \leqslant x < 45$	$45 \leqslant x < 60$
Frequency (f)	48	36	12	4

Calculate an estimate of the mean mass of a letter.

5 The table shows the prices paid for birthday cards sold in one day by a greetings card shop.

Calculate an estimate of the mean price paid for a birthday card that day.

Price of birthday card in pence (x)	Frequency (f)
$100 \leqslant x < 125$	18
$125 \leqslant x < 150$	36
$150 \leqslant x < 175$	45
$175 \leqslant x < 200$	31
$200 \leqslant x < 225$	17
$225 \leqslant x < 250$	9

Challenge 40.3

a) Design a data collection sheet, using appropriate groupings, and carry out one of the following tasks. Use the students in your class as the source of your data.
- Ask each person the amount of money they spent on lunch on a particular day.
- Obtain a piece of string, arrange it in a non-straight line and ask each person to estimate its length.

b) Calculate an estimate of   **(i)** the range of your data   **(ii)** the mean of your data.

- How to find the mean from a frequency table
- How to estimate the mean for grouped data
- How to estimate the range for grouped data
- How to find the class in which the median value lies in large data sets
- How to use a spreadsheet to calculate an estimate of the mean and range of grouped data

MIXED EXERCISE 40

1 For each of these sets of data
   (i) find the mode.   (ii) find the range.   (iii) calculate the mean.

a)

Score on octagonal dice	Number of times thrown
1	120
2	119
3	132
4	126
5	129
6	142
7	123
8	109
Total	1000

b)

Number of marbles in a bag	Number of bags
47	11
48	25
49	47
50	63
51	54
52	38
53	17
54	5
Total	260

c)

Number of pets per house	Frequency
0	64
1	87
2	41
3	26
4	17
5	4
6	1

d)

Number of broad beans in a pod	Frequency
4	17
5	36
6	58
7	49
8	27
9	13

e)

x	f
1	242
2	266
3	251
4	252
5	259
6	230

f)

x	f
15	9
16	13
17	18
18	27
19	16
20	7

2 Crazyphone top-ups cost £5, £10, £20 or £50 depending upon the amount of credit bought. The frequency table shows the number of each value of top-up sold in one shop on a Saturday.

Price of top-up (£)	5	10	20	50
Number of top-ups	34	63	26	2

Calculate the mean value of top-up bought in the shop that Saturday.

3 A sample of 350 people were asked how many magazines they had bought in September. The table below shows the data.

Number of magazines	0	1	2	3	4	5	6	7	8	9	10
Frequency	16	68	94	77	49	27	11	5	1	0	2

Calculate the mean number of magazines bought in September.

4 For each of these sets of data, calculate an estimate of
   (i) the range.
   (ii) the mean.

a)

Height of cactus in centimetres (x)	Number of plants (f)
$10 \leqslant x < 15$	17
$15 \leqslant x < 20$	49
$20 \leqslant x < 25$	66
$25 \leqslant x < 30$	38
$30 \leqslant x < 35$	15
Total	185

b)

Wind speed at noon in km/h (x)	Number of days (f)
$0 \leqslant x < 20$	164
$20 \leqslant x < 40$	98
$40 \leqslant x < 60$	57
$60 \leqslant x < 80$	32
$80 \leqslant x < 100$	11
$100 \leqslant x < 120$	3
Total	365

c)

Time holding breath in seconds (x)	Frequency (f)
$30 \leqslant x < 40$	6
$40 \leqslant x < 50$	29
$50 \leqslant x < 60$	48
$60 \leqslant x < 70$	36
$70 \leqslant x < 80$	23
$80 \leqslant x < 90$	8

d)

Mass of student in kilograms (x)	Frequency (f)
$40 \leqslant x < 45$	5
$45 \leqslant x < 50$	13
$50 \leqslant x < 55$	26
$55 \leqslant x < 60$	31
$60 \leqslant x < 65$	17
$65 \leqslant x < 70$	8

5 The table below shows the lengths, to the nearest minute, of 304 telephone calls.

Length in minutes (x)	$0 \leqslant x < 10$	$10 \leqslant x < 20$	$20 \leqslant x < 30$	$30 \leqslant x < 40$	$40 \leqslant x < 50$
Frequency (f)	53	124	81	35	11

a) What is the modal class?
b) In which class is the median length of call?
c) Calculate an estimate of the mean length of call.

6 The table shows the annual wages of the workers in a company.

Annual wage in £000	Frequency (f)
$10 \leqslant x < 15$	7
$15 \leqslant x < 20$	18
$20 \leqslant x < 25$	34
$25 \leqslant x < 30$	12
$30 \leqslant x < 35$	9
$35 \leqslant x < 40$	4
$40 \leqslant x < 45$	2
$45 \leqslant x < 50$	1
$50 \leqslant x < 55$	2
$55 \leqslant x < 60$	0
$60 \leqslant x < 65$	1

Calculate an estimate of the mean annual wage of these workers.

41 → PROBABILITY 1

THIS CHAPTER IS ABOUT

- Using the fact that the probability of an event not happening and the probability of the event happening add up to 1
- Calculating expected frequency
- Calculating relative frequency

YOU SHOULD ALREADY KNOW

- That probabilities are expressed as fractions or decimals
- All probabilities lie on a scale between 0 and 1
- How to find a probability from a set of equally likely outcomes
- How to add decimals and fractions
- How to subtract decimals and fractions from 1
- How to multiply decimals

The probability of an event not happening

Discovery 41.1

There are three pens and five pencils in a box.
One of these is picked at random.

a) What is the probability that it is a pen, P(pen)?
b) What is the probability that it is a pencil, P(pencil)?
c) What is the probability that it is not a pen, P(not a pen)?
d) What can you say about your answers to parts **b)** and **c)**?
e) What is P(pen) + P(not a pen)?

> **TIP**
> P( ) is often used when writing probabilities because it saves time and space.

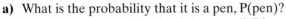

The probability of an event not happening = 1 − the probability of the event happening

If *p* is the probability of an event happening, then this can be written as

$$P(\text{not happening}) = 1 - p$$

EXAMPLE 41.1

a) The probability that it will rain tomorrow is $\frac{1}{5}$.
What is the probability that it will not rain tomorrow?

b) The probability that Phil scores a goal in the next match is 0.6.
What is the probability that Phil does not score a goal?

Solution

a) P(not rain) = 1 − P(rain)

$$= 1 - \frac{1}{5}$$

$$= \frac{4}{5}$$

b) P(not score) = 1 − P(score)

$$= 1 - 0.6$$

$$= 0.4$$

EXERCISE 41.1

1 The probability that Max will get to school late tomorrow is 0.1.
What is the probability that Max will not be late for school tomorrow?

2 The probability that Charlie has cheese sandwiches for his lunch is $\frac{1}{6}$.
What is the probability that Charlie does not have cheese sandwiches for lunch?

3 The probability that Ashley will pass her driving test is 0.85.
What is the probability that Ashley will fail her driving test?

4 The probability that Adam's mum will cook dinner tonight is $\frac{7}{10}$.
What is the probability that she will not cook dinner?

5 The probability that City will win their next game is 0.43.
What is the probability that City will not win their next game?

6 The probability that Alec will watch TV one night is $\frac{32}{49}$.
What is the probability that he will not watch TV that night?

Probability involving a given number of different outcomes

Often there are more than two possible outcomes.
If you know the probability of all but one of the outcomes, you can work
out the probability of the remaining outcome.

EXAMPLE 41.2

A bag contains only red, white and blue counters.

The probability of picking a red counter is $\frac{1}{12}$.

The probability of picking a white counter is $\frac{7}{12}$.

What is the probability of picking a blue counter?

Solution

You know that P(not happening) = 1 − P(happening).

So P(not happening) + P (happening) = 1

$$P(\text{not blue}) + P(\text{blue}) = 1$$
$$P(\text{red}) + P(\text{white}) + P(\text{blue}) = 1$$
$$P(\text{blue}) = 1 - [P(\text{red}) + P(\text{white})]$$
$$= 1 - (\tfrac{1}{12} + \tfrac{7}{12})$$
$$= 1 - \tfrac{8}{12}$$
$$= \tfrac{4}{12}$$
$$= \tfrac{1}{3}$$

There are only red, white and blue counters in the bag, so if a counter is not blue it must be red or white.

When there are a given number of possible outcomes, the sum of the probabilities is equal to 1.

For example, if there are four possible outcomes, A, B, C and D, then

$$P(A) + P(B) + P(C) + P(D) = 1$$

So, for example

$$P(B) = 1 - [P(A) + P(C) + P(D)]$$
$$\text{or}$$
$$P(B) = 1 - P(A) - P(C) - P(D)$$

EXERCISE 41.2

1 A shop has black, grey and blue dresses on a rail. Jen picks one at random.
The probability of picking a grey dress is 0.2 and the probability of picking a black dress is 0.1. What is the probability of picking a blue dress?

2 Heather always comes to school by car, bus or bike.
On any day, the probability that Heather will come by car is $\frac{3}{20}$ and the probability that she will come by bus is $\frac{11}{20}$.
What is the probability that Heather will come to school by bike?

3 The probability that the school hockey team will win their next match is 0.4.
The probability that they will lose is 0.25.
What is the probability that they will draw the match?

4 Pat has either boiled eggs, cereal or toast for breakfast.
The probability that she will have toast is $\frac{2}{11}$ and the probability that she will have cereal is $\frac{5}{11}$.
What is the probability that she will have boiled eggs?

5 The table shows the probability of getting some of the scores when a biased six-sided dice is thrown.

Score	1	2	3	4	5	6
Probability	0.27	0.16	0.14		0.22	0.1

What is the probability of getting 4?

6 When it is Jack's birthday, Aunty Chris gives him money or a voucher, or forgets altogether.
The probability that Aunty Chris will give Jack money for his birthday is $\frac{3}{4}$ and the probability that she will give him a voucher is $\frac{1}{5}$.
What is the probability that she forgets?

Challenge 41.1

The weather forecast says the probability that it will be sunny tomorrow is 0.4.
Terry says this means that the probability that it will rain is 0.6.
Is Terry correct? Why?

Challenge 41.2

A cash bag contains only 5p, 10p and 50p coins.
The total amount of money in the bag is £5.

A coin is chosen from the bag at random.

$P(5p) = \frac{1}{2}$
$P(10p) = \frac{3}{8}$

a) Work out $P(50p)$.
b) How many of each kind of coin is there in the bag?

Expected frequency

You can also use probability to predict how often an outcome will occur, or the **expected frequency** of the outcome.

EXAMPLE 41.3

Each time Ronnie plays a game of snooker, the probability that he will win is $\frac{7}{10}$.

In a season, Ronnie plays 30 games. How many of the games can he be expected to win?

Solution

The probability $P(\text{win}) = \frac{7}{10}$ tells us that Ronnie will win, on average, seven times in every ten games he plays. That is, he will win $\frac{7}{10}$ of the time.

In a season, he can be expected to win $\frac{7}{10}$ of 30 games.

$$\frac{7}{10} \times 30 = \frac{210}{10}$$
$$= 21$$

This is an example of an important result.

> Expected frequency = probability × number of trials

EXAMPLE 41.4

The probability of a child catching measles is 0.2. Out of the 400 children in a primary school, how many of them might you expect to catch measles?

Solution

Expected frequency = Probability × Number of trials
$$= 0.2 \times 400$$
$$= 80 \text{ children}$$

The number of 'trials' means the number of times the probability is tested. Here each of the 400 children has a 0.2 chance of catching measles. The number of trials is the same as the number of children: 400.

1. The probability that Beverley is late to work is 0.1.
   How many times would you expect her to be late in 40 working days?

2. The probability that it will be sunny on any day in April is $\frac{2}{5}$.
   On how many of April's 30 days would you expect it to be sunny?

3. The probability that United will win their next match is 0.85.
   How many of their next 20 games would you expect them to win?

4. When John is playing darts, the probability that he will hit the bull's eye is $\frac{3}{20}$.
   John takes part in a sponsored event and throws 400 darts.
   Each dart hitting the bull's eye earns £5 for charity.
   How much might he expect to earn for charity?

5. An ordinary six-sided dice is thrown 300 times.
   How many times might you expect to score

   a) 5?   b) an even number?

6. A box contains two yellow balls, three blue balls and five green balls.
   A ball is picked at random and its colour noted.
   The ball is then replaced. This is done 250 times.
   How many of each colour might you expect to get?

Challenge 41.3

a) If you roll a dice 30 times, how many times would you expect to get a 6?

b) Now roll a dice 30 times and see how many times you get a 6.
   Did a 6 come up as many times as you predicted?
   Do the same for another 30 throws of the dice.
   Compare and discuss your results with your classmates.

Relative frequency

You already know from Key Stage 3 how to **estimate** probabilities using **experimental evidence**.

$$\text{The experimental probability of an event} = \frac{\text{the number of times an event happens}}{\text{the total number of trials}}$$

The probability you estimate is known as **relative frequency**.

Discovery 41.2

Copy this table and complete it by following the instructions below.

Number of trials	20	40	60	80	100
Number of heads					
Relative frequency $= \dfrac{\text{Number of heads}}{\text{Number of trials}}$					

- Toss a coin 20 times and record, using tally marks, the number of times it lands on heads.
- Now toss the coin another 20 times and enter the number of heads for all 40 tosses.
- Continue in groups of 20 and record the number of heads for 60, 80 and 100 tosses.
- Calculate the relative frequency of heads for 20, 40, 60, 80 and 100 tosses.
  Give your answers to 2 decimal places.

a) What do you notice about the values of the relative frequencies?

b) The probability of getting a head with one toss of a coin is $\frac{1}{2}$ or 0.5. Why is this?

c) How does your final relative frequency value compare with this value of 0.5?

Relative frequency becomes more accurate the more trials you do.

When using experimental evidence to estimate probability, it is advisable to perform at least 100 trials.

EXERCISE 41.4

1 Ping rolls a dice 500 times and records the number of times each score appears.

Score	1	2	3	4	5	6
Frequency	69	44	85	112	54	136

a) Work out the relative frequency of each of the scores.
   Give your answers to 2 decimal places.

b) What is the probability of obtaining each score on an ordinary six-sided dice?

c) Do you think that Ping's dice is biased? Give a reason for your answer.

2 Rashid notices that 7 out of the 20 cars in the school car park are red.
   He says there is a probability of $\frac{7}{20}$ that the next car to come into the car park will be red.
   Explain what is wrong with this.

3 In a local election, 800 people were asked which party they would vote for.
The results are shown in the table.

Party	Labour	Conservative	Lib. Dem.	Other
Frequency	240	376	139	45

    a) Work out the relative frequency for each party.
        Give your answers to 2 decimal places.

    b) Estimate the probability that the next person to be asked will vote Labour.

4 Emma and Rebecca have a coin that they think is biased.
They decide to do an experiment to check.

    a) Rebecca tosses the coin 20 times and gets a head 10 times.
        She says that the coin is not biased.
        Why do you think she has come to this conclusion?

    b) Emma tosses the coin 300 times and gets a head 102 times.
        She says that the coin is biased.
        Why do you think she has come to this conclusion?

    c) Who do you think is correct?
        Give a reason for your answer.

5 Joe made a spinner numbered 1, 2, 3 and 4.
He tested the spinner to see if it was fair.
He spun it 600 times. The results are shown in the table.

Score	1	2	3	4
Frequency	160	136	158	146

    a) Work out the relative frequency of each of the scores.
        Give your answers to 2 decimal places.

    b) Do you think that the spinner is fair?
        Give a reason for your answer.

    c) If Joe were to test the spinner again and spin it 900 times, how many times would you
        expect each of the scores to appear?

6 Samantha carried out a survey into how students travel to school.
She asked 200 students. Here are her results.

Method of travel	Bus	Car	Bike	Walk
Number of students	49	48	23	80

    a) Explain why it is reasonable for Samantha to use these results to estimate the
        probabilities of students travelling by the various methods.

    b) Estimate the probability that a randomly selected student will use each of the various
        methods of getting to school.

Challenge 41.4

Work in pairs.

Put 10 counters, some red and the rest white, in a bag.

Challenge your partner to work out how many counters there are of each colour.

Hint: You need to devise an experiment with 100 trials.
At the start of each trial, all 10 counters must be in the bag.

WHAT YOU HAVE LEARNED

- **The probability of an event not happening = 1 − the probability of the event happening**
- **If three events, A, B and C, cover all possible outcomes then, for example,**
  P(A) = 1 − P(B) − P(C)
- **Expected frequency = Probability × Total number of trials**
- **Relative frequency =** $\dfrac{\text{Number of times an outcome occurs}}{\text{Total number of trials}}$
- **Relative frequency is a good estimate of probability if there are sufficient trials**

⊙ MIXED EXERCISE 41

1   The probability that Peter can score 20 with one dart is $\frac{2}{9}$.
    What is the probability of him not scoring 20 with one dart?

2   The probability that Carmen will go to the cinema during any week is 0.65.
    What is the probability that she will not go to the cinema during one week?

3   Some of the probabilities of the length of time that any car will stay in a car park are
    shown below.

Time	Up to 30 minutes	30 minutes to 1 hour	1 hour to 2 hours	Over 2 hours
Probability	0.15	0.32	0.4	

What is the probability that a car will stay in the car park for over 2 hours?

4   There are 20 counters in a bag. They are all red, white or blue in colour.
    A counter is chosen from the bag at random.
    The probability that it is red is $\frac{1}{4}$. The probability that it is white is $\frac{2}{5}$.
    a) What is the probability that it is blue?
    b) How many counters of each colour are there?

5   The probability that Robert goes swimming on any day is 0.4.
    There are 30 days in the month of June.
    On how many days in June might you expect Robert to go swimming?

6   Holly thinks that a coin may be biased.
    To test this, she tosses the coin 20 times. Heads turns up 10 times.
    Holly says 'The coin is fair.'
    a) Why does Holly say this?
    b) Is she correct? Give a reason for your answer.

7   In an experiment with a biased dice, the following results were obtained after 400 throws.

Score	1	2	3	4	5	6
Frequency	39	72	57	111	25	96

   a) If the dice was fair, what would you expect the frequency of each score to be?
   b) Use the results to estimate the probability of throwing this dice and getting
      (i)   a 1.
      (ii) an even number.
      (iii) a number greater than 4.

42 → PROBABILITY 2

THIS CHAPTER IS ABOUT	YOU SHOULD ALREADY KNOW

THIS CHAPTER IS ABOUT

- The addition rule for mutually exclusive events, P(A or B) = P(A) + P(B)
- The multiplication rule for independent events, P(A and B) = P(A) × P(B)
- Probability tree diagrams
- Finding probabilities of dependent events

YOU SHOULD ALREADY KNOW

- That a probability can be expressed as a fraction or as a decimal
- How to find probability from a set of equally likely outcomes
- P(an outcome not happening) = 1 − P (the outcome happening)
- How to add, subtract and multiply decimals
- How to add, subtract and multiply fractions

The addition rule for mutually exclusive events

Events A and B are **mutually exclusive** if A and B cannot both occur at the same time.

For mutually exclusive events A and B

$$P(A \text{ or } B) = P(A) + P(B)$$

The addition rule is sometimes also called the 'or' rule.

TIP

Remember: P(A) means the probability of A happening.

EXAMPLE 42.1

The probability that the school hockey team will win their next match is 0.4.
The probability that the school hockey team will draw their next match is 0.3.
What is the probability that they will win or draw their next match?

Solution

The results are mutually exclusive since they can't both win and draw their next match.

P(win or draw) = P(win) + P(draw)
            = 0.4 + 0.3 = 0.7

Sometimes two events are not completely separate so they are not mutually exclusive. In these cases you cannot just add the probabilities of the two events.

EXAMPLE 42.2

One card is drawn at random from an ordinary pack of 52.
What is the probability that it will be
a) a red card?
b) an ace?
c) a red card or an ace?

Solution

a) P(red card) $= \frac{26}{52} = \frac{1}{2}$

b) P(an ace) $= \frac{4}{52} = \frac{1}{13}$

c) There are 26 red cards (including 2 aces) and 2 black aces.
This makes 28 favourable outcomes.

P(red card or ace) $= \frac{28}{52} = \frac{7}{13}$

It would be wrong to use the addition rule in Example 42.2 because the two events, red card and ace, are not mutually exclusive. They can occur together (when the card is the ace of hearts or the ace of diamonds). Because of this you have to work out the required probability by working out the number of favourable outcomes.

Notice that $\frac{26}{52} + \frac{4}{52} \neq \frac{28}{52}$

> **TIP**
> It is important to note that many questions involving mutually exclusive events can be answered without using the addition rule. Sometimes a question just requires you to count up the number of favourable outcomes.

Check up 42.1

A fair dice is thrown.

Put the following events into pairs so that the two events in a pair are mutually exclusive.

A: Getting a 6   B: Getting a prime number
C: Getting a factor of 4   D: Getting a 2
E: Getting an odd number   F: Getting a number greater than 4

The multiplication rule for independent events

Two events are **independent** when the outcome of the second event is not affected by the outcome of the first.

For two independent events A and B

$$P(A \text{ and } B) = P(A) \times P(B)$$

The multiplication rule is sometimes also called the 'and' rule.

EXAMPLE 42.3

A dice is thrown twice.
What is the probability that a six occurs on both throws?

Solution

$P(\text{two sixes}) = P(\text{six and six}) = P(\text{six}) \times P(\text{six}) = \frac{1}{6} \times \frac{1}{6} = \frac{1}{36}$

EXAMPLE 42.4

It is known that the probability that a certain type of seed will germinate is 0.8.
What is the probability that four seeds sown will all germinate?

Solution

Since the outcome of the second throw is unaffected by the outcome of the first throw, these are independent events.

$P(4 \text{ seeds germinate}) = P(\text{seed 1 and seed 2 and seed 3 and seed 4 germinate})$
$= 0.8 \times 0.8 \times 0.8 \times 0.8$
$= 0.4096$

EXAMPLE 42.5

There are six black counters and four white counters in a bag.
Elaine picks a counter at random, notes its colour and replaces it.
She then picks another counter.
What is the probability that she chooses one counter of each colour?

Solution

Since the counter is replaced, the second choice is independent of the first. The probabilities stay the same each time she takes a counter.

One counter of each colour could arise in two different ways:

        black then white

or   white then black.

P(one counter of each colour) = P(black **and** white) **or** P(white **and** black)

$$= P(black) \times P(white) + P(white) \times P(black)$$

$$= \frac{6}{10} \times \frac{4}{10} + \frac{4}{10} \times \frac{6}{10}$$

$$= \frac{24}{100} + \frac{24}{100}$$

$$= \frac{48}{100} = \frac{12}{25}$$

Check up 42.2

This is the answer sheet to the multi-choice section of a pub quiz.

Bill's team are not very good and decide to guess the answers at random.

What is the probability that they guess all five answers correctly?

ANSWER SHEET

1. (a)	(b)	(c)	
2. (a)	(b)		
3. (a)	(b)	(c)	
4. (a)	(b)	(c)	(d)
5. (a)	(b)		

EXERCISE 42.1

1 A bag contains ten red balls, five blue balls and eight green balls.
   What is the probability of selecting

   a) a red ball or a blue ball?   **b)** a green ball or a red ball?

2 When Mrs Smith goes to town the probability that she goes by bus is 0.5, the probability that she goes in a taxi is 0.35 and the probability that she goes on foot is 0.15.
   What is the probability that she goes to town

   a) by bus or taxi?   **b)** by bus or on foot?

3 In any batch of computers made by a company, the probabilities of the number of faults per computer are as follows.

Number of faults	0	1	2	3	4	5
Probability	0.44	0.39	0.14	0.02	0.008	0.002

   What is the probability that any given computer will have

   a) 1 or 2 faults?   **b)** 2, 3 or 4 faults?   **c)** an odd number of faults?
   d) fewer than 2 faults?   **e)** at least 1 fault?

4 Channel 10 weather report says

> The probability of rain on Saturday is $\frac{3}{5}$ and the probability of rain on Sunday is $\frac{1}{2}$.
> This means it is certain to rain on Saturday or Sunday.

Explain why the report is wrong. What mistake have they made?

5 A coin is tossed and a dice is thrown.
What is the probability of getting a head on the coin and an odd number on the dice?

6 A fair spinner is numbered 1 to 5.
The spinner is spun three times.
What is the probability that the spinner lands on
1 each time?

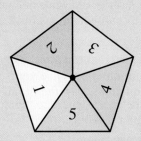

7 The probability that Holly wins the 100-metre race is 0.4.
The probability that Robert wins the 400-metre race is 0.3.
What is the probability that

a) Holly and Robert both win their race?

b) neither of them wins their race?

8 Each of the letters of the word INDEPENDENT is written on a card.
The cards are shuffled and one is selected.
This card is returned to the pack which is again shuffled.
A second card is selected.
What is the probability that the two cards are

a) both P?   **b)** both N?   **c)** both a vowel?

9 A box contains a large number of red counters and a large number of black counters.
25% of the counters are red.
A counter is taken out of the box, its colour is noted and it is replaced.
A second counter is then taken.
What is the probability that

a) both counters are red?

b) both counters are black?

c) one counter of each colour is chosen?

10 The probability that a darts player hits the bull's eye is 0.6.
He has three throws at the dartboard.
What is the probability that

a) he hits the bull's eye each time?

b) he misses the bull's eye each time?

c) he hits the bull's eye on two of his three throws?

Probability tree diagrams

When you are dealing with two or more events occurring, you can show the possible outcomes and their probabilities on a tree diagram. 'Branches' are drawn to show the possibilities for each event. Some of the questions in Exercise 42.1 could have been answered using a probability tree diagram.

This method can be a useful way of setting out the different possible outcomes of a series of events and can be extended, if necessary, to take account of any subsequent events.

> A tree diagram can look very messy if you are not careful with your presentation. Give yourself plenty of space and make sure that you don't draw the branches too close together.

EXAMPLE 42.6

The probability that I am woken by my alarm on any morning is 0.7.

a) Draw a probability tree diagram to show whether I am woken by my alarm on two days.

b) Use the tree diagram to work out the probability that
   (i) I am woken by my alarm on both days.
   (ii) I am woken by my alarm on just one of the two days.

Solution

a)

First day	Second day	Outcome	Probability
	Woken	WW	$0.7 \times 0.7 = 0.49$
Woken 0.7	Not woken	WN	$0.7 \times 0.3 = 0.21$
Not woken 0.3	Woken	NW	$0.3 \times 0.7 = 0.21$
	Not woken	NN	$0.3 \times 0.3 = 0.09$

(0.7 Woken, 0.3 Not woken branches from each node)

> Notice how you can use letters to make the work less cluttered.

You *multiply* the probabilities when you follow each path *along* the branches of the tree.

You *add* the probabilities *down* when you are interested in more than one possible outcome.

> **TIP** A good check to see if you have completed the tree correctly is to add the final probabilities together. The total should be 1.

b) (i) P(woken both days) $= 0.7 \times 0.7 = 0.49$
(ii) P(woken on just one of the days) $= 0.7 \times 0.3 + 0.3 \times 0.7$
$$= 0.21 + 0.21 = 0.42$$

EXAMPLE 42.7

A bag contains five red balls and three green balls.
A ball is selected from the bag at random. Its colour is checked and it is returned to the bag.
A second ball is then selected.
What is the probability that
a) both balls are green?
b) both balls are the same colour?
c) at least one of the balls is red?

Solution

First ball	Second ball	Outcome	Probability
Red $\frac{5}{8}$	Red $\frac{5}{8}$	RR	$\frac{5}{8} \times \frac{5}{8} = \frac{25}{64}$
	Green $\frac{3}{8}$	RG	$\frac{5}{8} \times \frac{3}{8} = \frac{15}{64}$
Green $\frac{3}{8}$	Red $\frac{5}{8}$	GR	$\frac{3}{8} \times \frac{5}{8} = \frac{15}{64}$
	Green $\frac{3}{8}$	GG	$\frac{3}{8} \times \frac{3}{8} = \frac{9}{64}$

a) P(both green) $= \frac{9}{64}$
b) P(both same colour) $=$ P(both green) $+$ P(both red)
$$= \frac{9}{64} + \frac{25}{64}$$
$$= \frac{34}{64} = \frac{17}{32}$$
c) P(at least one red) $=$ P(one red) $+$ P(two reds)
$$= \frac{15}{64} + \frac{15}{64} + \frac{25}{64} = \frac{55}{64}$$
Or
P(at least one red) $= 1 -$ P(no red)
$$= 1 - \frac{9}{64} = \frac{55}{64}$$

> **TIP** Do not cancel fractions until your final answer. You will find it easier to add them when they have the same denominator.

1   Terry is playing a board game where two dice are thrown each turn.
    To start the game he must throw the two dice and get two sixes.

    a) Copy and complete the tree diagram.

First dice	Second dice	Outcome	Probability

Six
   Six   SS
   Not a six   SN
Not a six
   Six   NS
   Not a six   NN

    b) Use the tree diagram to work out the probability that
       (i) Terry gets two sixes.
       (ii) Terry gets just one six.

2   The probability that the bus to school is late on any day is 0.2.

    a) Draw a tree diagram to show the bus being late or not late on two days.

    b) Work out the probability that the school bus is
       (i) not late on either of the days.
       (ii) late on one of the two days.

3   There are seven black discs and three white discs in a bag.
    A disc is selected, its colour is noted and it is then replaced in the bag.
    A second disc is then selected.

    a) Draw a probability tree diagram to show the outcomes of the two selections.

    b) Use the tree diagram to find the probability that
       (i)   both discs are black.
       (ii) both discs are the same colour.
       (iii) at least one disc is white.

4 The probabilities that the school rugby team win, draw or lose any match are 0.5, 0.3 and 0.2 respectively.

   a) Copy and complete the tree diagram to show the outcomes of the next two matches.

First match	Second match	Outcome	Probability

```
         W  <  W
               D
               L

         D  <  W
               D
               L

         L  <  W
               D
               L
```

   b) Work out the probability that
      (i)   the team win both of their matches.
      (ii) the team win one of the two matches.
      (iii) the results of the two matches are the same.

5 A coin is tossed three times.

   a) Draw a tree diagram with three sets of branches to show all the possible outcomes.

   b) Work out the probability that
      (i)   the result is three heads.
      (ii) exactly two heads are obtained.
      (iii) at least two tails are obtained.

Challenge 42.1

There are x red balls and y blue balls in a bag.
A ball is selected at random from the bag.
The ball is replaced and another is selected.
What is the probability of choosing

a) one ball of each colour?

b) two balls of the same colour?

Give your answers in terms of x and y as a single fraction in its simplest form.

Dependent events

So far in this chapter, the outcome of the second event was not influenced by the result of the first event. There are, however, many situations where the outcome of the second event is dependent on what happened in the first event. These are called **dependent events**. Since the probability of the second event is influenced by the outcome of the first event, it is called a **conditional probability**.

Tree diagrams can be used to represent dependent events, as is illustrated in the following two examples.

EXAMPLE 42.8

There are 10 boys and 15 girls in a Maths class.
Two students are to be selected at random to represent the class in a competition.
What is the probability that the team will consist of one boy and one girl?

Solution

The probabilities for the second event are out of 24 since one of the students has already been selected in the first choice and so only 24 remain to be chosen from.

First choice	Second choice	Outcome	Probability
	Boy $\frac{9}{24}$	BB	$\frac{10}{25} \times \frac{9}{24} = \frac{90}{600}$
Boy $\frac{10}{25}$	Girl $\frac{15}{24}$	BG	$\frac{10}{25} \times \frac{15}{24} = \frac{150}{600}$
Girl $\frac{15}{25}$	Boy $\frac{10}{24}$	GB	$\frac{15}{25} \times \frac{10}{24} = \frac{150}{600}$
	Girl $\frac{14}{24}$	GG	$\frac{15}{25} \times \frac{14}{24} = \frac{210}{600}$

P(one boy and one girl) $= \frac{150}{600} + \frac{150}{600} = \frac{300}{600} = \frac{1}{2}$

EXAMPLE 42.9

Whether I catch the bus to work or not depends on the weather.
The probability that it rains on any day is 0.3.
When it rains, the probability that I catch the bus to work is 0.9.
When it doesn't rain, the probability that I catch the bus is 0.2.
What is the probability that I catch the bus on any given day?

Solution

Weather	Transport	Outcome	Probability

$$\text{P(catch the bus)} = 0.27 + 0.14 = 0.41$$

EXERCISE 42.3

1 There are seven yellow balls and four red balls in a box.
A ball is selected at random and not replaced.
A second ball is then selected.

   a) Draw a probability tree diagram to show all the possible outcomes.

   b) Find the probability that the two balls are
     (i) both yellow.
     (ii) one of each colour.

2 There are 500 electronic components in a box.
It is known that 100 are defective.
Two components are selected at random without replacement.
What is the probability that

   a) both are defective?

   b) at least one is defective?

3 Judith and Suki play two games of tennis.
The probability of Judith winning the first game is 0.7.
If she wins the first game, the probability of her winning the second game is 0.7. If she
loses the first game, however, the probability of her winning the second game is 0.5.

   a) Draw a probability tree diagram to show the possible outcomes.

   b) What is the probability of
     (i) Judith winning both games?
     (ii) Suki winning both games?
     (iii) Judith and Suki winning one game each?

4 The probability that the school basketball team will win their next match is 0.6.
If they win the next match, the probability that they will win the following match is 0.7,
otherwise it is 0.4.
What is the probability that

    a) they win their next two matches?

    b) they win one of their next two matches?

5 At a school fair, thirty tickets numbered 1 to 30 are placed in a box.
People pay to pick a ticket from the box.
A ticket drawn from the box wins a prize if its number is a multiple of 5.
Once drawn, a ticket is not replaced.
John picks three tickets, one after the other, from the box.
What is the probability that

    a) none of his tickets win?

    b) at least one of his tickets wins?

    c) two of his tickets win a prize?

Challenge 42.2

Frodo is searching a castle for the treasure room.
He has a bunch of five keys, only one of which will open the door to the treasure.
He does not know which key will open the door.
Frodo tries the keys in turn.
What is the probability that

a) the first key he tries opens the door?

b) the first key does not open the door but the second key does?

c) the third key he tries opens the door?

d) the fourth key he tries opens the door?

e) the fifth key he tries opens the door?

Challenge 42.3

There are x red balls and y blue balls in a bag.
A ball is chosen at random from the bag.
The ball is replaced and another is chosen.
What is the probability of choosing

a) one ball of each colour?

b) two balls of the same colour?

Give your answers in terms of x and y as a single fraction in its simplest form.

◎ MIXED EXERCISE 42

1 In a school, every student has to take either art, music or drama.
60% of the students study art and 25% of the students study music.
A student is selected at random.
What is the probability that they study

a) art or music? **b)** drama or music?

2 A tin of biscuits contains ten chocolate biscuits, eight cream biscuits and seven plain biscuits.
A biscuit is selected at random from the tin.
What is the probability that it is

a) a chocolate or a plain biscuit? **b)** a cream or a plain biscuit?

3 During any week, I take sandwiches for my school lunch, have a meal in the school
canteen or go home for my lunch.
The probability that I take sandwiches for my school lunch is 0.4 and the probability that
I have a meal in the school canteen is 0.1.
What is the probability that, for my lunch one day

a) I either take sandwiches or go to the school canteen?

b) I either take sandwiches or go home?

4 A fair spinner is numbered 1 to 5.
In two spins, what is the probability that I get

a) a 1 on the first spin and a 5 on the second spin?

b) a 3 or a 4 on each of the two spins?

c) an even number on the first spin and an odd number
on the second spin?

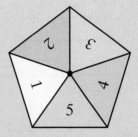

5 One bag contains two red sweets, one orange sweet and four yellow sweets.
A second bag contains one green sweet, two black sweets and three purple sweets.
One sweet is taken at random from each bag.
What is the probability that

a) one is orange and one is green?

b) one is red and one is purple?

c) one is either red or yellow and one is either black or purple?

6 An ordinary dice is thrown three times.
What is the probability of obtaining three sixes?

7 There are two black discs and three white discs in a bag.
A disc is selected, its colour noted and it is then replaced in the bag.
A second disc is then selected.

   a) Draw a probability tree diagram to show the outcomes of the two selections.

   b) Use the tree diagram to find the probability that
       (i) both discs are black.
       (ii) both discs are the same colour.
       (iii) there is one disc of each colour.

8 70% of the plants grown from a large batch of seeds will have a red flower.
Two plants grown from this batch are selected at random and the colour of the flowers produced is noted.

   a) Copy and complete the tree diagram.

First plant	Second plant	Outcome	Probability
	Red	RR	
Red			
	Not red	RN	
	Red	NR	
Not red			
	Not red	NN	

   b) Use the tree diagram to work out the probability that
       (i) neither plant has red flowers.
       (ii) one of the plants has red flowers.

9 A small box of chocolates contains four chocolates with hard centres and five chocolates with soft centres.
One chocolate is taken at random from the box and eaten.
Then a second one is taken.
What is the probability that

   a) both centres are of the same type?   **b)** there is one of each type of centre?

10 There are two fuses in a circuit.
The probability that the first fuse will blow is 0.2.
If the first fuse blows, the probability that the other fuse will blow is 0.7, otherwise it is 0.1.
What is the probability that

   a) both fuses will blow?   **b)** one of the two fuses will blow?

THIS CHAPTER IS ABOUT

- Posing statistical questions and planning how to answer them
- Primary and secondary data
- Choosing a sample and eliminating bias
- The advantages and problems of random samples
- Designing a questionnaire
- Collecting data
- Writing a statistical report

YOU SHOULD ALREADY KNOW

- How to make and use tally charts
- How to calculate the mean, median, mode and range
- How to draw diagrams to represent data, such as bar charts, pie charts and frequency diagrams

Statistical questions

Discovery 43.1

Are boys taller than girls?

Discuss how you could begin to answer this question.

- What information would you need to collect?
- How would you collect it?
- How would you analyse the results?
- How would you present the information in your report?

To answer a question using statistical methods, the first thing you need to do is make a written plan.

You need to decide which statistical calculations and diagrams are relevant to the problem. You should think about this before you start collecting data, so that you can collect it in a useful form.

It is a good idea to rewrite the question as a **hypothesis**. This is a statement such as 'boys are taller than girls'. Your report should present evidence either for or against your hypothesis.

Different types of data

When you investigate a statistical problem such as 'boys are taller than girls', there are two types of data which you can use.

- **Primary data** are data which you collect yourself. For example, you could measure the heights of a group of girls and boys.
- **Secondary data** are data which someone else has already collected. For example, you could use the internet database CensusAtSchool, which has a large number of students' heights already collected. Other sources of secondary data include books and newspapers.

Data samples

Most statistical investigations do not have a definite, obvious answer. For example, some girls are taller than some boys, and some boys are taller than some girls. What you are trying to find out is whether girls or boys are taller most often. You cannot measure the heights of all boys and all girls, but you can measure the heights of a group of boys and girls and answer the question for that group. In statistics, a group like this is called a **sample**.

The size of your sample is important. If your sample is too small, the results may not be very reliable. In general, the sample size needs to be at least 30. If your sample is too large, the data may take a long time to collect and analyse. You need to decide what is a reasonable sample size for the hypothesis you are investigating.

You also need to eliminate **bias**. A biased sample is unreliable because it means that certain results are more likely. For example, if all the boys in your sample were members of a basketball team, your data might appear to show that boys are taller than girls but the results would be unreliable because basketball players are often of above-average height.

It is often a good idea to use a **random** sample, where every person or piece of data has an equal chance of being selected. You may want, however, to make sure that your sample has certain characteristics. For example, random sampling within the whole school could mean that all the boys selected happen to be in Year 7 and all the girls in Year 11: this would be a biased sample, as older children tend to be taller. So you may instead want to use random sampling to select five girls and five boys from within each year group.

Random numbers can be generated by your calculator or a spreadsheet. To select a random sample of five girls from Year 7, for example, you could allocate a random number to each Year 7 girl and then select the five girls with the smallest random numbers.

When you write your report, include reasons for your choice of sample.

EXAMPLE 43.1

Candace is doing a survey about school meals. She asks every tenth person going into lunch.

Why may this not be a good method of sampling?

Solution

She will not get the opinions of those who dislike school meals and have stopped having them.

Discovery 43.2

A borough council wants to survey public opinion about its library facilities.

How should it choose a sample of people to ask?

Discuss the advantages and disadvantages of each method you suggest.

When you collect large amounts of data, you may need to group it in order to analyse it or to present it clearly. It is usually best to use equal class widths for this. Tally charts are a good way of obtaining a frequency table, or you can use a spreadsheet or other statistics program to help you. Before you collect your data, make sure you design a suitable data collection sheet or spreadsheet. Two-way tables are often useful for both recording and presenting your data.

Designing a questionnaire

A **questionnaire** is often a good way of collecting data.

You need to think carefully about what information you need and how you will analyse the answers to each question. This will help you get the data in the form you need.

For instance, if you are investigating the hypothesis 'boys are taller than girls', you need to know a person's sex as well as their height. If you know their age as well, you can find out whether your hypothesis is true for all ages of boys and girls. The data can be organised in a two-way table. However, asking people their height would probably not be the best way of finding this information – you would be more likely to get reliable results if you asked whether you could measure their height.

Here are some points to bear in mind when you design a questionnaire.

- Make the questions short, clear and relevant to your task.
- Only ask one thing at a time.
- Make sure your questions are not 'leading' questions. Leading questions show bias. They 'lead' the person answering them towards a particular answer: for example, 'do you agree that the cruel sport of fox-hunting should be illegal?'
- If you give a choice of answers, make sure there are neither too few nor too many.

EXAMPLE 43.2

Suggest a sensible way of asking an adult their age.

Solution

Please tick your age-group:

☐ 18–25 years   ☐ 26–30 years   ☐ 31–40 years
☐ 41–50 years   ☐ 51–60 years   ☐ Over 60 years

This means that the person does not have to tell you their exact age, which many adults don't like doing.

When you have written your questionnaire, test it out on a few people. This is called doing a **pilot survey**. Try also to analyse the data from this pilot, so that you can check whether it is possible. You may then wish to reword one or two questions, regroup your data, or change your method of sampling, before you do the proper survey.

If you encounter practical problems in collecting your data, describe them in your report.

Discovery 43.3

- Think of a survey topic about school lunches. Make sure that it is relevant to your own school or college. For instance, you may wish to test the hypothesis 'fish and chips is the favourite meal'.
- Write some suitable questions for a survey to test your hypothesis.
- Try them out in a pilot survey. Discuss the results and how you could improve your questions.

Writing up your report

Your report should begin with a clear statement of your aims and end with a conclusion. Your conclusion will depend on the results of the statistical calculations you have done with your data and on any differences or similarities illustrated by your statistical diagrams. Throughout the report, you should give your reasons for what you have done and describe any difficulties you encountered, and how you dealt with them.

Use this checklist to make sure that the whole project is clear.

- Use statistical terms whenever possible.
- Make sure you include a written plan.
- Explain how you selected your sample and why you chose to select it that way.

- Show how you found your data.
- Say why you have chosen to draw a particular diagram or table, and what it shows.
- Relate your findings back to the original problem. Have you proved or disproved your hypothesis?
- Aim to extend the original problem, using ideas of your own.

EXERCISE 43.1

1 State whether the following are primary data or secondary data.

   a) Measuring people's foot length

   b) Using school records of students' ages

   c) A librarian using a library catalogue to enter new books on the system

   d) A borrower using a library catalogue

2 A borough council wants to survey public opinion about the local swimming pool. Give one disadvantage of each of the following sampling situations.

   a) Selecting people to ring at random from the local phone directory.

   b) Asking people who are shopping on Saturday morning.

3 Paul plans to ask 50 people at random how long they spent doing homework yesterday evening.
Here is the first draft of his data collection sheet.

Time spent	Tally	Frequency
Up to 1 hour		
1–2 hours		
2–3 hours		

Give two ways in which Paul could improve his collection sheet.

4 For each of these survey questions
   - state what is wrong with it.
   - write a better version.

   a) What is your favourite sport: cricket, tennis or athletics?

   b) Do you do lots of exercise each week?

   c) Don't you think this government should encourage more people to recycle waste?

5 Janine is doing a survey about how often people have a meal out in a restaurant. Here are two questions she has written.

   Q1. How often do you eat out?
   □ A lot   □ Sometimes   □ Never

   Q2. What food did you eat the last time you ate out?

   a) Give a reason why each of these questions is unsuitable.

   b) Write a better version of Q1.

6 Design a questionnaire to investigate use of the school library or resource centre.
You need to know
- which year group the student is in.
- how often they use the library.
- how many books they usually borrow on each visit.

Design two-way tables to show how you could organise the data.

WHAT YOU HAVE LEARNED

- **Primary data are data which you collect yourself**
- **Secondary data are data which someone else has already collected, and are found in books or on the internet, for example**
- **You need to plan how to find evidence for or against your hypothesis, giving evidence of your planning**
- **Avoid bias when sampling**
- **In a random sample, every member of the population being considered has an equal chance of being selected**
- **Make sure the size of the sample is sensible**
- **In a questionnaire, questions should be short, clear and relevant to your task**
- **You can do a pilot survey to test out a questionnaire or data collection sheet**
- **In your report, you should give reasons for what you have done and relate your conclusions back to the original problem, saying whether your hypothesis has been shown to be correct or not**

◎ MIXED EXERCISE 43

1 Jan uses train times from the internet.
Are these data primary or secondary? Give a reason for your answer.

2 Ali wants to test the hypothesis 'older students at secondary school are better at estimating angles than younger ones'.
- **a)** What could he ask people to do in order to test this hypothesis?
- **b)** How should he choose a suitable sample of people?
- **c)** Design a collection sheet for Ali to record his data.

3 Write three suitable questions for a questionnaire asking a sample of people about their favourite music or musicians.
If you use questions without given categories for responses, show also how you would group the responses to the questions when analysing the data.

44 → REPRESENTING AND INTERPRETING DATA

Cumulative frequency diagrams

Check up 44.1

The heights of 60 plants were measured.
Look at these data and calculate an estimate of the mean height of the plants.

Height (h cm)	$0 < h \leqslant 10$	$10 < h \leqslant 20$	$20 < h \leqslant 30$	$30 < h \leqslant 40$	$40 < h \leqslant 50$
Frequency	15	31	8	2	4

Why may the mean not be a good average to use here?

When you have a set of grouped continuous data, you can analyse it by calculating an estimate of the mean. However, sometimes the median is a more appropriate measure of average to use. For this and other purposes, you can use a **cumulative frequency** graph.

The data in Check up 44.1 can be combined to give this table.

Height (h cm)	$h \leqslant 0$	$h \leqslant 10$	$h \leqslant 20$	$h \leqslant 30$	$h \leqslant 40$	$h \leqslant 50$
Cumulative frequency	0	15	46	54	56	60

This tells you that, for instance, there are 54 plants which are 30 cm or less in height, obtained from 15 + 31 + 8.

> **TIP**
> When you look at a cumulative frequency table, make sure that you understand what information is given by the cumulative frequencies in the context of the data.

You use the table to draw a cumulative frequency graph, plotting $(0, 0)$, $(10, 15)$, $(20, 46)$, $(30, 54)$, $(40, 56)$ and $(50, 60)$.

Make sure that you use the top end of the class to plot a cumulative frequency graph and not the midpoints as you would for a frequency polygon.

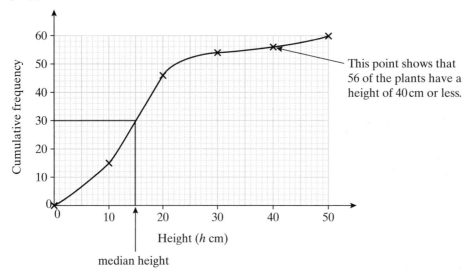

This point shows that 56 of the plants have a height of 40 cm or less.

median height

With 60 plants, the median is the average of the heights of the 30th and 31st plants. From the graph, find the value of the height at a cumulative frequency of 30.5. With a large set of data, the approximation $\frac{60}{2} = 30$ may be used. Cumulative frequency graphs are only suitable for large sets of data, so you should be able to use this approximation.

The graph shows that the median is about 15 cm.

Some statisticians prefer to join the points of the graph with straight lines rather than curves, showing that the graph is used for estimates. For GCSE, either method is accepted.

The cumulative frequency graph may also be used to find other information. For instance, reading the frequency corresponding to 25 on the horizontal axis, you can see that 52 plants are less than or equal to 25 cm in height.

The graph may also be used to find the **interquartile range**. This measures the spread of the middle 50% of the plants.

You need to read values for the lower quartile and the upper quartile, $\frac{1}{4}$ and $\frac{3}{4}$ of the way through the data respectively. Here, $\frac{1}{4}$ of 61 = 15.25, and $\frac{3}{4}$ of 61 = 45.75, so these are at the 15th and the 45th or 46th plants approximately.

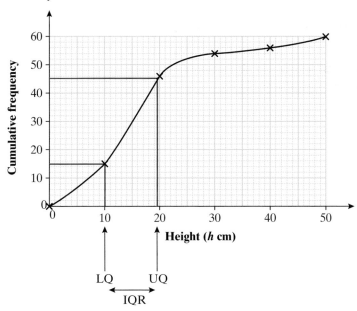

Lower quartile (LQ) = 10 cm

Upper quartile (UQ) = 19.5 cm

Interquartile range (IQR) = UQ − LQ
= 19.5 − 10 cm
= 9.5 cm

You can see here that the interquartile range is small, compared to the whole range. This tells you that the middle 50% of the data are closely bunched together. The steep curve between the quartiles also shows this.

Since the data are grouped and you do not have the original data, you cannot tell what the smallest and largest values of the data are, so you can only estimate the range. It could be as large as 50 − 0 = 50 cm and its lower bound is 40 − 10 = 30 cm.

The following notation is sometimes useful.
The lower quartile is Q_1 ($\frac{1}{4}$ of the way through the data).
The median is Q_2 ($\frac{2}{4}$ or $\frac{1}{2}$ way through the data).
The upper quartile is Q_3 ($\frac{3}{4}$ through the data).
The interquartile range is $Q_3 − Q_1$.

Note: Mathematicians and software packages vary as to the cumulative frequencies they use for the quartiles. In this chapter $\frac{n+1}{2}$ is used for the median and $\frac{n+1}{4}$ and $\frac{3(n+1)}{4}$ are used for the lower and upper quartiles as the most accurate values to be used, with $\frac{n}{2}, \frac{n}{4}$ and $\frac{3n}{4}$ being used as approximations when n is large. When interpreting a cumulative frequency graph for a large data set, there is unlikely to be much difference between the values for these definitions, and the acceptable values in GCSE questions will make allowance for these differences. For small data sets, the quartiles have only limited value and you are advised to use them with caution in these circumstances.

Cumulative frequency graphs can only be used with continuous data. Check up 44.2 illustrates the use of the interquartile range for a small set of continuous data.

Check up 44.2

Collect data from at least 50 students. (Or use data that has already been collected.) For instance, you could find the heights of students or the weight of their school bags. Group the data suitably into equal classes and find the frequencies. Make a cumulative frequency table and draw the cumulative frequency graph.

From your graph, find the median and the quartiles, the range and the interquartile range.

Outliers

Outliers are extreme values which occur in a distribution. There may be good reason for them to occur; however, sometimes outliers may be errors in recording results and you may wish to discount them.

A rule for outliers is that they are values whose distance from the nearest quartile is more than $1.5 \times$ interquartile range. If you discount outliers when doing a statistics project, make sure you mention them and give your reasons for ignoring them.

1 This cumulative frequency graph shows the heights of 200 girls and 200 boys.

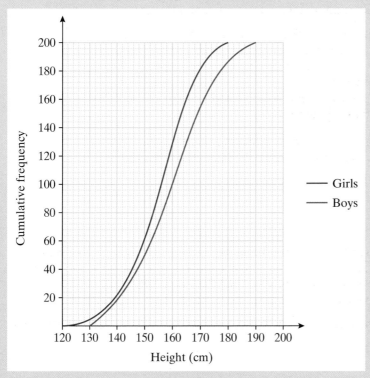

Find the median, the quartiles and the interquartile range for the girls and for the boys.

2 The left-hand table shows information about the masses of 200 potatoes.

   a) Copy and complete the cumulative frequency table on the right.

Mass (*m* grams)	Frequency
$0 < m \leqslant 50$	16
$50 < m \leqslant 100$	22
$100 < m \leqslant 150$	43
$150 < m \leqslant 200$	62
$200 < m \leqslant 250$	40
$250 < m \leqslant 300$	13
$300 < m \leqslant 350$	4

Mass (*m* grams)	Cumulative frequency
$m \leqslant 50$	16
$m \leqslant 100$	38
$m \leqslant 150$	
$m \leqslant 200$	
$m \leqslant 250$	
$m \leqslant 300$	
$m \leqslant 350$	

   b) Draw the cumulative frequency graph.

   c) Use your graph to find the median and interquartile range of these masses.

3 The left-hand table shows the ages of people in a netball club.

   a) Copy and complete the cumulative frequency table on the right.

Age (years)	Frequency
11–15	7
16–18	10
19–24	15
25–34	20
35–49	12
50–64	7

Age (*y* years)	Cumulative frequency
$y < 11$	0
$y < 16$	7
$y < 19$	17
$y <$	
$y <$	

Note: the upper boundary of the 11–15 age group is the 16th birthday.

   b) Draw the cumulative frequency graph.

   c) How many people in this club are aged under 30?

   d) How many people in this club are aged 40 or over?

   e) Find the median and the quartiles.

4 The cumulative frequency graph shows the head circumferences of a sample of 50 girls and 50 boys.

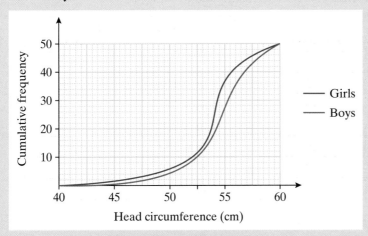

Make two comparisons between these distributions.

5 This table shows the earnings for a group of students one week.

Earnings (£w)	$0 < w \leqslant 20$	$20 < w \leqslant 40$	$40 < w \leqslant 60$	$60 < w \leqslant 80$	$80 < w \leqslant 100$
Frequency	5	15	26	30	6

Draw a cumulative frequency to represent this distribution.

6 A company tested a sample of 500 light bulbs of each of two types it produces. The table summarises the results, showing the time, in hours, that each light bulb lasted.

Time (t hours)	Frequency for type A	Frequency for type B
$0 < t \leqslant 250$	12	2
$250 < t \leqslant 500$	88	58
$500 < t \leqslant 750$	146	185
$750 < t \leqslant 1000$	184	223
$1000 < t \leqslant 1250$	63	29
$1250 < t \leqslant 1500$	7	3

a) On the same axes, draw cumulative frequency graphs to represent these distributions.

b) Which of the two types of light bulb is more reliable?

Histograms

This frequency table summarises the data for the earnings one week of a group of students (from question **5** in Exercise 44.1).

Earnings (£w)	$0 < w \leqslant 20$	$20 < w \leqslant 40$	$40 < w \leqslant 60$	$60 < w \leqslant 80$	$80 < w \leqslant 100$
Frequency	5	15	26	30	6

Work in pairs, with one drawing a frequency diagram to represent these data and the other drawing a frequency polygon.
Compare your graphs and discuss the advantages and disadvantages of each type of graph.

The data may be grouped differently and presented like this.

Earnings (£w)	$0 < w \leqslant 10$	$10 < w \leqslant 30$	$30 < w \leqslant 50$	$50 < w \leqslant 70$	$70 < w \leqslant 100$
Frequency	2	8	24	32	16

How could you present this information on a graph?
Try out your ideas and discuss the results.

You may have realised in the task above that our eyes take area into account when judging relative size. So, when the width of the groups is unequal, you need to consider the impact of the area of the bars in a bar chart.

A histogram is a frequency diagram which uses the *area* of each bar to represent frequency. In doing this it represents fairly the frequencies of groups of unequal widths.

The table shows the ages of people in the netball club from question **3** of Exercise 44.1.

Age (years)	11–15	16–18	19–24	25–34	35–49	50–64
Frequency	7	10	15	20	12	7

The histogram below shows the distribution of the ages of the members. Remember that the upper boundary of the 11–15 age group is the 16th birthday. So the boundaries of the bars in the histogram are at 11, 16, 19, 25, 35, 50 and 65.

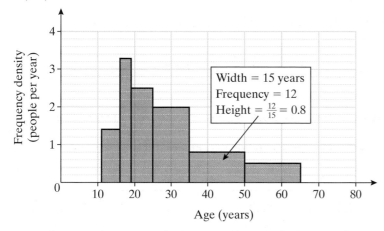

The calculation of the frequency density for one group has been added to the histogram to show you how it was done. Sometimes, instead of frequency densities being shown, a key is given, showing what each unit of area represents. On this histogram, one rectangle of the grid represents two people.

Usually the calculations for the frequency densities are done in a table, as in the next example.

EXAMPLE 44.1

Draw a histogram to represent this distribution for the money raised for charity by runners in a sponsored race.

Amount raised (£x)	Frequency
$0 < x \leqslant 50$	6
$50 < x \leqslant 100$	22
$100 < x \leqslant 200$	31
$200 < x \leqslant 500$	42
$500 < x \leqslant 1000$	15

Solution

You add two columns to the table, one to work out the width of each group and the other to work out the frequency density.

You calculate the frequency density using

$$\text{Frequency density} = \frac{\text{frequency}}{\text{width of group}}$$

Amount raised (£x)	Frequency	Group width	Frequency density (people per £)
$0 < x \leqslant 50$	6	50	$6 \div 50 = 0.12$
$50 < x \leqslant 100$	22	50	$22 \div 50 = 0.44$
$100 < x \leqslant 200$	31	100	$31 \div 100 = 0.31$
$200 < x \leqslant 500$	42	300	$42 \div 300 = 0.14$
$500 < x \leqslant 1000$	15	500	$15 \div 500 = 0.03$

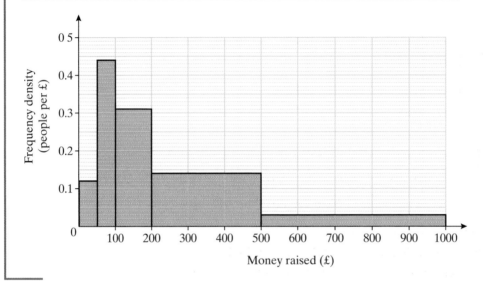

TIP

An alternative method is to calculate the group widths as multiples of £50 (or £10).

Using £50 would give group widths (×£50) of 1, 1, 2, 6 and 10.

The frequency densities (people per £50) would be 6, 22, 15.5, 7 and 1.5.

This avoids the need for a decimal scale.

You may wish to draw the histogram using this scale and compare it with the one drawn in the solution above.

EXAMPLE 44.2

The histogram shows the distribution of money raised in another sponsored race.

The key has not been shown, but you are given that 15 people raised £50 or less.

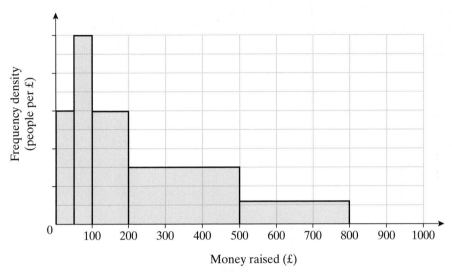

a) Calculate the frequency for each group.

b) Calculate also an estimate of the total amount raised.

c) Make two comparisons with the distribution for the sponsored race in Example 44.1.

Solution

a) The bar for the first group has a width of $\frac{1}{2}$ a rectangle on the grid and a height of 6 rectangles.

Its area is $\frac{1}{2} \times 6 = 3$ grid rectangles.
So 1 grid rectangle represents 5 people.

The 50–100 group has an area of 5 rectangles, which represents 25 people.
The 100–200 group has an area of 6 rectangles, which represents 30 people.
The 200–500 group has an area of 9 rectangles, which represents 45 people.
The 500–800 group has an area of $1.2 \times 3 = 3.6$ rectangles, which represents 18 people.

Alternatively, frequency density could be used.
The frequency density for the first group $= 15 \div 50 = 0.3$, so now label the scale on the frequency density axis accordingly.
For the second group, the frequency density is 0.5, so the frequency $= 0.5 \times 50 = 25$.
For the third group, the frequency density is 0.3, so the frequency $= 0.3 \times 100 = 30$, etc.

b)

Amount raised (£x)	Frequency	Midpoint	Midpoint × Frequency
$0 < x \leqslant 50$	15	25	375
$50 < x \leqslant 100$	25	75	1 875
$100 < x \leqslant 200$	30	150	4 500
$200 < x \leqslant 500$	45	350	15 750
$500 < x \leqslant 800$	18	650	11 700
Total	133		34 200

So a total of approximately £34 200 was raised.

c) More people took part in the second race: 133 in the second race compared with 116 in the first.
Both groups had a positively skewed distribution, with the majority of people raising less than £200, but with a few people raising much larger amounts.

Challenge 44.1

Discuss other comparisons between these two distributions.

EXERCISE 44.2

1 The table shows the earnings for a group of students one week.

Earnings (£w)	$0 < w \leqslant 20$	$20 < w \leqslant 40$	$40 < w \leqslant 70$	$70 < w \leqslant 100$	$100 < w \leqslant 150$
Frequency	5	15	26	30	6

Draw a histogram to represent this distribution.
Label your vertical scale or key clearly.

2 This distribution shows the ages of people watching a local football team one week.

Age (years)	Under 10	10–19	20–29	30–49	50–89
Frequency	24	46	81	252	288

a) Explain why the boundary of the 10–19 group is 20 years.

b) Calculate the frequency densities and draw a histogram to represent this distribution.

3 This histogram represents a distribution of waiting times in an outpatients department one day.

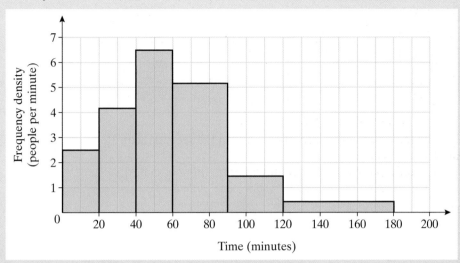

Time (minutes)

a) Make a frequency table for this distribution.

b) Calculate an estimate of the mean waiting time.

4 The histograms represent the time spent on mobile phone calls for a sample of girls and boys one week.

Histogram to show the time spent on mobile phone calls by girls

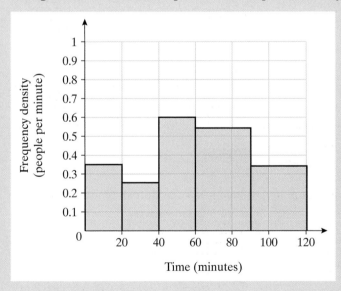

Time (minutes)

Histogram to show the time spent on mobile phone calls by boys

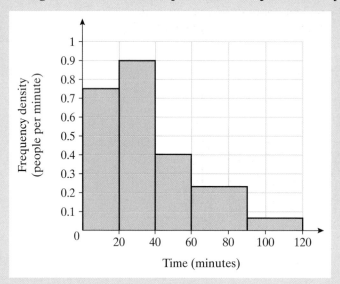

a) Find how many girls and how many boys spent between 20 and 40 minutes on the phone.

b) Compare the distributions.

5　The members of a gym were weighed. The histogram represents their masses.

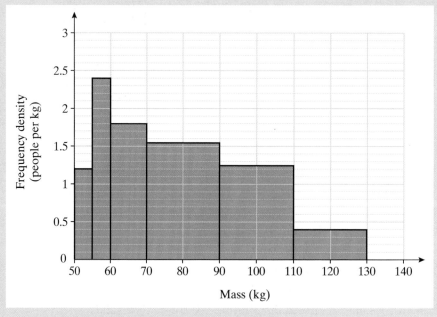

a) How many members of the gym were weighed?

b) Calculate an estimate of their mean mass.

Misleading graphs and diagrams

Statistical information is often illustrated using graphs in newspapers, reports and other publications. Sometimes the graphs may be misleading, as in this example.

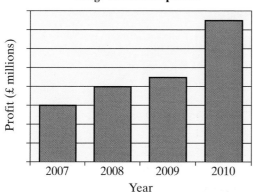

With no vertical scale given, the increase looks impressive. However, compare this with the two graphs below, which both represent the same information.

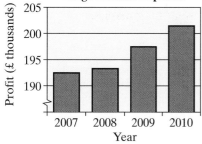

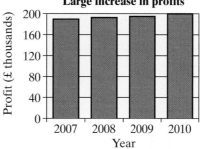

Discovery 44.2

Look at some newspapers and find some graphs and diagrams representing statistical information.

Discuss how clearly they show the information.

Discuss how fairly they show the information.

Read the accompanying newspaper articles.

Do they interpret the statistics accurately?

Are there comparisons made between distributions? If so, are they valid?

Share your results.

- Frequencies can be combined to give cumulative frequencies, showing how many there are in that group or below
- Cumulative frequency graphs can be drawn by plotting the cumulative frequency at the top end of each group
- The values of the median, the upper quartile and the lower quartiles can be obtained from a cumulative frequency graph
- Interquartile range = Upper quartile − Lower quartile
- A histogram is a frequency diagram which uses the area of each bar to represent frequency
- The vertical axis of a histogram is frequency density
- Frequency density = $\dfrac{\text{frequency}}{\text{width of group}}$
- How to compare distributions
- How to identify misleading graphs

MIXED EXERCISE 44

1 The table on the left-hand side shows information about the heights of 200 children in a primary school.

a) Copy and complete the cumulative frequency table on the right.

Height (h cm)	Frequency
$120 < h \leqslant 130$	5
$130 < h \leqslant 140$	27
$140 < h \leqslant 150$	39
$150 < h \leqslant 160$	62
$160 < h \leqslant 170$	45
$170 < h \leqslant 180$	18
$180 < h \leqslant 190$	4

Height (h cm)	Cumulative frequency
$h \leqslant 120$	0
$h \leqslant 130$	5
$h \leqslant 140$	
$h \leqslant 150$	
$h \leqslant 160$	
$h \leqslant 170$	
$h \leqslant 180$	
$h \leqslant 190$	

b) Draw the cumulative frequency graph.

c) Use your graph to find the median and the interquartile range of these heights.

2 The cumulative frequency graph shows how far a group of people, each using a pedometer, walked one day.

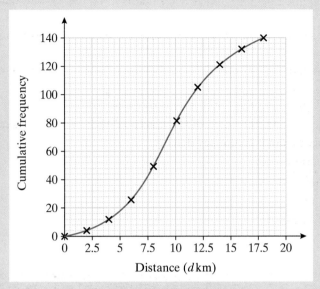

a) How many people walked less than 4 kilometres?

b) How many people walked more than 12 km?

c) What was the median distance?

3 The distribution shows the ages of people watching a rugby match one week.

Age (years)	Under 10	10–19	20–29	30–49	50–89
Frequency	240	1460	2080	4950	6120

Calculate the frequency densities and draw a histogram to represent this distribution.

4 The histogram shows the distance run one week by some people in training for a charity race.

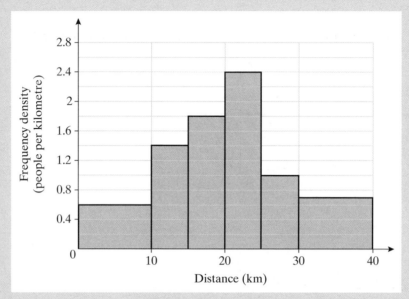

Calculate the frequency for each group and hence calculate an estimate of the mean distance run.

5 The histogram shows the distance run one week by another group of people.

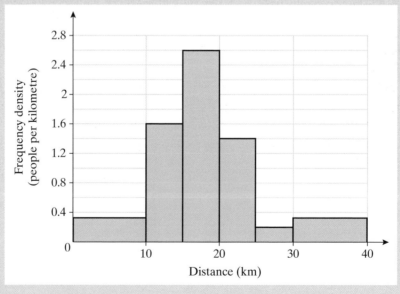

Compare this distribution with that in question **4**.
Make at least two comparisons.

6 The graphs show the percentiles for body mass index (BMI) for children.
For instance they show that 75% of girls aged 13 years have a body mass index of 21 or less.

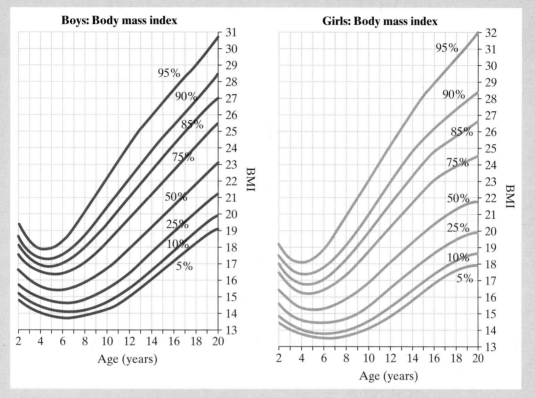

Graphics courtesy of Kidshealth.org/The Nemours Foundation.
All rights reserved. © 2005

a) What is the 75th percentile BMI for boys aged 14 years?

b) Peter is 9 years old. He has a BMI of 18.
Between which two percentile lines on the graph is his BMI?

c) Sumita is 11 years old. Her BMI is on the lower quartile.
What is her BMI?

1 A gardener wishes to create a lawn. The shape of the lawn is shown below. All the angles in the diagram are right angles.

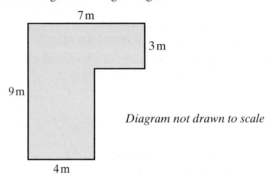

Diagram not drawn to scale

a) Find the area of the lawn in square metres.

b) He must decide whether he will use grass seed or turf to create the lawn.
A box of seed costs £6 and will cover 12 square metres.
Turf is sold at £1.55 per square metre.

Which is the cheaper option and by how much?

c) He is told that the seed may be bought in a large box or a small box.
A small box costs £6 and covers 12 square metres and a large box costs £8 and covers 20 square metres.

(i) What is the cost of seed per square metre for each box?
(ii) What is the cheapest way of buying enough seed to create the lawn?

2 Susan and Amir visit a high street department store in order to buy small gifts for their friends.
Two special offers on display in the store are shown below.

> **KEY RINGS**
> Usually £1.25 each
>
> **SPECIAL OFFER**
> 3 for the price of 2

> **BATH OILS**
> Buy 3 and get the cheapest
> **FREE**
> Large bottle £3.25
> Medium bottle £2.75
> Small bottle £2.25

a) Susan wants to buy 10 key rings as gifts for her friends.
How much money does she save by making full use of the offer?

b) Amir wants to buy 7 large, 4 medium, and 4 small bottles of bath oils.
Investigate how he can buy these 15 bottles in order to maximise his saving.
How much money does he save?

3 Four people volunteered to collect money in a town centre for a local charity over one weekend.
The organiser drew the following two pie charts to compare the shares of the money each volunteer collected on Saturday and on Sunday.

Saturday

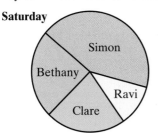

Sunday

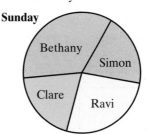

a) The total collected on Saturday was £520.
Estimate the amount of money collected by Bethany on Saturday.

b) Can you tell from the pie charts on which of the two days Ravi collected more money? Give a reason for your answer.

4 A survey was carried out in order to obtain data about how often people went to watch their local rugby team play.
People were asked the following question outside a supermarket one Saturday afternoon in October.

> How many times do you go to watch your local rugby team play?
> Please tick one box
>
> 1 or 2 times ☐   2 or 3 times ☐   3 or more times ☐

a) Give **two** reasons why the question is not suitable.

b) Give **one** criticism of how the survey was carried out.

5   Michael is planning a trip to the USA and Mexico.
    He notes that the current exchange rates are

        £1 = $2.05 $1 = 9.20 pesos   £1 = 20.25 pesos

    a) Michael has £600 to spend on his trip which he converts into
       US dollars ($). He spends half of these dollars in the USA and
       converts the remaining dollars into pesos.
       Would Michael have been better off if he had converted £300 into
       dollars and £300 into pesos at the start of his trip?

    b) Two similar fishing trips are advertised: one in the USA costing
       $100 and one in Mexico costing 750 pesos. Michael decides to
       book one of these trips before he leaves the UK.
       What is the difference in cost to the nearest pound?

6   Sarah and Mandy live in Nottingham and are planning a trip to
    Liverpool. They need to be in Liverpool by 2 p.m. They can travel by
    train, coach or in Sarah's car.

    Showing all your reasoning, how do you recommend they travel from
    Nottingham to Liverpool?
    Give **one** advantage and **one** disadvantage for your choice of
    transport.

Train timetable

Nottingham	09:27	10:52	11:44	12:53
Chesterfield	10:20	11:31	12:32	13:32
Manchester	11:37	12:37	13:37	14:37
Warrington	11:57	12:57	13:57	14:57
Liverpool	12:27	13:27	14:28	15:29

The train fare from Nottingham to Liverpool is £39.50 return **each**.

Coach timetable

Depart Nottingham	07:15	07:50	09:00
via	Sheffield	Birmingham	Leeds
Arrival Liverpool	11:55	13:35	14:40

The coach fare from Nottingham to Liverpool is £32 return for **two people**
travelling together.

Travelling by car
The distance from Nottingham to Liverpool is 105 miles.
The car's expected average speed on this journey is 35 mph.
The cost of running Sarah's car is 30p per mile.

7 Mr and Mrs Crawford and their three children aged 15, 13 and 8 are planning a holiday in the Peak District. They require accommodation for six nights.
They are looking at the following three options.

RISE HOTEL
****
Restaurant 12 noon – 10 p.m.
Pool, Sauna & Gym
6 a.m. – 9 p.m.
Double room (sleeps 2) – £110 per night
Single room (sleeps 1) – £70 per night
Breakfast included

THE MOUNT
Guest House
Evening meal (7 p.m.),
Bed & Breakfast
SPECIAL OFFER
2 adults, 2 children – £150 per night
(each additional child £40 per night)

FERN COTTAGE
Holiday Cottage to Let (Sleeps 6 people)
Fee
£160 per day for the first 4 days
£120 per day for each additional day

Investigate the cost of **each** option and discuss the advantages and disadvantages of each one.

8 A person's taxable income is calculated using the formula shown below.

Taxable income = gross income − pension contribution − other allowances

Sunita has a gross income of £52 500 and pays a pension contribution of £4725. Her other allowances total £5435.
Sunita pays tax at the rate of 20% on the first £34 600 of her taxable income and at a rate of 40% on the rest of her taxable income.

Calculate the total amount of tax Sunita pays.

9 Elizabeth has arranged a two-week holiday. The first week is in the USA, and the second week is in Canada.
Her father has given her 920 US dollars he had left from a previous holiday and she has used £450 to buy Canadian dollars.
The exchange rates were £1 = 1.84 US dollars and £1 = 2.24 Canadian dollars.

a) Calculate the number of Canadian dollars Elizabeth bought.

b) Find the total value, in pounds, of the US dollars and the Canadian dollars that Elizabeth had for her holiday.

10   Sally decides to buy a new freezer. She compares the costs in two stores.

Electric Superstore

SALE NOW ON

Super Slim Freezer
Normal selling price £320 + VAT at 17.5%

Sale   30% off selling price

Discount Electrics

SALE OFFERS

Super Slim Freezer
Normal selling price
        £450 (including VAT)

Sale offer   40% discount

a) Which of these stores has the lower price during the sale?

b) Calculate the difference between the sale prices in the two stores.

11 a) Two brothers share £80 in the ratio 1 : 4.
        How much does each brother receive?

    b) To make pink paint, red paint and white paint are mixed together in
        the ratio 2 : 5.
        Mike has 600 ml of white paint and an unlimited quantity of red paint.

        Find the largest quantity of pink paint that Mike can make.

12   A train takes 3 hours 30 minutes to travel a distance of 168 miles.
    Calculate the average speed of the train.

13 Mrs Williams received an electricity bill from Wales Electricity Company.
The bill, with some of the entries removed, is shown below.

Wales Electricity Company

Bill period 24th February 2009 to 7th May 2009

Mrs S Williams
22 Southdown Road
Aberfelin

INVOICE

Meter reading Last time	This time	Tariff C–Customer reading E–Estimated reading	Units used	Price of each unit in pence	Amount £
75692C	79148C	Units used	….…………	11.9	….……………
		Quarterly charge			13.55
		Total charge			….……………
		VAT at 5% of the total charge			….……………
		Total charge including VAT			….……………
		Previous balance carried forward. Credit (CR)			29.37 (CR)
		Amount to pay			….……………

Use the information given on the bill to complete all of the missing
entries and calculate the total amount that Mrs Williams has to pay.

14 A contractor has been asked to give a time scale for completing a
contract for surface-treating 8 miles of road.
Last year the contractor surface-treated a similar road in 18 days.
At that time he had a workforce of 20 and treated 12 miles of road.
Now his workforce is only 16.

Showing all your calculations, how long should he expect the work to
take, assuming that all other conditions will be similar to last year?

15 Mr Thomas is planning a holiday for himself, his wife and their three children. The holiday will be in Majorca at the Hotel Golden Sands. The brochure costs are shown in the table below.

Hotel Golden Sands					
Number of nights	7	10	11	14	All holidays
Starting date	Adult	Adult	Adult	Adult	First child
15 Apr – 30 Apr	£325	£355	£375	£405	£79
01 May – 11 May	£369	£399	£405	£485	£89
12 May – 19 May	£359	£385	£395	£495	£99
20 May – 01 June	£459	£465	£469	£475	£199
02 June – 22 June	£409	£435	£445	£449	£149
23 June – 15 July	£455	£479	£489	£509	£199

Children
On all holidays first child as above, additional children £50 each.

Supplements per person per night
Balcony £5
Balcony with sea view £12
Half board £17

Mr Thomas decides to book an 11-night holiday for himself and his family departing on 14 May.
He wants a room with a balcony and sea view, and meals on a half-board basis.

Find the total cost of the holiday for Mr Thomas and his family.

16 Tim decides to work out some costs of running his car.

a) He bought the car for £6500 on 1 January 2009 and sold it on 1 January 2010. During that period the car had decreased in value by 15%.

Calculate the decrease in the value of the car during the year.

b) His total mileage for the year was 12 425 miles, and the car had averaged 35 miles to the gallon.

In order to work out the cost of the petrol he had used, Tim used the fact that 1 gallon is approximately 4.55 litres. He used 85p per litre as the average cost of the petrol he had bought during the year.

Calculate the cost of the petrol Tim bought for the car in 2009.

17 Mr Adams is planning a holiday for himself, his wife and their two children. The holiday will be in Tenerife in the Hotel Sun Beach, and the brochure costs are shown in the table below.

Hotel Sun Beach					
Number of nights	7	10	11	14	All holidays
Starting date	Adult	Adult	Adult	Adult	First child
17 Apr – 30 Apr	£325	£355	£375	£405	£79
01 May – 10 May	£369	£399	£405	£495	£79
11 May – 18 May	£359	£385	£395	£495	£79
19 May – 29 May	£459	£465	£469	£475	£199
30 May – 21 June	£409	£435	£445	£449	£149
22 June – 12 July	£455	£479	£489	£509	£199

Children
Second child pays first child price plus £40 per week

Supplements per person per night
Balcony £2
Bed and breakfast £5
Half board £11

Mr Adams decides to book a 14-night holiday for himself and his family departing on 17 May.
He wants a room with a balcony, and meals on a half-board basis.

Find the total cost of the holiday for Mr Adams and his family.

18 Three batsmen are being considered for the one remaining place on the county's cricket team.
The numbers of runs each cricketer scored in their last five games are shown below.

Player A	49	55	48	51	47
Player B	7	15	113	21	9
Player C	102	3	5	60	80

a) For each player, calculate the range and mean of the number of runs scored per match in their last five games.

b) Using your calculations, decide which player you would choose. Give a reason for your choice.

19 A company has an order to produce plastic cubes of side length 6 cm. The company distributes the cubes to its customers in cardboard boxes sized 70 cm by 50 cm by 45 cm.

6 cm

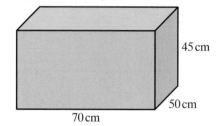

45 cm

50 cm

70 cm

a) Find the maximum number of cubes that can be fitted into the cardboard box and calculate the percentage of the volume of the box that is not used.

b) Find the dimensions of a smaller cardboard box that could contain the same number of cubes found in part **a)** but have no wasted volume.

c) What is the length, h, of each side of the cardboard box shown below, which can contain exactly 125 of the plastic cubes with no wasted space?

h

h

h

20 Most people have to pay income tax on part of their income.
Each person is given an annual tax-free allowance on which no tax is paid.
Once a person's taxable income has been calculated, they have to pay tax at a rate of 20% of the first £37 400 of taxable income, and then 40% on any amount above £37 400.
Siân's annual income is £54 000 and she has a tax-free allowance of £6475.

Calculate the tax that Siân should pay.

21 Halima wishes to cover her kitchen floor with a vinyl covering costing £8.34 per square metre.
The vinyl is sold in rolls that are 4 m wide.
Her kitchen floor is rectangular and measures 11 m by 7 m.

Showing all your calculations, find the cheapest way of covering the kitchen floor with the vinyl.

22   The diagram below shows the positions of two houses, P and Q.
House P is situated on Moss Road and house Q is situated on Brook
Road. The two roads meet at a junction, J.
It is possible to walk in a straight line between P and Q across open fields.
You can travel from P to Q either by walking across the fields at an
average speed of 6 km per hour, or by cycling on the roads at an average
speed of 25 km per hour.

Which would take less time, and by how many minutes?

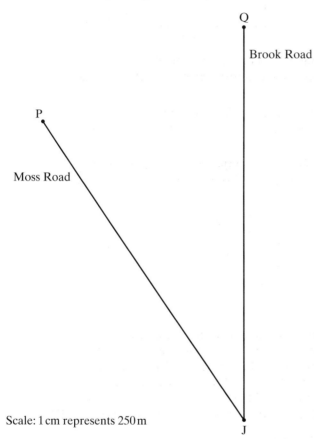

Scale: 1 cm represents 250 m

23   The owner of a holiday cottage decides to increase the weekly hire price
by 20% for the summer months.
Towards the end of the summer she reduces this new price by 20%.

What percentage of the original price does it now cost to hire the cottage
for one week?

24 A catering company charges for providing buffet meals at events using the following rates.

> ## Catering for your BIG DAY
> **Buffet meals for any event**
>
> | | |
> |---|---|
> | Up to 20 people | £8.50 per person |
> | 21 to 50 people | £7.75 per person |
> | Over 50 people | £7.25 per person |

a) What is the cheapest way to order meals for 48 people?
Show all your working.

b) The method the company uses to calculate the cost of providing meals is

£4.30 per person for ingredients plus a basic cost.

The basic cost is £23 for up to 20 people, increasing by £7 for every additional 30 (or part of 30) people.

Show that the company makes over 50% profit when catering for 100 people.

c) What is the smallest number of people that the company can cater for and still make a profit?

25 Alan and Barbara have decided to use some time during their holiday to collect money for a local charity.

a) Alan visited 45 houses near his home asking for donations. Four-fifths of the houses he visited gave an average of £2 per house. There was no reply from the other houses. It took him, on average, 15 minutes to visit each house.

Calculate the amount Alan collected and the time it took him.

b) Barbara did some part-time work at a local business for £6 per hour and gave the money she made to the charity.

Compare the two methods of collecting money, giving **one** advantage for each method.

26 A store is reducing the price on all of its goods by 15%.
Find the selling price of a computer that was originally priced at £460.

27 Find, to the nearest penny, the compound interest earned when £3500 is invested for 3 years at 5% per annum.

28 A leaflet called 'Pricing made easy' shows the cost of sending items
through the post to any destination within the United Kingdom.

Pricing made easy			
	Weight up to	First class	Second class
Letter			
Length: 240 mm maximum	100 g	£0.36	£0.27
Width: 165 mm maximum			
Thickness: 5 mm maximum			
Large letter			
Length: 353 mm maximum	100 g	£0.52	£0.42
Width: 250 mm maximum	250 g	£0.78	£0.66
Thickness: 25 mm maximum	500 g	£1.08	£0.90
	750 g	£1.57	£1.31
Packet			
Length: over 353 mm	100 g	£1.14	£0.95
Width: over 250 mm	250 g	£1.45	£1.24
Thickness : over 25 mm	500 g	£1.94	£1.63
	750 g	£2.51	£2.08
	1 kg	£3.08	£2.49
	1.25 kg	£4.30	Second class
	1.5 kg	£5.00	is not available
	1.75 kg	£5.70	for items
	2 kg	£6.40	weighing over
	4 kg	£8.22	1 kg
	Each additional	£2.80	
	2 kg or part thereof		

a) You have four items, described below, to send through the post to
different addresses within the United Kingdom.

Item	Dimensions			Weight
	Length (L)	Width (W)	Depth (D)	
1	21 cm	10 cm	4 mm	70 g
2	32 cm	21 cm	2 cm	400 g
3	40 cm	30 cm	20 cm	1.6 kg
4	50 cm	40 cm	20 cm	7 kg

Item 2 is to be sent second class. The others are all to be sent first class.

How much will it cost to send each item?

b) Your friend is intending to send four separate packets to the same
address in the United Kingdom. Each packet weighs 1.3 kg.

How could you could save him the most money on his postage?
Calculate the exact amount that would be saved.

29 You have ordered a circular fruit cake which is 8 cm high and 30 cm in diameter.
The volume of the fruit cake can be found using the formula

$$\textbf{Volume} = \pi \times \textbf{(Radius)}^2 \times \textbf{Height}$$

a) Calculate the volume of the cake, in cm^3, to the nearest $100\,cm^3$.

b) You need to choose a box to carry the cake home.
The shop sells boxes in the three sizes shown below.
The price of a box is proportional to its volume.

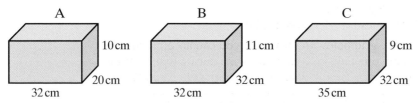

Boxes are not drawn to scale

Which box would you choose? Show your calculations and reasoning.

c) Your cake weighed 4 kg.
The following quantities of ingredients were used to make it.

600 g of flour   2.2 kg of fruit   9 eggs

Calculate the amount of each of these ingredients that would be needed to make a similar cake weighing 3 kg.

30 The Jones family is considering buying a new car.
They wish to investigate the cost difference between buying and running a car with a diesel engine and a car with a petrol engine.

a) Using the information given below, calculate an estimate for the cost of fuel used in one year by **each** type of car.

Average mileage:	800 miles per month
Fuel consumption:	Diesel engine – 55 miles per gallon
	Petrol engine – 38 miles per gallon
Fuel price:	Diesel – 123.9p per litre
	Petrol – 112.9p per litre

1 gallon = 4.55 litres

b) The new diesel engine car costs £21 000.
The new petrol engine car costs £18 400.
The value of the diesel engine car depreciates by 20% in the first year.
The value of the petrol engine car depreciates by 25% in the first year.

Calculate the difference in their values at the end of the first year.

31 A small guesthouse has four single bedrooms, numbered 1, 2, 3 and 4, available for guests.
Seven people wish to stay at the guesthouse on the days shown below.

Week 9 – 15 August
Person A: Sunday, Monday, Tuesday
Person B: Tuesday, Wednesday, Thursday, Friday
Person C: Monday, Tuesday, Wednesday
Person D: Monday, Tuesday, Wednesday, Thursday, Friday
Person E: Sunday, Monday
Person F: Friday, Saturday
Person G: Wednesday, Thursday, Friday

a) Use a copy of the grid below to show how you can accommodate them without anyone having to change rooms during their stay. Fit in as many as you can. Person A has already been allocated Room 1.

	Room 1	Room 2	Room 3	Room 4
Sunday	A			
Monday	A			
Tuesday	A			
Wednesday				
Thursday				
Friday				
Saturday				

b) The price charged is £48 per person per night.
An assistant is paid £7.50 per hour for 4 hours' work each day.
Other costs total £6.42 per guest per night.
Calculate how much profit will be made during this week.

c) What percentage of the available accommodation for the week was taken by the guests?
Give your answer to the nearest whole number.

32 a) George wants to make a rectangular patio, with an area between $34 \, m^2$ and $36 \, m^2$.
He uses square concrete slabs of side 60 cm. He does not have the equipment to cut the slabs.
His first plan is to make the patio 7 m long and 5 m wide.
What is the smallest number of slabs required to completely cover a rectangle 7 m by 5 m?

b) George realises that his first plan covers too large an area so he decides to alter one or both dimensions of the patio.
Investigating all possible solutions, calculate how many slabs are required so that the area of the patio is between $34 \, m^2$ and $36 \, m^2$.
(Ignore any extra length resulting from the joins between the slabs.)

1   What is the smallest number of addition signs you would need to insert into 987654321 to make a total of 99?
Show your answer.

2   Jane is older than Kim. Kim is older than Shaun.
Shaun is younger than Jane. Rachel is older than Jane.

Arrange the children in order of age, starting with the oldest.

3   I start with a one-digit whole number, multiply it by 3, add 8, divide by 2 and subtract 6.
I now have the same one-digit number that I started with.

What is my starting number?

4   Place the digit cards 9, 7, 4, 5, 6 and 1 in the boxes to make the largest possible answer.

$$\square\square \times \square\square + \square \times \square = ?$$

5   In school, Susan has been learning about time, fractions and decimals.
She makes mistakes in her work as shown below.

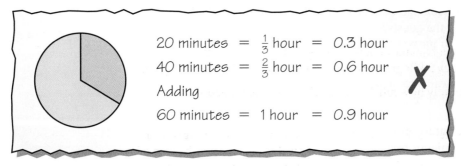

Explain the mistakes that Susan has made.

6   Debbie opens her maths textbook and finds that the sum of the two consecutive page numbers is 245.

What are the numbers on the pages?

7   Zen's maths textbook has a page missing. Where the page is missing, the sum of the page numbers on the facing pages is 127.

What are the page numbers on the missing page?

8   David earns £960 per month and spends $\frac{2}{3}$ of this amount on rent, electricity and rail fares. He spends £200 per month on food and entertainment and saves what he has left each month.

What fraction of his earnings does David save each month? Give your answer in its simplest form.

9   What is the sum of all the **digits** in the sequence

$$1, 2, 3, 4, 5, \ldots, 97, 98, 99, 100?$$

10 The sum of two numbers is 13.
The product of the same two numbers is 36.

What is the difference between the squares of these numbers?

11 The usual time taken to complete a task is 2 hours 20 minutes.
A target is set to decrease this time by 15%.

Calculate the target time in hours and minutes.

12 In 2010 the price of a gemstone was £8.
A jeweller predicts that the value of the gemstone will rise by 25p per year.

In which year is the gemstone predicted to be worth £9.75?

13 The peel of a banana weighs about $\frac{1}{8}$ of the total weight of the banana. I buy 3 kilograms of bananas at a cost of 60p per kilogram.

Approximately how much am I paying for banana peel?

14 The terms $2x + 1, 2x - 3, x + 2, x + 5$ and $x - 3$ are arranged in a different order. When rearranged in this different order, the sum of the first three terms is $4x + 3$ and the sum of the last three terms is $4x + 4$.

Find the third term.

15 Which is the first term in the sequence $1002, 999, 996, 993, 990, \ldots$ that will be less than zero?

16 Copy and complete the following fraction sum.

$$\frac{1}{\square} + \frac{1}{\square} = \frac{19}{72}$$

17 Tony and Bob share prize money in the ratio 7 : 4.
Tony spends 10% of his prize money immediately, leaving him with £1953.

Find the total amount of prize money won by Tony and Bob.

18 David gave Jenny half of his CDs.
Jenny gave Henry half of the CDs David gave her.
Henry kept 8 of the CDs and gave the remaining 10 to Faryl.

How many CDs did David give Jenny?

19 A ball is dropped from a height of 100 metres.
Each time it hits the ground it bounces $\frac{3}{5}$ of the height it fell.

How far has the ball travelled in total by the fifth bounce?

20 A rectangular table is three times as long as it is wide.
If it was 3 metres shorter and 3 metres wider it would be a square.

Calculate the length and width of the table.

21 Bacteria in a Petri dish double the area they cover each day.
It takes 16 days for the bacteria to cover the dish.

After how many days is $\frac{1}{4}$ of the dish covered in bacteria?

22 In a mental maths test I get eight times as many answers correct as
incorrect. There are 315 questions in the test.

How many do I get correct?

23 A ferry can carry 10 cars or 6 trucks across a lake at one time.
They never load both cars and trucks on the ferry for safety reasons.
The ferry made five trips across the lake and was full each time.
A total of 42 vehicles were carried across the lake.

How many cars crossed the lake on the ferry?

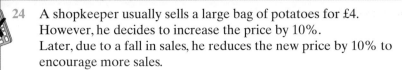

24 A shopkeeper usually sells a large bag of potatoes for £4.
    However, he decides to increase the price by 10%.
    Later, due to a fall in sales, he reduces the new price by 10% to
    encourage more sales.

    What percentage of the original price does a customer now pay for a
    large bag of potatoes?

25 A manufacturer claims that

    '*a low-energy light bulb lasts 20 times longer and uses $\frac{2}{3}$ of the
    electricity used by an ordinary light bulb*'.

    An ordinary light bulb costs 49p and uses £3.30 of electricity over its
    lifetime.
    A low-energy light bulb costs £15.

    Which type of light bulb offers better value for money and by how
    much?
    Hint: Consider the period of the lifetime of one low-energy light bulb.

26 A net of a cube is cut from a rectangular piece of card x cm by y cm
    as shown in the diagram.

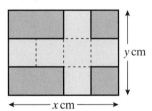

    a) Find, in terms of x, an expression for the surface area of the cube.

    b) Write down an expression for y in terms of x.

27 Five times one number is equal to twice a second number.
    The sum of the two numbers is 0.7.

    Find the two numbers.

28 A farm has pigs and hens. There are 38 heads and 100 feet.

    How many hens and pigs are there on the farm?

29   A rectangular shape is made using 12 square tiles placed with equal gaps between them.
The overall length of the rectangle is 645 mm and the overall width is 475 mm.

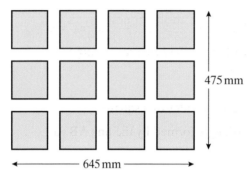

Find the dimensions of the tiles and the width of the gap in millimetres.

30   Whole numbers greater than 1 are arranged in five columns like this.

	2	3	4	5
9	8	7	6	
	10	11	12	13
17	16	15	14	
	…	…	…	…
…	…	…	…	
	…	…	…	…

Which column will contain the number 1000?
Give a reason for your answer.

31   Is it possible to pour a half pint of milk into a 250 ml glass?

32   A turntable rotates 720° in one second.

Calculate the number of revolutions in one minute.

33   John walks at 4 km/h and Sue walks at 5 km/h.
They have agreed to meet halfway along a 12-kilometre track.
If they both start walking at the same time from different ends, after how long will they meet?

34 The coordinates of the points A and B are $(4, 7)$ and $(-2, -1)$ respectively.

Calculate the length of the line AB.

35 It takes one man one day to dig a 2 metre by 2 metre by 2 metre hole.

How long does it take three men working at the same rate to dig a hole 4 metres by 4 metres by 4 metres?

36 A and B are the midpoints of two adjacent sides of a square.

What is the ratio of the area of the triangle formed by the line AB to the area of the square?

37 A cyclist rides from Cornhill to Dinas and back again without stopping. This is a **total** distance of 20 km. The return journey from Dinas to Cornhill is uphill and takes 16 minutes longer than the journey from Cornhill to Dinas. The average speed for the entire journey is 10 km/h.

Find the average speed for the first part of the journey, from Cornhill to Dinas.
Give your answer in km/h correct to 1 decimal place.

38 A regular pentagon of side length 5.88 metres is drawn inside a circle such that each vertex of the pentagon lies on the circumference of the circle. The radius of the circle is 5 metres.

Calculate the area of the pentagon.

39 A six-sided polygon is to be drawn using a computer program. The designer has stated that three of the internal angles should be $140°$ each and the remaining three angles should all be acute angles.

Explain whether or not this design is possible.
Show your working and give a reason for your answer.

40 The diagram shows a concrete paving slab.

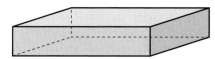

A garden designer suggests that all the dimensions should be increased by 20%.

Find the percentage increase in the volume of concrete required.

41 A square of width x metres is increased in size to form a rectangle. The length of the rectangle is 5 metres greater than the width of the square and the width of the rectangle is 3 metres greater than the width of the square.

Write down an expression for the area of the rectangle.

42 An open box is made from a square piece of card. The length of the card is y cm. Squares of side length x cm are cut from the corners of the card, leaving the net of an open box.

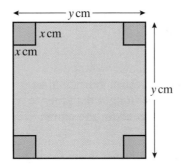

a) Write down an expression in terms of x and y for the volume of the open box.

b) Write down an expression in terms of x and y for the percentage of the original square piece of card wasted in making the open box.

43 The length of the hour hand on a clock is 4 cm and the length of the minute hand is 6 cm.

Calculate the distance between the tips of the hands at 2 o'clock.

44 The diagram shows a rectangle between two parallel lines.

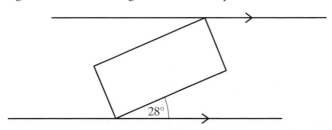

The rectangle is of length 7.1 cm and width 3.4 cm.

Calculate the perpendicular distance between the parallel lines.

45 The diagram shows a pyramid with a square base.
The length of each of the edges is x.

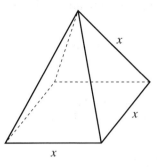

Show that the surface area of the pyramid is $x^2(1 + \sqrt{3})$.

46 A new logo is being designed for an airline company. The graphic designer starts with a circle surrounded by a square, with each side of the square being a tangent to the circle. Each side of the square is of length $2x$ cm.

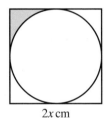

$2x$ cm

Find an expression, in terms of x and π, for the shaded area of the logo.

47 The diagram shows two vertical walls EF and HG.
The point P is on horizontal ground between the bases of the two walls.

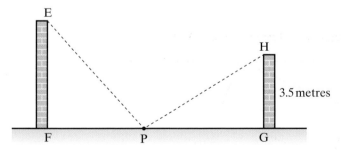

3.5 metres

The angle of elevation of E from P is 48° and the angle of elevation of H from P is 33°.

Given that HG = 3.5 m and 3FP = 2PG, find the height of the wall EF.

48 The diagram shows a vertical pole PQ standing on horizontal level ground.

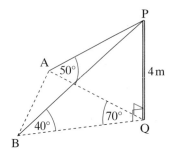

Straight wires PA and PB are attached to the pole at P and to the horizontal ground at A and B respectively.

Find AB, given that PQ = 4 m, $P\hat{A}Q = 50°$, $P\hat{B}Q = 40°$ and $A\hat{Q}B = 70°$.

49 Eliza visits her friend Abdul and then returns home by the same route. She walks at 2 km/h when going uphill, 6 km/h when going downhill and 3 km/h on level ground. Her total walking time to and from Abdul's house is 6 hours.

What is the distance between Eliza's and Abdul's houses?

50 The first part of a journey is x km and the time taken is t hours. The second part of the journey is twice as far but takes half the time.

Write down an expression in terms of x and t for the average speed of the entire journey. Give your answer in its simplest terms.

51 Eleanor finds an old pencil box, 2 inches by 2 inches by 8 inches. Inside it is the longest pencil that could possibly be fitted inside it.

a) Given that 12 inches is approximately 30 cm, calculate the maximum possible length of the pencil in centimetres.

b) Other than the use of 12 inches to be approximately 30 cm, state an assumption made that may affect the accuracy of the answer.

52 A right-angled triangle in the first quadrant is formed by the lines with equations $y = 0$, $y = x$ and $x + y = 5$.

Find the area of the triangle.

53 The diagram shows quadrilateral ABCD with $B\hat{C}A = 46°$,
$D\hat{A}C = 127°$, BC = 5.8 cm and AD = 15.2 cm.

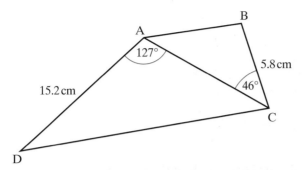

Given that the area of triangle ABC is 20.4 cm², find the area of the
quadrilateral.

54 The diagram shows a flowerpot.

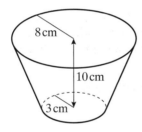

The height of the flowerpot is 10 cm, the radius of the base is 3 cm
and the radius of the top is 8 cm.

Find the volume of the flowerpot.

55 The diagram shows two circles. OA is the radius of the larger circle
and the diameter of the smaller circle.

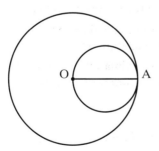

Find, in its simplest form, the area of the smaller circle as a fraction
of the area of the larger circle.

56 The diagram shows an equilateral triangle, ABC, with sides of length 2 cm.
The lines AD and CE are perpendicular to the sides BC and AB respectively, and P is the point of intersection of AD and CE.

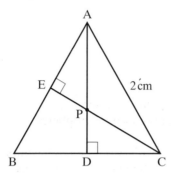

a) Calculate the length of AP.

b) Find the area of the quadrilateral BDPE.

57 The diagram shows a triangle ABC and a point D on BC.

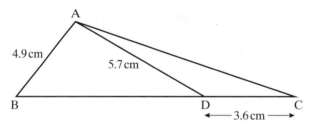

Given that $A\hat{B}D = 32°$, AB = 4.9 cm, AD = 5.7 cm and DC = 3.6 cm, calculate the area of triangle ADC.

58 The bearing of a cruise liner C from a lighthouse L is 056° and the distance CL is 4.2 km.
The bearing of the cruise liner from a tanker T is 318° and the distance CT is 9.6 km.

Calculate the bearing of the tanker from the lighthouse.

59 There are 12 people in a room. Six of them are wearing socks, four are wearing shoes and three are wearing both shoes and socks.

How many people have bare feet?

60 The mean monthly rainfall for 6 months was 23.8 mm.

If it had rained 1 mm more each month, what would the mean monthly rainfall have been?

61 Five numbers are written in ascending order.
You are given that
- the second number is four times the first number
- the fourth number is six greater than the second number
- the range of the five numbers is 62
- the median is 47
- the mean is 45.

Find the five numbers.

62 An ice-cream shop sells nine different flavours of ice-cream.
The flavours are vanilla, chocolate, mint, butterscotch, raspberry, strawberry, coffee, cherry and pineapple.
A group of children all buy double scoops with two different flavours of ice-cream. No children buy the same combination of flavours and every different combination of flavours is chosen.

How many children are in the group?

63 Ilyas had a mean of 56% for his first seven maths tests.

What percentage would Ilyas have to get in his eighth test to raise his overall mean to 60%?

64 Four unbiased coins are tossed.

Find the probability that there will be at least three heads.
Give your answer as a fraction.

65 You and your friend are playing a game which involves spinning a coin. Your friend gets five heads in his first five spins. Your friend assures you that the coin is a fair coin.

What is the probability that your friend will throw heads on the next spin of the coin?
Explain your answer.

66 Lisa travels to work each day by bus.
The probability of the bus being late is 0.3.
The probability of the bus being late and Lisa not getting a seat on the bus is 0.18.

Given that the events 'the bus being late and 'Lisa not getting a seat on the bus' are independent, find

a) the probability of Lisa not getting a seat on the bus

b) the probability of the bus not being late and Lisa getting a seat on the bus.

67 The histogram represents the lengths of anchor lines for small boats as found in a survey.

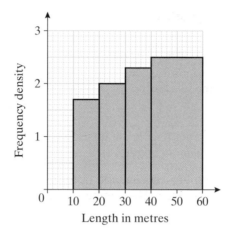

Use the histogram to find an estimate for the 70th percentile.

68 The histogram represents the lengths of leaves as found in a survey.

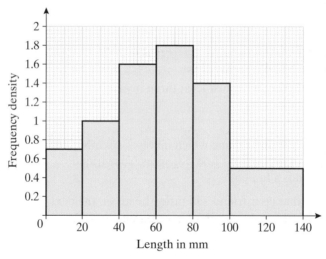

Use the histogram to calculate an estimate for the median.

69 Jack eats toast for breakfast.
Sometimes he eats it without butter or marmalade, sometimes he has either butter or marmalade, but when he is really hungry he has both butter and marmalade. The probability that he eats his toast without butter or marmalade is 0.2, and the probability that he eats it with both is 0.3. He is nine times more likely to eat his toast with only butter than he is to eat it with only marmalade.

Calculate the probability that he eats his toast with only butter.

For an extra challenge, now try these questions.

1   There are six sacks of grain.
    They weigh 54 kg, 30 kg, 49 kg, 53 kg, 62 kg and 50 kg.
    Each sack contains either wheat or barley or oats.
    There is twice as much wheat as there is barley.
    Only one sack contains oats.

    Work out what each sack contains.

2   I am thinking of two numbers. Their highest common factor is 6.
    Their lowest common multiple is 36. One of the numbers is 12.

    What is the other number?

3   Find the area of a parallelogram with base $(12 + 5\sqrt{3})$ cm and
    corresponding perpendicular height $(12 - 5\sqrt{3})$ cm.
    Simplify your answer.

4   The first term in a sequence of consecutive integers is $4a + 7$.
    Prove that the sum of the first five terms of the sequence is five times
    the third term of the sequence.

5   Is it possible to connect the points with coordinates $(-1, 2), (8, 47)$ and
    $(-10, -43)$ with a single straight line?
    You must show your working and state a reason for your answer.

6   The diagram shows a triangle ABC with RS drawn parallel to BC.

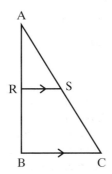

    Given that RB = 5.2 cm, BC = 9.3 cm and RS = 5.4 cm, calculate AR.

7 A two-dimensional structure is made using six rods of equal length and five joints as shown in the diagram. The joints connecting the rods allow the structure to move so that the total area contained within the structure changes.

The structure lies flat on a horizontal surface.

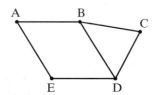

a) Given that each rod is of length 4 cm and $B\hat{A}E = y°$, find, in terms of y, an expression for the area of the structure.

b) Find the minimum and maximum areas of the structure.

Explain the reasoning for both of your answers.

→ INDEX